Frontline

CCW

THE UNIVERSITY OF OXFORD

CHANGING CHARACTER OF WAR PROGRAMME

The Changing Character of War Programme is an inter-disciplinary research group at the University of Oxford, and was funded by the Leverhulme Trust between 2003 and 2009.

Frontline

Combat and Cohesion in the Twenty-First Century

Edited by
Anthony King

OXFORD
UNIVERSITY PRESS

OXFORD
UNIVERSITY PRESS

Great Clarendon Street, Oxford, OX2 6DP,
United Kingdom

Oxford University Press is a department of the University of Oxford.
It furthers the University's objective of excellence in research, scholarship,
and education by publishing worldwide. Oxford is a registered trade mark of
Oxford University Press in the UK and in certain other countries

First Edition published in 2015

Published in the United States of America by Oxford University Press
198 Madison Avenue, New York, NY 10016, United States of America

British Library Cataloguing in Publication Data
Data available

Library of Congress Control Number: 2014959967

ISBN 978–0–19–871966–3

Acknowledgements

This volume is based on a series of papers which were originally presented at a mixed academic-practitioner conference 'Frontline: combat and cohesion in Iraq and Afghanistan', held at All Souls College Oxford on 18–19 March 2013. The conference was part of an ESRC-funded research project 'Combat, Cohesion and Gender: the elementary forms of the military life' (ES/J006645/1) which provided the financial support for the event. The conference could not have taken place without the support and guidance of Sir Hew Strachan, Chichele Professor of the History of War, to whom I am personally indebted; Rob Johnson, Director of the Changing Character of War Programme; and Ruth Murray, the administrator of the Changing Character of War Programme. I would also like to thank the Warden and Fellows and the staff of All Souls College not only for allowing the use of the College for the conference but also for their hospitality throughout 2012–13 when I was a Visiting Fellow at the College. I am deeply indebted to the Manciple of All Souls, Paul Gardner, Lorraine Scaysbrooke, Helen Green, and John Ward who were immensely helpful in assisting me in organizing the conference and especially the formal dinner at the College on the night of 18 March. I would also like to thank Rosemary Strawson and John Keeling of Trinity College who were responsible for the accommodation of delegates and the final dinner. Finally, it is important to acknowledge all the participants in the conference. Even though not all of them were able to appear in this volume, they made a vital contribution to the discussions which informed many of the following chapters: Peter Bradley, Katherine Brown, Neil Brown, Helena Carreiras, Sergio Catignani, Doug Chalmers, Jack Clark, Randall Collins, Bill Combes, James Cowan, Ulrike Franke, Lucy Giles, Todd Greentree, Andrew Harrison, Bernd Horn, Tom Hughes, Charlotte Isaakson, Anna Krylova, John Keeling, Ewen McLay, David Segal, Mady Wechsler Segal, Sibylle Scheipers, Sean Scott, Deividas Slekys, Justin Stenhouse, Martin Todd, Eleanor Taylor, Nicholas Welch, and Karl Yden. At the same time, we did not forget to have some fun.

Acknowledgements

Contents

List of Figures ix
List of Tables xi
List of Contributors xiii

Part I. Concepts

1. On Cohesion 3
Anthony King

2. Subaltern Soldiers: Eurocentricism and the Nation-State in the Combat Motivation Debates 24
Tarak Barkawi

3. Killing and Cohesion in Close Combat: Contexts and Concepts from the First World War to the Present 46
Rob Johnson

Part II. Current Debates

4. From a Sociology of Units to a Sociology of Combat Formations: Militaries and Cohesion in Urban Combat 73
Eyal Ben-Ari

5. Discipline and Punish: Encouraging Combat Performance in the Citizen and Professional Army 93
Anthony King

Part III. Historical Experience

6. Combat and Cohesion in the Early Twentieth Century: Strengthening the Inverted Pyramid 119
Antulio J. Echevarria II

7. Re-evaluating Combat Cohesion: The British Second Army in the Northwest Europe Campaign of the Second World War 134
Jonathan Fennell

Part IV. Case Studies

8. New Model Diggers: Australian Identity, Motivation, and Cohesion in Afghanistan 167
Garth Pratten

9. A Profession of Love: Cohesion in a British Platoon in Afghanistan 200
Patrick Bury and Anthony King

10. Hierarchy, Status, and Combat Motivation in the French Foreign Legion 216
Mikaela Sundberg

11. Cohesion and Misconduct: The French Army and the Mahé Affair 234
Pascal Vennesson

12. Cohesion, Political Motivation, and Military Performance in the Italian Alpini 250
Chiara Ruffa

Part V. The Home Front

13. Of FOBs and Families 269
Leonard Wong and Stephen Gerras

14. Beyond the Battlefield 291
Christopher Dandeker and Simon Wessely

15. The Future of Cohesion 309
Anthony King

Bibliography 325
Index 349

List of Figures

7.1. Second Army, monthly admissions per 1000 to general hospitals and casualty clearing stations for sickness, battle exhaustion (BE), desertion, AWOL, and SIW as compared with the morale score derived from the censorship summaries, June 1944 to April 1945. 148

7.2. Second Army, weekly admissions per 1000 to general hospitals and casualty clearing stations for sickness, battle exhaustion, and SIW, 11 June 1944 to 5 May 1945. Morale scale equivalents are presented on the right-hand Y-axis. 150

7.3. Second Army, rate of sick, battle exhaustion, and SIW per 1000 and number of prisoners captured per week. 151

7.4. 21st Army Group, area of advance in ten-day periods to ammunition expended (1 July 1944 to 28 March 1945). 152

13.1. Father's role by age. Percentage of each age who say it is 'extremely important' for a father to provide this for his children. 284

List of Tables

7.1. Strength of the fighting portion of Second Army plus weekly admissions for casualties (cas)*, sick, battle exhaustion (BE), and self-inflicted wounds (SIW), 6 June 1944 to 26 May 1945. 145

7.2. Number of convictions by courts martial for desertion and AWOL, Second Army, July 1944 to April 1945. 146

7.3. Morale descriptions and scores based on the censorship summaries, Second Army, June 1944 to April 1945. 146

7.4. Monthly sick, battle exhaustion (BE), desertion/AWOL, and SIW rates per 1000 in Second Army, June 1944 to April 1945. 147

List of Contributors

Tarak Barkawi is Reader in International Relations at the London School of Economics. He earned his doctorate at the University of Minnesota and specializes in the study of war, armed forces, and society with a focus on conflict between the West and the global South. He has written on colonial armies, 'small wars' and imperial warfare, the Cold War in the Third World, and on counter-insurgency and the War on Terror. More generally, he is interested in the place of armed force in histories and theories of globalization, modernization, and imperialism, especially from a postcolonial perspective.

Eyal Ben-Ari is Director of the Centre for Society, Security and Peace at Kinneret Academic College and has carried out research in Israel, Japan, Singapore, and Hong Kong. In Israel he has studied various aspects of the Israel Defense Forces and has now (with colleagues) completed a book on the ways in which the ground forces of the Israeli military have waged combat in the Al-Aqsa Intifada. In Japan he has carried out research on white-collar communities, early childhood education, and the Japanese community in Singapore. He is currently carrying out research on the Japanese Self-Defense Forces and the armed forces of the industrial democracies. Among his recent books are (with Zev Lehrer, Uzi Ben-Shalom, and Ariel Vainer) *Rethinking the Sociology of Warfare: A Sociological View of the Al-Aqsa Intifada* (2010), (with Kobi Michael and David Kellen) *The Transformation of the World of Warfare and Peace Support Operations* (2008), and (with Nissim Otmazgin) *The State and Popular Culture in East Asia* (2012).

Patrick Bury is an ESRC-funded PhD student studying the British Army Reserves at Exeter University. He served as a Captain in the Royal Irish Regiment between 2005 and 2009. *Callsign Hades*, an account of his platoon's tour in Sangin, Afghanistan, was published in 2010. Patrick has also worked for the NATO Parliamentary Assembly and as a senior analyst for a private security company.

Christopher Dandeker is Professor of Military Sociology in the Department of War Studies, King's College London and (with Professor Simon Wessely), Co-Director of the King's Centre for Military Health Research (KCMHR). He joined King's from the University of Leicester as a Senior Lecturer in 1990, where he studied for his BSc [Soc] and PhD. He studies all aspects of civil–military relations. Professor Dandeker's publications include *The Structure of Social Theory* (1984), *Surveillance Power and Modernity* (1990), *Les Armées en Europe* (1998) (edited with Bernard Boene), *Nationalism and Violence* (1998), *Facing Uncertainty: Sweden in international* *perspective* (2000), and contributions to *Israel's Armed Forces in Comparative Perspective* (2009) (edited by Stuart Cohen) and to *Modern Warfare and the Utility of Force* (2010) (edited by Jan Angstrom and Isabelle Duveysteyn).

Antulio J. Echevarria II is the editor of the US Army's War College Quarterly, *Parameters*, and holds the Elihu Root Chair of Military Studies. He is a former US Army officer with more than twenty years' military service, and holds a doctorate in modern history from Princeton University. He is the author of four books, the most recent of which is *Reconsidering the American Way of War: US Military Practice from the Revolution to Afghanistan*, Georgetown University Press, 2014; and numerous articles on strategy and military history.

Jonathan Fennell is a Lecturer in Defence Studies at King's College London. He was awarded a Doctorate from the University of Oxford in 2008 and his first book, *Combat and Morale in the North African Campaign*, was published in 2011 by Cambridge University Press. He has addressed conferences in the UK, Europe, and North America and has had papers published in *The Journal of Military and Strategic Studies, Air Power Review, War in History and The Journal of Strategic Studies*. His current research, which has as its main objective the delivery of a second book for Cambridge University Press, investigates how not only economic, technological, and logistical factors, but also political, social, institutional, and cultural ones, affected the performance and effectiveness of the British and Commonwealth armies in the Second World War.

Stephen Gerras is a Professor of Behavioral Sciences in the Department of Command, Leadership, and Management at the US Army War College. He served in the Army for over twenty-five years, including commanding a light infantry company and a transportation battalion, teaching leadership at West Point, and serving as the Chief of Operations and Agreements for the Office of Defense Cooperation in Ankara, Turkey. Dr Gerras holds a BS from the US Military Academy and an MS and PhD in industrial and organizational psychology from Penn State University.

Rob Johnson is the Director of the Changing Character of War Research Programme at the University of Oxford and Senior Research Fellow at Pembroke College. As a Lecturer in the History of War, his research interests are conflicts 'amongst the people' in Afghanistan, Pakistan, and the Middle East, and he is frequently invited to act as a specialist advisor to the British, American, and Afghan armed forces on security, strategy, and defence. He is the author of *The Afghan Way of War* (2011), 'Managing Helmand' in the *International Journal of Area Studies*, and 'The Taliban' in Beatrice Heuser (ed.), *National Experiences of Insurgencies and Counter-insurgencies* (Proceedings of the British Academy, 2013). He is currently working on the use of partnered auxiliary forces in counter-insurgency.

Anthony King (editor) is a Professor of Sociology at the University of Exeter. He has written on sport, social theory, and the armed forces, including his most recent books *The Transformation of Europe's Armed Forces* (Cambridge University Press, 2011) and *The Combat Soldier* (Oxford University Press, 2013). He is currently working on a book on divisional headquarters. He has acted as a mentor and adviser to the armed forces for a number of years, including membership of the ISAF RC (South) Prism Cell in 2009–10.

Garth Pratten is a Senior Lecturer in Military Operations at the Strategic and Defence Studies Centre at the Australian National University. An historian by training, he has had a varied career having worked for the Australian Army's Training Command and the Australian War Memorial, and taught at Deakin University and in the War Studies Department at the Royal Military Academy Sandhurst. In 2010, Dr Pratten deployed to Afghanistan as part of the team compiling the war diary for ISAF's Regional Command South. In April 2006, Dr Pratten was awarded the Australian Army's CEW Bean prize for his PhD thesis, the book of which, *Australian Battalion Commanders in the Second World War,* was runner up for the Templer Medal in 2010. He is currently working on a project to examine the conduct and experience of unit-level command in the ground forces of Australia, Canada, the United Kingdom, and the United States in southern Afghanistan.

Chiara Ruffa received her PhD from the European University Institute in 2010. Having been a Research Fellow at the Belfer Center for Science and International Affairs, Harvard Kennedy School of Government, she was appointed as an Assistant Professor at Uppsala University, Department of Peace and Conflict Research in January 2012. She is currently a senior lecturer at the Swedish Defense University, Stockholm. Dr Ruffa's research interests are International Relations at large and security studies in particular with a strong focus on humanitarian, development, and military actors in complex humanitarian emergencies. She combines it with an expertise in qualitative empirical research: mainly ethnography and comparative case study research. For her PhD thesis, which she is currently turning into a book manuscript, Dr Ruffa has conducted extensive ethnographic research in Southern Lebanon and Afghanistan embedded with several different NGOs, civil society actors, and Western and non-Western armies. Her work has been published in *Security Studies, Comparative European Politics, Security and Defense Analysis, Armed Forces and Society, Small Wars and Insurgencies, The New York Times,* and book chapters in several edited volumes.

Mikaela Sundberg is a Senior Lecturer at the Department of Sociology, Stockholm University. She has published many articles within the field of sociology of science, notably on the role and use of numerical simulations. Some of her recent publications include 'Creating Convincing Simulations in Astrophysics' (*Science Technology and Human Values*, 2012) and 'The Dynamics of Coordinated Comparisons' (*Social Studies of Science*, 2011). As a part of the research project 'Social Relations within a Total and Greedy Institution: Brotherhood among French Foreign Legionnaires', funded by the Swedish Research Council, she has conducted ethnographic work within the Foreign Legion.

Pascal Vennesson is Professor of Political Science at the S. Rajaratnam School of International Studies (RSIS), Nanyang Technological University, Singapore. His research and teaching lie at the intersection of the fields of international relations and strategic studies. He is finishing a book entitled *War in the Global Village: Transnational Challenges and the Struggle for Freedom of Action*. He recently published 'Sanctions and Embargoes in EU-Asia Relations' (with Clara Portela) in Thomas Christiansen, Emil Kirchner, Philomena B. Murray, eds, *The Palgrave Handbook of EU-Asia Relations*

(Palgrave, 2013); 'The Transnational Politics of Warfare Accountability: Human Rights Watch versus the Israel Defense Forces' (with Nikolas Rajkovic) *International Relations* 26, no. 4 (2012); 'War Without the People' in: Hew Strachan, Sibylle Scheipers, eds., *The Changing Character of War* (Oxford University Press, 2011). He is the author, co-author, and editor of five books and his refereed articles have been notably published in *Armed Forces and Society, International Relations, The Journal of Strategic Studies, The Review of International Studies*, and *La Revue Française de Science Politique* (*The French Political Science Review*), and *Security Studies*. He is a member of the editorial boards of *La Revue Française de Science Politique* (*The French Political Science Review*), and *Security Studies*.

Simon Wessely is Vice Dean, Chair, and Head of the Department of Psychological Medicine at the Institute of Psychiatry and a Consultant Liaison Psychiatrist at King's College Hospital. His clinical work is in general hospital psychiatry, with a particular interest in unexplained syndromes including chronic fatigue syndrome (CFS), and he continues to work in the dedicated service that he set up twenty years ago. His research interests are in the grey areas between medicine and psychiatry, clinical epidemiology, psychiatric injury, and military health. His first paper was called 'Dementia and Mrs Thatcher', whilst his doctoral thesis was on crime and schizophrenia. He has written over 650 papers on various aspects of psychiatry and military health, which is his main current research focus. He is co-director of the King's Centre for Military Health Research and Civilian Consultant Advisor in psychiatry for the British Army. He cycles each year, albeit slowly, to Paris to raise funds for the Royal British Legion. He was knighted for services to military health and psychological medicine in 2013.

Leonard Wong is a Research Professor in the Strategic Studies Institute at the US Army War College. His research focuses on the human and organizational dimensions of the military and includes topics such as combat motivation, the military profession, and leadership development. His research has led him to locations such as Iraq, Afghanistan, Kosovo, Bosnia, and Vietnam. He has testified before Congress and his work has been highlighted in media such as the *New York Times, Wall Street Journal, Washington Post, New Yorker*, CNN, NPR, PBS, and 60 Minutes. He is a retired US Army Lieutenant Colonel whose career has included teaching leadership at West Point and serving as an analyst in the Pentagon. He is a Professional Engineer and holds a BS from the US Military Academy and an MS and PhD in Business Administration from Texas Tech University.

Part I
Concepts

1

On Cohesion

Anthony King

Introduction

Why *Frontline*? The title of this volume seems to be drawn from a now obsolete military lexicon. It is, after all, commonly assumed that frontlines no longer exist. For instance, Western troops in Iraq and Afghanistan fought a 360 degree battle without obvious front and rear areas. Hybrid counter-insurgency warfare seems to have led to a radical revision of the geometry of the battle-space in which concepts applicable to interstate warfare of the twentieth century seem to be palpable archaisms. It is certainly true that operations in Iraq and Afghanistan or those in Mali, Sierra Leone, or Libya have not involved fronts of the kind which were evident in the First and Second World Wars, Korea, or the Cold War. Forces have not deployed along clear territorial boundaries. It is also the case that support troops, such as logisticians, have been subject to similar risks as combat troops because the enemy operates everywhere; rear-area troops have been subject to close quarters engagement by insurgent snipers and gunmen and to the ubiquitous improvised explosive device (IED). Consequently, they have often had to fight like infantry soldiers.

The distinctive geography of the conflicts, and especially in Iraq and Afghanistan where the fighting has been most intense, has to be recognized. Nevertheless, while grand fronts have disappeared, it is an exaggeration to claim that no territorial distinctions could be drawn during these campaigns. On the contrary, while the geometry of the battle-space has become more complex; it is palpably false to claim that the combat zone is without distinction or gradation today. Counter-insurgency campaigns may not have fronts, as contiguous lines along which battle is drawn, but they do clearly involve controlled, contested, and enemy territories, each clearly demarcated. These areas appear as concentric rings or lozenges on operating maps, located in particular cities, districts, villages, or valleys. Soldiers, operating out of exposed patrol bases at the very edge of contested areas have been only too well-aware of a 'forward line of enemy troops'. In Iraq and Afghanistan, Western troops would typically be engaged by small-arms fire or IEDs on crossing that clearly defined line of control, often a distance of 200 metres from their patrol bases. In these contested areas, micro-fronts developed and Western forces and insurgents concentrated to fight for control of these spaces. These points, distributed at critical zones in theatres from West Africa to Afghanistan, constitute the new frontlines of the twenty-first century.

Indeed, soldiers and civilians in conflict zones today have been eminently aware of the existence of these micro-fronts. In the summer of 2010, NATO launched a major offensive, Operation Moshtarrak III, against insurgent positions in the Arghandab Valley in Panjwai and Zhari outside Kandahar City in Afghanistan. US forces fighting through enemy-held villages were engaged in intense and brutal combat for weeks during this period. Yet, less than ten miles south at Kandahar Airfield, soldiers and civilians could enjoy pizza and fresh coffee, provided by the Canadian franchise, Tim Hortons, on the Boardwalk, in almost total safety. Occasionally, they were forced to take cover as a result of rocket attacks, which caused a (very small) number of casualties, including a US female soldier who lost her legs when one struck the Boardwalk. Taliban fighters once made a suicidal assault on the main entrance. Yet the difference between the two zones was profound; Kandahar Airfield was plainly a rear-area, despised as such by combat troops, Panjwai, the front. The same observation would be equally valid in conflicts in Mali, the Central African Republic, Syria, Libya, or Somalia.

Indeed, notwithstanding their relative safety, the vulnerability of support troops in today's battle-space is in no way a new phenomenon. During the era of interstate wars, when grand fronts plainly existed, rear areas were continually subject to raids, artillery, and air attack. It is possibly true that support troops have had to be more proficient at close combat because the chances of them being directly engaged are higher than was the case for logistics units in interstate warfare, yet this difference does not seem to warrant the claim that there are no longer any fronts at all. Rather, because Western forces have often operated on exterior lines of communication, support troops have frequently had to run a gauntlet across a domain of semi-permissive or contested territory as they advance from controlled, safe areas to genuinely contested districts. The geography of combat has changed. However, micro-fronts, where most of the combat occurs, still exist as insurgents and Western troops struggle for control of critical areas. These are today's frontlines.

This volume explores the experience of soldiers at those points in contemporary warfare; it examines the frontline and the special conditions which pertain there in order to contribute to current debates about the armed forces and above all, the question of cohesion. Together, the chapters record and analyse the character of combat in the first two decades of the twenty-first century, where possible highlighting continuities and discontinuities between the today's frontline and the frontline of the twentieth century. It seems an opportune moment to conduct such an enterprise. The war in Afghanistan, which followed the terrorist attacks of 11 September 2011, has drawn to a close: for Western forces at least. By the end of 2014, conventional combat troops had left Afghanistan leaving only mentors, Special Operations Forces, and drones to continue military operations in this theatre. The war in Iraq seemed to have ended some time ago in 2011 when American ground troops finally withdrew. The recent assaults by ISIS have radically altered the situation and it is not inconceivable that Western and, above all, American combat troops will once again be deployed to that theatre. Nevertheless, a possible re-intervention into Iraq does not invalidate the attempt to take some account of Western military operations in the first fifteen years of the twenty-first century. In stark contrast to the conflicts of the 1990s, these wars have provided a rich and extensive archive of material about close quarters ground combat. Indeed, an evidential base of this quality has not existed since Vietnam. Consequently, an analytical retrospective on the last decade and more of conflict seems not only appropriate but necessary in order to begin to take some account of the plethora of new data.

This volume seeks to contribute to this process; it aims to analyse the question of combat and cohesion with a general but not exclusive focus on Iraq and Afghanistan. It explores the dynamics of today's micro-fronts, spreading rhizomatically across theatres and, perhaps, the globe, in order to examine the changing character of Western military forces from the perspective of history, the social sciences, and psychology. Specifically, it addresses the long-standing question of cohesion and combat performance. It explores how and why professional Western soldiers fought or have prepared to fight in the early twenty-first century.[1] Consequently, although many of the contributors to this volume are not sociologists, but historians or political scientists, by addressing the question of cohesion, this volume is avowedly sociological in orientation. The chapters, in very different ways, explore the problem of collective action. Each of the contributors seeks to highlight the distinctive ways in which the armed forces have engendered cohesion among their troops and encouraged performance from them.

Classical Cohesion

There is, of course, a long precedent for this endeavour which should be recognized from the outset and which has provided the framework for the volume as a whole and for each of the chapters. Following each of the major conflicts of the mid- to late-twentieth century, the Second World War, Korea, and Vietnam, social scientists, especially in the United States, sought to provide some systematic and objective analysis of the wars which they had experienced. There have been many dimensions to these interventions, but one of the most obviously recurring themes in terms of combat troops has been the question of cohesion. Cohesion has fascinated social scientists precisely because, as an extreme form of human activity, combat represents an almost ideal exemplification of the collective action problem which is so central to the social sciences and to social philosophy. The ability of soldiers to create and sustain group solidarity in combat is a specific example of the fundamental phenomenon of human social cooperation. Distinctively, in combat, soldiers are often willing to cooperate with each other even at great, even ultimate, personal risk to themselves. In this case, they contribute to the collective goods of their group even though they might never enjoy the benefit of those goods. Indeed, in order to prevail in battle, armies typically rely on the willingness of some individuals to die, so that the majority of soldiers may survive. It is a surprising fact that even in this situation, individuals are often still willing to cooperate and to contribute to group goals, the benefit of which they will never experience. Social scientists have rightly recognized that the central problem of cooperation and group cohesion is illustrated with particular clarity in combat. Combat serves a useful methodological function; it distills the collective action problem into a pure and brutal form. The frontline is a unique space. Yet it illustrates the general dynamics of group formation with a rare lucidity.

Of course, Janowitz and Shils' celebrated article on the Wehrmacht was decisive here; it was the first attempt to articulate the primary group thesis and remains a critical reference point to this day. Like many others during the Second World War, Janowitz and Shils were struck by the 'extraordinary tenacity of the German Army' which continued to fight well even when defeat was inevitable.[2] For Janowitz and Shils, the performance of the Wehrmacht was ultimately explicable only by reference to the cohesiveness of its primary groups: 'it is the main hypothesis of this paper, however, that the unity of the German Army was in fact sustained only to a very slight extent by the National Socialist political convictions of its members, and that more important in the motivation of the determined resistance of the German soldier was the steady satisfaction of certain *primary* personality demands afforded by the social organization of the army'.[3] Primary groups, consisting of small groups of soldiers of up to perhaps fifty individuals, were held together by bonds of comradeship produced by 'spatial proximity, the capacity for intimate communications, the provision of paternal protectiveness by NCOs and junior officers, and the gratification of certain personality needs, for example, manliness, by the military organization and its activities'.[4] Janowitz and Shils identified the interpersonal relations within the primary groups as critical to the fighting effectiveness of the Wehrmacht. As long as his primary group fulfilled the 'major primary needs' of the individual soldier, it possessed a 'leadership with which he could identify himself', and the soldier was able to give 'affection to and received affection from the other members of his squad and platoon', the German soldier would be willing to perform in combat.[5] The German soldier would continue to fight 'as long as he felt himself to be a member of his primary group and therefore bound by the expectations and demands of its other members'.[6]

In the following year, in a magisterial survey of US soldiers in the Second World War which was deeply important to the development of both sociology and social psychology as academic disciplines, Sam Stouffer also stressed the importance of the primary group in combat. Thus, he concluded that 'combat posed a challenge for a man to prove himself to himself and others'.[7] Masculinity was a key motivating factor used to encourage solidarity on the line and 'the man who lived up to the code of the combat soldier had proved his manhood'.[8] Bound by their masculinity, comrades would rather fight and die than be shamed. Like Janowitz and Shils, Stouffer et al. highlighted the importance of interpersonal relations between male soldiers to their subsequent performance on the battlefield. Following the Korean War, Roger Little extended the primary group approach, first outlined by Janowitz and Shils and Stouffer et al. Little explored the 'buddy system' within the US infantry platoon in order to establish the mechanisms by which cohesion was established.[9] Being a man in the buddy system was behaving in a buddy-like way, helping comrades, performing common duties, and sharing danger. Crucially, Little noted that soldiers believed that assisting their 'buddy' away from combat had highly beneficial effects on the battlefield: 'You don't always have a chance to do a favour in combat, but if you share everything, you can be pretty sure that your buddy will remember it if you need help.'[10] Buddies, who had demonstrated their commitment to each other, would not betray each other in combat. For Little, interpersonal masculine bonds between soldiers were critical to the cohesion and combat performance of the platoon.

Predictably, the Vietnam War engendered intense interest from the academic community in the question of warfare. There were a number of pieces published in the 1970s which directly addressed and extended the primary group thesis. Ronald Eisenhart provides a shocking account of how the US Marines—and by implication the Army—actually built men for the Vietnam War. He recorded a degrading system of brutalization in which violence was routine and every transgression was punished by a physical assault. The conscripts were constantly accused of being inadequate as men; they were women, girls, or homosexuals. 'You can't hack it, you goddamned faggot' was a favoured form of abuse.[11] Eisenhart's experiences were, perhaps, extreme. However, responding to Vietnam, other sociologists also emphasized the critical role of masculinity in uniting the US conscript forces. Gideon Aran and William Arkin and Lynne Dobrofsky produced significant pieces in prominent sociology journals which addressed the issue of cohesion. Arkin and Dobrofsky's piece most explicitly identified masculinity as central to solidarity in the armed forces. Closely echoing Eisenhart's work, they argued that the process of military training aims to turn 'the boy into a man'.[12] Aran's analysis of parachuting similarly emphasized the role of masculinity. In that piece, Aran described in Durkheimian fashion how military parachuting constituted a 'rites of passage' which affirmed social solidarity.[13] Paratroopers are part of a highly cohesive social group, but, as they launch themselves from the plane and in the brief moments as they hang beneath their parachutes, they are free of all social ties. They become individuals. For Aran, this affirmation of the individual ultimately united paratroopers even more closely once they had landed. Anomic individuality promoted mechanical solidarity because airborne units were filled only with individuals who had proved their bravery and self-discipline; 'parachuting is a sort of test' which 'helps the organization distinguish between the action seekers from those who are not'.[14] In each of these works, following the line that Janowitz and Shils had taken after the Second World War, the formation of masculine primary groups was taken to be essential to military performance. The interpersonal bonds between soldiers was regarded as critical to the cohesion they demonstrated or were likely to demonstrate in combat.

However, some of the most important post-Vietnam interventions seemed to be directly or implicitly opposed to the approach originally advocated by Janowitz and Shils. For instance, Charles Moskos was sceptical about how much weight could be put upon the concept of the primary group. 'Rather than viewing soldiers' primary groups as some kind of semi-mystical bond of comradeship, they can be better understood as pragmatic and situational responses.'[15] Against the sentimental view of cohesion as deep personal attraction, he has emphasized the temporary and contingent nature of the solidarity of the primary group in relation to his work on Vietnam. In the face of danger, it was rational for the individual members of combat units to unite in collective self-protection; 'primary group ties are best viewed as mandatory necessities arising from immediate life and death exigencies'. Moreover, 'if the individual soldier is realistically to improve his survival chances, he must necessarily develop and take part in primary-group relations'.[16] For Moskos, soldiers in Vietnam were rational individuals who cooperated with each other for their own calculated benefit; a soldier 'gets such support largely to the degree that he reciprocates'.[17] At the same time, and substantially contradicting his thesis of rational self-preservation, Charles Moskos also observed the appearance of deviant cohesion among black American soldiers in Vietnam who, developing their own subcultures, eventually owed 'higher fealty to the black community than to the United States Army'.[18] The intense bonds between these groups of black soldiers did not result, as the primary group thesis would have it, in excellent combat performance, but its opposite; disobedience, mutiny, and, ultimately, the murder of superiors.

The question of cohesion remained central to the post-Vietnam and post-Cold War literature in the 1980s and 1990s, when the darker aspects of the phenomenon began to be appreciated more fully. Charles Moskos was once again a major figure. Having reduced primary cohesion in the 1970s to rational self-interest, by the 1980s, Charles Moskos was ironically concerned that venal selfishness was undermining the newly professionalized, all-voluntary US military. Cohesion was evaporating. The armed forces had become 'institutional', with soldiers motivated only by their individual careers.[19] Similarly, Darryl Henderson feared the collapse of cohesion in the professional US armed forces, which he also ascribed to careerism.[20] However, perhaps the most important contribution to deviant cohesion literature in the last decades of the twentieth century was Donna Winslow's work on the Canadian Airborne in Somalia. Winslow has provided a vivid ethnographic account of how the hyper-masculine bonds between soldiers undermined the military effectiveness of the Canadian Airborne in Somalia in 1993 when members of the Regiment tortured and murdered a local. Winslow rightly identifies failings of command, training, and recruitment as well as alcohol abuse as factors. However, she also maintains that the distinctive cohesiveness based on an 'exaggerated masculine ethic' developed by the Canadian paratroopers provided a social context in which violence and torture might occur.[21] The intense interpersonal solidarity of paratroopers could not be doubted, but, in Somalia, airborne cohesiveness produced a deviant and criminal performance. Despite the recognition of the potential pathologies of the primary group, all this scholarship from Janowitz and Shils to Winslow was united by the presumption that interpersonal relations within these small groups had an overwhelmingly influence on military performance. The social cohesion of the primary group was critical to the cohesion which units displayed in combat. The primary group thesis continues to be advocated even up to the present.[22]

Beyond the Primary Group Thesis

There was then a clear research programme from the Second World War to the end of the twentieth century and beyond which explored how the social cohesion of the primary group contributed to (or undermined) combat performance. There is no denying the importance, richness, and validity of this literature. It remains a key reference point for the analysis of the armed forces today and is central to any attempt to understand the frontline experience of the twenty-first century. However, the primary group thesis did not go unchallenged. On the contrary, from the 1970s, a number of scholars criticized its failure to recognize political and ideological factors. Thus, following his research on Vietnam, Charles Moskos emphasized the role which political ideology played for US conscripts throughout the twentieth century: 'That the American soldier has a general aversion to overt ideological symbols and patriotic appeals should not obscure latent ideological factors which serve as preconditions in supporting the soldier to exert himself under dangerous conditions'.[23] For Moskos, the 'salient' values of this latent ideology were materialism and a belief in the 'worthwhileness of US society'.[24] For Moskos some commitment to worthwhile national goals was essential to the performance of citizen soldiers, as the collapse of the US Army in Vietnam demonstrated. He was not alone in emphasizing political factors. Explicitly rejecting Janowitz and Shils' primary group thesis, Stephen Wesbrook and Omer Bartov similarly argued that political motivation was essential to combat performance.[25] These scholars did not dismiss the primary group thesis entirely but maintained that it had to be augmented by political factors.

The question of political motivation remains a central concern today, as Alexander Watson's work shows.[26] However, in the last twenty years, an alternative cohesion paradigm has emerged which has reoriented the debate away from the interpersonal bonds of social cohesion towards practical military teamwork. In this way, scholars have explicitly sought to transcend the primary group thesis by emphasizing what has sometimes been called 'task cohesion'. Elizabeth Kier[27] and Robert MacCoun[28] were among the first in this emergent tradition. They argued that the classical literature was based on the presumption that citizen soldiers unite because of their mutual 'likeness'; they are socially *alike*, with homogeneous ethnic, racial, class, and gender backgrounds, and, therefore, they *like* one another. As a result of this 'likeness', deep bonds of interpersonal obligation develop between them; they are able to generate 'social cohesion', from which the special motivation, encouraging them to fight for each other, grows. MacCoun and Kier reject the primary group thesis outright, concluding that there is no necessary relationship between social cohesion, as a set of interpersonal relations, and subsequent military performance, especially in a professional force. By contrast, Kier prioritizes teamwork—or task cohesion—based on individual and collective competence over comradeship: 'the sense of group cohesion based on "teamwork" has little to do with whether members enjoy one another's company, share an emotional bond or feel part of some "brotherhood of soldiers"'.[29]

Although Kier does not discuss it, training was implicitly vital to the generation of her definition of task cohesion or teamwork. As long as teams trained together sufficiently, they could perform adequately no matter what their personal relations with each other.

More recently, a number of scholars have similarly sought to demonstrate the signal importance of training to the generation of cohesion in combat in order to displace the primary group thesis (Ben-Shalom et al. 2005; King 2006; Strachan 2006; Smith 2008; Ben-Ari et al. 2010; Coss 2010). Social scientists have increasingly observed that professional soldiers unite around quite impersonal procedures and drills, which they have learnt in training, whatever their social backgrounds. In his 2006 article on the cohesion debate, Hew Strachan argues that training is particularly interesting, since not only does it enable soldiers to perform together on the battlefield, but it also provides a self-referential form of motivation for them: 'Training creates the psychological capacity to elongate the peak phase [of combat effectiveness] and to surmount the low points of the later phases. It does so in part through the inculcation of battle drills, a set of procedures, so that when exhaustion makes rational thought impossible, or when fear has taken over, individuals react without thinking' (Strachan 2006: 217). With immediate relevance to the question of cohesion among the professional soldiers who fought in Iraq and Afghanistan, Strachan concludes by drawing a connection between the special cohesion generated by intense training and professionalism: 'A professional soldiers may pass his whole career without ever confronting it [combat]. But that does not relieve him of the need to prepare for it, nor erode the centrality of training for all he does' (Strachan 2006: 227).

Other scholars have similarly sought to explain the connection between training and professionalism and combat performance or cohesion (e.g. King 2006, 2007, 2011, 2013). Indeed, a year before Strachan's paper was published, Ben-Uzi et al. (2005) and, later, Ben-Ari et al. (2010) developed the concept of 'swift trust' to describe the alternative form of solidarity which emerges in the highly trained Israel Defense Forces. They observed how Israeli soldiers did not necessarily need to know each other or to have personal relations in order to cooperate; they could perform with each other by reference to common drills and adherence to shared doctrine. In this context, personal affection or interpersonal attraction, 'likeness', between males in dense primary groups, once regarded as essential to cohesion, became less relevant or even supererogatory. The social identity of soldiers and especially their social homogeneity—their 'likeness'—was less important than whether each one could fulfil their allotted roles, which they have been trained to perform (MacCoun et al. 2006).

The rise of professionalism has encouraged a profound reorientation of scholarship away from interpersonal relations towards military performance and teamwork; scholars are not so interested in what soldiers individually feel about their comrades (or profess to feeling in research interviews) but in what they collectively do together in training and on operations. Scholars, writing in this emergent tradition, are far more concerned with the details of battlefield performance and the practices, procedures, and drills which soldiers are trained to perform. The analysis of personal friendships has been partially displaced by an investigation of specific forms of practice. Moreover, scholars have been less impressed by appeals to apparently overarching factors like political ideology or the primary group and more interested in the specific social dynamics on the battlefield or within military units. Randall Collins' recent work on violence exemplifies this approach very well.[30] In that work, Collins dismisses various presumptions and myths about violence and seeks, instead, to demonstrate the social dynamics which recurrently structure violent interaction.

A practical or performative turn is observable in military scholarship in the last in two decades. It is worth noting that this approach is an expression of a longer current of scholarship which was often submerged in the second half of the twentieth century. Thus, it would be possible to argue that the work of William Cockerham and John Keegan in the 1970s constituted an early example of the practical task-cohesion approach. Against most of the literature emerging from Vietnam which focuses on the primary group, Cockerham, based on his experiences as a paratrooper in Vietnam, prioritized drills and training as the key variables in explaining combat performance. He accepted Moskos's arguments about the importance of latent ideology and comradeship to military performance but argued that 'theories of primary group relations and latent ideology are not in themselves all-inclusive explanations of combat motivation'.[31] Rather, he describes 'high level of identification with immediate superiors' and above all the 'strong value of teamwork' as critical to performance.[32] Cockerham maintained that in combat, soldiers were united through drills: 'one of the most efficient techniques which allows soldiers generally to adjust to combat is to ignore the danger by interpreting combat not as a threat to life but as a sequence of requirements to be met by an effective technical performance'; he records 'only knowing almost from rote what to do, can men carry out their tasks come what may'.[33]

He was not alone. In his seminal exploration of the experience of combat, John Keegan showed how situational processes could generate specific kinds of activity.[34] Thus, famously, Keegan observed that the flight of the Imperial Guard on the evening of Waterloo was not precipitated at the front of the column, even though troops in the front ranks were in the most danger; they were committed to the fight. Rather, the column stalled from the rear, precisely where soldiers could disengage themselves, seeing the effects of fire from a distance and, unlike their comrades, having the opportunity to retreat.

Indeed, it might be possible to trace this approach back to the Second World War itself. Although it has now become conventional to dismiss his work as an example of the primary group thesis and as methodologically unsound,[35] S.L.A. Marshall's work is better categorized as an early attempt to analyse combat performance. Marshall was in the unique position of having been a combat historian in the Second World War, Korea, and Vietnam. In each case, Marshall did not attribute combat performance primarily to the interpersonal relations between soldiers but to the specific dynamics within small units in combat. For instance, in his famous work on US combat soldiers in the Second World War, *Men against Fire*, S.L.A. Marshall explained the poor performance of American rifleman partly by reference to the morals of Western culture.[36] However, although Marshall regarded morality as a significant problem for the Western soldier, his fundamental explanation of non-firing was sociological; it was a situational product of the peculiar environment of the twentieth-century battlefield. The modern battlefield had become an empty space and, paradoxically, despite the increasingly large numbers of troops involved in ever larger encounters, the experience for the soldier was of isolation and loneliness.[37] In moral isolation, it was natural and understandable that riflemen froze and were unable to fire their weapons. Alone, they felt weak and exposed to the attention of the enemy. Instructively, Marshall noted the higher performance of those crews working machine guns, bazookas, or mortars, which he specifically related to the social context of their action. These soldiers were no longer isolated; they were part of a team, and the mutual support and pressure which crew members offered each other encouraged more active performance. Marshall provided a genuinely sociological explanation of the battlefield performance of US soldiers not in terms of their interpersonal social bonds but by reference to the actual small unit dynamics in combat.

The Historic Origins of the Performative Paradigm

The performative task-cohesion paradigm, then, can be traced back to the 1970s and, indeed, to the Second World War. However, throughout that period, it remained the subordinate approach to the study of cohesion; the primary group thesis, prioritizing social cohesion, remained dominant. There seems to be two reasons for the resurgence of the practical approach in the early twenty-first century: one military and one academic. In the first instance, the re-appearance of professional forces in the late twentieth century seems to have been a significant factor in encouraging this scholarly reorientation towards task cohesion. For most of the twentieth century, the dominant mode of military participation was conscription; most states relied on national service and mass citizen armies. In the latter part of the century, initially after the 1970s and especially following the end of the Cold War, Western militaries moved to the all-volunteer force. Anglophone countries took the lead here. Canada abolished conscription in 1945 immediately after the Second World War, while the United Kingdom introduced also all-volunteer force in 1960, the United States and Australia in 1973. Although European powers continued to retain conscription, they also reduced their national service requirements in the 1970s, to such a level that scholars could start to discuss the general end of the mass army.[38] After the end of the Cold War, all major European powers abandoned conscription and professionalized their forces, including, finally, Germany in 2010. All major Western power now field professional forces. The emergence of an academic paradigm which prioritizes training and skill as the primary explanatory variables just as Western forces themselves professionalized does not seem to be coincidental.

Yet there is an anomaly which needs to be explained here. All Anglophone forces (including bilingual Canada) have been professional since at least 1973 and, yet, until the 1990s and especially the last ten years, the scholarship (most of which was written in English) still prioritized the primary group. A time lag seems to be evident when the scholarship trailed behind the institutional conditions which it was nominally analysing. It is possible to explain this hysteresis effect. Although Anglophone forces have been all-volunteer for decades, they were not always particularly professional in terms of their performance. Indeed, in the 1970s and early 1980s, as Stanley McChrystal's recent autobiography illustrates, the American army was professional only in name; morale was still affected by Vietnam and training and equipment were often poor: 'Pride in craft was an elusive trait in the post-Vietnam Army.'[39] The observation of Charles Moskos and Darryl Henderson, that there was little cohesion in the American army at this time confirmed McChrystal's own experience. Although professionalized from 1973, the American Army only began to become *professional* towards the end of the 1980s and particularly during the 1990s, as the Revolution of Military Affairs began to be implemented.

The Canadian Army experienced a similar trajectory. Some commentators have observed that from the 1945 until 2000 Canada was a professional force in name only. Indeed, Jack Granatstein and John English have both argued that the abuses perpetrated by Canadian airborne soldiers in Somalia in 1993 were the product not so much of hyper-masculinity, as Winslow suggests, but of inadequate professionalism.[40] They reflected the poor standard of training, discipline, and morale in the Canadian army, which was, they claim, typical throughout the entire post-war period. The Canadian army was only nominally professional throughout this time. Only the pressure of operations in Iraq and especially Afghanistan since 2006 has transformed the professionalism of

Canada's forces. The British are a partial exception to this process, demonstrating reasonable levels of proficiency through the Cold War period. There seem to have been a number of institutional factors which explain their generally higher level of professionalism. Not only were the British forces professional for most of their history, but they were engaged in a series of counter-insurgency campaigns almost continually throughout the Cold War, including the long war in Northern Ireland. In addition, these campaigns were punctuated by more dramatic episodes such as the Suez Crisis and the Falklands War. These operations necessitated higher levels of professionalism than were displayed by Canada, which had no external commitments, or the United States which only professionalized in the 1970s. However, even in the British case, the severe reduction of the forces from the 1990s and the campaigns in Iraq and Afghanistan have intensified its professionalism.

This is a widespread experience. As a result of recent campaigns, Western forces have not only abandoned conscription and adopted the volunteer model but they have become far more competent, capable, and experienced than their Cold War predecessors (even if they were all-volunteer forces). They have professionalized in both senses of the word. This professionalism, advanced by the operational demands of Iraq and Afghanistan, has provided a context in which scholars have begun to reconsider the issue of cohesion and to question the primary group thesis. Precisely because training, preparation, and drills are so important to today's all-volunteer forces, scholars have similarly paid more attention to these activities as the principal explanation of cohesion in combat rather than to interpersonal relations. The scholarship has reflected the transformation of the forces themselves.

Yet other influences from within the academy do not seem to have been irrelevant here and, specifically, a change of paradigm in the social sciences which was observable from the 1960s. During the middle decades of the twentieth century and especially after the Second World War, Anglophone and Francophone social sciences were dominated by structuralist or functionalist traditions which prioritized objective social factors as central to their explanatory accounts; functionalists explained the working of the social system by reference to institutional needs and norms, while structuralists highlighted underlying cultural codes which patterned thought and action. In France, Claude Levi-Strauss was the prime representative of the latter approach, while Talcott Parsons dominated American sociology in the 1940s and 1950s.

There was a strong reaction to these approaches from the 1960s, substantially influenced by the later philosophy of Ludwig Wittgenstein. Famously, in the *Philosophical Investigations*, Wittgenstein had reacted against the linguistic philosophy of the Vienna School and other analytical traditions, including his own earlier work, to argue that language was not to be comprehended in purely logical terms but rather as a practice. Against the a priori analysis of how language ought to work, he advocated the examination of particular types of usage; 'Don't think, but look'.[41] For Wittgenstein, the apparently mysterious character of language became explicable only by looking at its application in particular 'forms of life'.[42] The social context, the *practice*, not prior linguistic structures, explained how words gained their meanings and could be used to do things.

Accordingly, during the 1960s and into the 1970s, a series of research programmes emerged in the social sciences which reflected this new interest in practice; Erving Goffman's interactionism, Harold Garfinkel's ethnomethodology, Clifford Geertz's interpretative anthropology, Pierre Bourdieu's practical theory, and Anthony Giddens' structuration theory might all be seen in different and complex ways as part of this process. In history, the Annales School, especially Emanuel Le Roy Ladurie and Philippe Ariès, and the work of E.P. Thompson, might be seen as part of this reorientation towards everyday social practice as a critical domain of investigation. Natalie Zemon Davis's celebrated paper 'The Rites of Violence' is a particularly pertinent example of this approach.[43] In that piece, she showed that much of the mob violence during the Wars of Religion in France in the late sixteenth century was not the structural effect of socio-economic conditions, as social historians tended to presume at that time. On the contrary, the practice of violence was patterned by two critical dynamics. Firstly, it arose at heightened ritualistic moments when large numbers of Protestants or Catholics were already gathered to celebrate religious festivals; the crowd was already sensitized to its own collective identity and its power. Secondly, the practice of violence was informed by the collective understandings of these religious sects: Catholic crowds burned the corpses of their victims or disposed of them in rivers in acts of ritual purification, while Protestants targeted priests, friars, and monks and desecrated Catholic relics, buildings, and images in order to demonstrate their disdain for superstition and its purveyors. The paradigm shift from social to task cohesion or from interpersonal relations to drills and training reflects fundamental changes in both the armed forces and the academy in the late twentieth and early twenty-first centuries.

Historical Cohesion

In the last twenty years, scholars have begun to consider the details of military performance and the actual dynamics of combat itself, partially because professional armies themselves emphasize training and drills. Ironically, this attention to training, professionalism, and performance—to the interactional realities of the battlefield—increasingly evident in a contemporary literature, represents something of a return to the past and specifically to the military literature of the early modern era. Almost entirely ignored by social scientists, military writers from the fifteenth to nineteenth centuries focused on specific tactical formations, drills, and particular combat situations to develop an understanding of how cohesion might be encouraged on the battlefield. The fact that in both early modern and contemporary eras, Western states have fought with small professional armies does not seem to be coincidental here. Precisely because early modern armies—like today's all-volunteer forces—were small and had the opportunity and requirement to train extensively, technical questions of formation, tactics, and drills were highly pertinent.

Of course, one of the earliest examples of what might be called a practical orientation to the question of cohesion is to be found in Niccolò Machiavelli's work. In *The Prince* and *The Art of War*, Machiavelli disdained mercenaries who, fighting for money, refused to endanger themselves for the state. He blamed them for the catastrophes which Florence and Italy generally suffered in the early sixteenth century when Charles V invaded. Machiavelli idealized the citizen soldier who, together with his comrades, fought to protect civic freedom. Machiavelli's citizen soldier was the embodiment of civic virtue, and consequently, seems to have resonances with Janowitz and Shils' and Stouffer's ideal soldier. Yet Machiavelli in no way believed that the civic virtue of the soldier was in itself adequate to guarantee military performance. On the contrary, most of *The Art of War* is involved in a technical discussion of the specific formations which the army of the city-state should optimally assume in battle. In Book III, Machiavelli advocated a mixture of pikes and muskets which imitated Roman manipular tactics in order to generate high levels of performance in combat.[44] For Machiavelli, combat performance was not reducible to virtue—morale—alone; tactical doctrine was critical in sustaining cohesion among troops in the line of battle. He was deeply sensitive to the fact that different formations generated alternative performances from troops.

Machiavelli was in no way unique in relating combat performance to specific tactical formations and practices. On the contrary, seventeenth- and eighteenth-century practitioners and commentators were eminently aware of the great influence which tactical formation exerted on battlefield cohesion and combat performance; the works of Condé, Turenne, Montecuculli, Chevalier Folard, Maizeroy, du Teil, Pirch, Puysegar, Mesnil-Durand, Guibert, and Bourcet speak directly to the question of military practice and specifically investigate how tactical formations could maximize firepower, thereby engendering confidence and solidarity among the troops. Indeed, the question of tactical formation generated a huge literature, especially in France after its humiliations during the Seven Years War.[45] In *Reveries on War*, Marshall de Saxe explicitly highlighted the ways in which different tactical formations affected the morale of the troops within them. De Saxe expressed his admiration for the career and writings of Chevalier Folard, but he profoundly disagreed with an essential element in Folard's thought: 'he supposes all men to be brave at all times and does not realize that the courage of the troops must be reborn daily, that nothing is so variable, and that the true skill of a general consists in knowing how to guarantee it by his dispositions, his positions, and those traits of genius that characterize great captains'.[46] It is easy to overlook the significance of this statement as it appears merely to replace Folard's assertion about the enduring individual bravery of soldiers with its opposite; that individual bravery is uncertain and has to be actively generated by a general. For De Saxe, the performance of an army was not dependent on the individual bravery of its soldiers but their collective solidarity, which he termed 'the human heart'.[47] Crucially, that solidarity—or heart—is not just an aggregate of their individual dispositions. Rather, it is generated by the specific disposition of military units as they deployed for battle. Generals could actually increase the bravery of their troops not by addressing individual psychology but rather by deploying their troops in formations which engender collective confidence. Tactical and psychological dispositions were interrelated.

De Saxe gives a number of examples of the way in which combat formation explains the collective behaviour of troops. Thus, during the pursuit at Friedelingen in 1702, the panic of victorious French troops, who had only moments before shown 'incomparable valour', occurred because, having defeated an enemy in ordered formation, they had disintegrated into a mob.[48] When troops imagined they were being counter-attacked, out of formation, they felt utterly exposed; consequently, soldiers who had shown enormous bravery in formation in the initial assault, fled ignominiously in 'frightful disorganization'.[49] The courage and cowardice of these French troops was not individual but was dependent upon the situational context. By organizing soldiers in particular formations and drilling them thoroughly, it was possible for generals to embolden their troops, irrespective of whether they were individually cowardly or heroic.

De Saxe only began an argument in the French Army. From the 1750s to the beginning of the Napoleonic wars there was a prodigious literature on tactics, focusing on the issue of whether the French army should adopt *ordre mince* or *profonde* (thin or deep order, i.e., line or column)—or a mixture of the two.[50] One of the key figures here was Comte Jacques de Guibert, who sought to lay out a science of tactics where pure mass was replaced with the skilful use of military formations.[51] Guibert, for instance, noted that in the seventeenth century when modern tactics were still in their 'infancy', 'one continued to believe that the force of the infantry consisted in the density of its order and in its impetus'.[52] Guibert's chief opponent was Jean de Mesnil-Durand, who advocated a complex order of battle, relying primarily on dense columns of infantry. Each of these writers recognized that alternative combat formations encouraged or undermined solidarity, decisively influencing soldiers' performance in battle. The column was more manoeuvrable than the line and also, crucially, generated confidence and morale; surrounded by large numbers of other soldiers all in close proximity with each other, individuals were emboldened in the column. Yet the advantages of morale and manoeuvrability were vitiated by the column's vulnerability to fire and its own inability to generate firepower. Guibert rejected Mesnil-Durand precisely because so few of the soldiers in his columns could bring their weapons to bear:

> let us begin by destroying the old prejudice, according to which it is believed that to increase the force of a unit, one augments its depth. All physical laws about the movement and shock of bodies become illusions, when one wants to adapt them to tactics. For firstly, a unit cannot be compared to a mass, since it is not a compact body without interstices. Secondly, in a troop which approaches the enemy, it is only the men in the rank which closes with it which have the power of shock. All those, who are behind them and cannot close up and unite with the compactness and pressure which exists in a physical body, are useless and often only cause disorder and tumult.[53]

The thin line, by contrast, was potentially more fragile, but, ironically, precisely because the soldiers in it were manifestly responsible for their own protection, the line might engender higher levels of performance, especially since soldiers could fire more easily from it. Soldiers in the line would be more active firers not only because they could be but also because, in this formation, they had to be. In the end, Guibert recommended an *ordre mixte*, which for him combined the morale effects of the column with the firepower of the line.

In the nineteenth century in a different tactical context, Ardant du Picq took up many of these concerns. He was interested in showing how the situational dynamics of combat, in which tactics and formation played a crucial role, were decisive to performance. The word 'morale' in du Picq's work closely resembles the concept of 'the heart' in de Saxe's work and might be understood as 'cohesion'. It refers to the willingness and ability of soldiers to fight together. For Du Picq, like de Saxe, military formations were critical in engendering morale, and he noted situations when the morale of troops collapsed. In particular, when soldiers were no longer in formations which put them under the physical observation of their peers, their natural fears overtook them: 'The bewildered men, even the officers, have no longer the eyes of their comrades or of their commanders upon them, sustaining them. Self-esteem [i.e., recognition by one's peers] no longer impels them.'[54] As a result of the importance which du Picq invested in mutual social pressure, he was a fierce proponent of the bayonet charge (despite its inevitable casualties) precisely because individuals were always under the observation of their comrades. At the same time, the bayonet charge demoralized defenders not primarily because they feared the wounds the bayonet inflicted (bullets and grenades caused worse injuries) but because a force charging with bayonets demonstrate its unshakable collective resolve; they would close and kill together as a unit. In the light of this display of ineluctable collective determination, Du Picq plausibly claimed that no force ever withstood a bayonet charge.

Clearly, none of these military writers could be defined as a sociologist; their concerns were entirely practical, and sociology emerged as an academic discipline only much later in the middle decades of the nineteenth century. Yet, in highlighting the way in which the physical disposition of soldiers in military formations engendered different kinds of solidarity among them to have a decisive effect on their battlefield performance, these writers offered sociological insights into combat. Like Emile Durkheim, they tried to show how individual and collective action was a product of the specific solidarity of the social milieu. Like him, they sought to demonstrate an internal connection between social and physical disposition and individual and collective psychological disposition. Interestingly, this is precisely what contemporary scholars are seeking to do when they explain combat performance by reference to particular military practices and specific social dynamics. The appearance of a new cohesion paradigm in the twenty-first century represents, then, a return to the past; facilitated by the re-emergence of the professional force.

Conclusion

Contemporary scholars, who have prioritized training, drills, and situational dynamics to explain cohesion in combat, are part of a very long tradition of military science and a more recent but still well-established 'practical' intellectual paradigm. Scholars working within this framework seek to situate military performance within its specific historical and micro-context, highlighting the organizational and social dynamics which generate particular kinds of collective action. The current volume consciously locates itself in this emergent tradition. It seeks to understand combat performance and cohesion by examining the specific practices of today's professional soldiers within the historical and social context in which they occurred; it examines their training, their drills, and specific examples of their performance on operations and in combat. At the same time, and highlighting its divide from classical primary group theory of the twentieth century, the volume seeks to explore the differences—and continuities—between the citizen armies of the twentieth century and the professional force of the twenty-first century. To this end, contributors are drawn from history, sociology, political science, and psychology. Focusing on a variety of case studies including the First World War, the French Foreign Legion, the Italian Alpini, Sikh regiments in the Indian Army, the Australian, British, and American Armies in Afghanistan, and the post-operational effects of combat, the volume is diverse. However, united by an interest in cohesion, it seeks to provide both a resource for understanding contemporary issues about combat and advancing the performative paradigm promoted here in a historical and comparative framework. Above all, the volume, like Janowitz and Shils' before it, is a response to and recognition of the profound importance which the conflicts that Western forces have fought since 2000 are likely to have in the coming decades. No one can know where or when Western forces will have to fight again in the future, but it is important to provide some audit of what combat has been like for those involved in it in the last decade and a half. This is what this volume seeks to do.

Notes

1. One chapter in the volume considers the case of the Israel Defense Forces. They are included because, notwithstanding the controversies surrounding some of their activities, they represent one of the most professional and operationally experienced forces in the world today and they are closely allied to the United States. Accordingly, they typically fight like a professional Western force.
2. Morris Janowitz and Edward Shils, 'Cohesion and Disintegration in the Wehrmacht in World War II', *Public Opinion Quarterly* (Summer 1948): 281.
3. Janowitz and Shils, 'Cohesion and Disintegration in the Wehrmacht in World War II', 281.
4. Janowitz and Shils, 'Cohesion and Disintegration in the Wehrmacht in World War II', 315.
5. Janowitz and Shils, 'Cohesion and Disintegration in the Wehrmacht in World War II', 284.
6. Janowitz and Shils, 'Cohesion and Disintegration in the Wehrmacht in World War II', 284.
7. Samuel Stouffer, Arthur Lumsdaine, Marion Harper Lumsdaine, Robin Williams, M. Brewster Smith, Irving Janis, Shirley Star, and Leonard Cottrell, *The American Soldier*, vol. 2: *Combat and Its Aftermath* (Princeton, NJ: Princeton University Press, 1949), 131, 134.
8. Stouffer et al., *The American Soldier*, 134.
9. Roger Little, 'Buddy Relations and Combat Performance', in Morris Janowitz (ed.) *The New Military: Changing Patterns of Organization* (New York: Russell Sage Foundation, 1964), 195–224.
10. Little, 'Buddy Relations and Combat Performance', 217.
11. Ronald Eisenhart, 'You can't Hack it Little Girl: A Discussion of the Covert Psychological Agenda of Modern Combat Training', *Journal of Social Issues* 31(4) (1975): 15.
12. William Arkin and Lynne Dobrofksy, 'Military Socialization and Masculinity', *Journal of Social Issues* 34(1) (1978): 154.

13. Arkin and Dobrofksy, 'Military Socialization and Masculinity', 150.
14. Arkin and Dobrofksy, 'Military Socialization and Masculinity', 148.
15. Charles Moskos, 'The American Combat Soldier in Vietnam', *Journal of Social Issues* 31(4) (1978): 37.
16. Charles Moskos, *The American Enlisted Man. The Rank and File in Today's Military.* (New York: Russell Sage Foundation, 1970), 145.
17. Moskos, *The American Enlisted Man*, 145.
18. Moskos, *The American Enlisted Man*, 128.
19. Charles Moskos, *Soldiers and Sociology* (United States Army Research Institute for Behaviour and Social Sciences, 1988).
20. Darryl Henderson, *Cohesion: the Human Element* (Washington, DC: National Defence University Press, 1985).
21. Donna Winslow, *The Canadian Airborne Regiment: A Socio-Cultural Inquiry* (Ottawa, Canada: Minister of Public Works and Government Services, 1997), 123–4.
22. Guy Siebold, 'The Essence of Military Cohesion', *Armed Forces and Society* 33(2) (2007): 286–95; Leonard Wong, Thomas Kolditz, Raymond Millen, and Terrence Potter, *Why They Fight: Combat Motivation in the Iraq War* (Carlisle, PA: Strategic Studies Institute, 2003).
23. Charles Moskos, *The American Enlisted Man. The Rank and File in Today's Military* (New York: Russell Sage Foundation, 1970), 155.
24. Moskos, *The American Enlisted Man*, 152, 155.
25. Omer Bartov, *Hitler's Army* (Oxford: Oxford University Press, 1992); Omer Bartov, 'The Conduct of War: Soldiers and the Barbarization of Warfare', *Journal of Modern History* 64, Supplement (December 1992): S32–45; Stephen Wesbrook, 'The Potential for Military Disintegration', in S. Sarkesian (ed.) *Combat Effectiveness and Cohesion* (London: Sage, 1980).
26. Alexander Watson, 'Fighting for Another Fatherland: The Polish Minority in the German Army, 1914–18', *English Historical Review* CXXVI No. 522 (October 2011): 1137–66.
27. Elizabeth Kier, 'Homosexuality in the US Military: Open Integration and Combat Effectiveness', *International Security* 23(2) (1998): 19.
28. Robert MacCoun. 'What is Known About Unit Cohesion and Military Performance' in National Defence Research Institute *Sexual Orientation and US Military Personnel Policy: Options and Assessment* (Washington, DC: RAND, 1993); Robert MacCoun, Elizabeth Kier and Aaron Belkin. 'Does Social Cohesion Determine Motivation in Combat? An Old Question with an Old Answer', *Armed Forces and Society* 32(4) (July 2006): 646–54; MacCoun, Robert and Hix, William. "Cohesion and performance', in *Sexual Orientation and U.S. Military Policy: An Update of RAND's 1993 Study*, (Santa Monica: RAND, 2010).
29. Kier, 'Homosexuality in the US Military', 19.
30. Randall Collins, *Violence: A Microsociological Theory* (Princeton, NJ: Princeton University Press, 2008).
31. William Cockerham, 'Attitudes Towards Combat Among US Army Paratroopers', *Journal of Political and Military Sociology* 6 (Spring 1978): 12.
32. Cockerham, 'Attitudes Towards Combat among US Army Paratroopers', 13.

33. Cockerham, 'Attitudes Towards Combat Among US Army Paratroopers', 13.
34. John Keegan, *The Face of Battle* (London: Penguin, 1976).
35. For a full discussion of the Marshall debate and a rehabilitation of this work, see Anthony King, *The Combat Soldier: Infantry Tactics and Cohesion in the Twentieth and Twenty-First Centuries* (Oxford: Oxford University Press, 2013), Chapter 3.
36. Dave Grossman, *On Killing: The Psychological Costs of Learning to Kill in War and Society* (Boston, MA: Back Bay Books, 1996).
37. Marshall, *Bird* (New York: Warner Books, 1968), 44–7.
38. Michel Martin, 'Conscription and the Decline of the Mass Army in France, 1960–75', *Armed Forces & Society* 3(3) (1977): 355–406; Catherine Kelleher, 'Mass Armies in the 1970s: The Debate in Western Europe', *Armed Forces & Society* 5(1) (1978): 3–30.
39. Stanley McChrystal, *My Share of the Task* (London: Penguin, 2014), 29.
40. Jack Granatstein, *Canada's Army: Waging War and Keeping the Peace* (Toronto: University of Toronto Press, 2004), 406; John English, *Lament for an Army* (Toronto: Canadian Institute of International Affairs, 1998).
41. Ludwig Wittgenstein, *Philosophical Investigations*, trans. G. Anscombe (Oxford: Blackwell, 1976), §66.
42. Wittgenstein, *Philosophical Investigations*, 226.
43. Natalie Zemon Davis, 'Rites of Violence: Religious Riot in Sixteenth Century France', *Past and Present* 59 (May 1973): 51–91.
44. Niccolò Machiavelli, *The Art of War*, trans. Neal Wood (New York: Bobbs Merrill, 1965), 83–92.
45. Christopher Duffy, *The Military Experience in the Age of Reason* (London: Routledge and Kegan Paul, 1987); Spenser Wilkinson, *The French Army before Napoleon* (Oxford: Clarendon Press, 1915); Jean Colin, *L'Éducation Militaire de Napoleon* (Paris: Librairie Militaire R. Capelet et Cie, 1900).
46. Maurice du Saxe, *Reveries on the Art of War*, trans. Thomas Phillips (New York: Dover Publications, 2007), 18.
47. du Saxe, *Reveries on the Art of War*, 18.
48. du Saxe, *Reveries on the Art of War*, 18.
49. du Saxe, *Reveries on the Art of War*, 19.
50. Jean Colin, *The Transformation of War*, trans. L. Pope-Hennessy (London; Hugh Rees, 1912); Jean Colin, *L'Infanterie au XVIIIe Siecle: la tactique* (Paris, Berger-Levrault, 1907); Robert Quimby, *The Background of Napoleonic Warfare: the theory of Military Tactics in 18th Century France* (New York: Columbia University Press, 1957).
51. Comte Jacques de Guibert, *Essai Général de Tactique* (Paris: Economica, 2004).
52. de Guibert, *Essai Général de Tactique*, 29.
53. de Guibert, *Essai Général de Tactique*, 71.
54. Ardant du Picq, *Battle Studies*, trans. J. Greely and R. Cotton (Charleston, SC: Bibliobazaar, 2006), 136.

2

Subaltern Soldiers

Eurocentricism and the Nation-State in the Combat Motivation Debates

Tarak Barkawi

> "Since the military profession is an expression of national identity, only those with strong feelings of nationalism tend to enter."
>
> Morris Janowitz[1]

> "[M]any battles and campaigns have been won by men who had little idea of why they were fighting, and, perhaps cared less."
>
> Lord Wavell[2]

Morris Janowitz was an American sociologist who served in the US Army and helped found the modern discipline of military sociology. Archibald Wavell, among other things, was the penultimate Viceroy of British India and one of the last British Commanders-in-Chief of the Indian Army. Janowitz spent the Second World War in Washington, DC, London, and France, where he interviewed German PoWs. Wavell spent his war commanding the diverse and multicultural legions of the British Empire in North Africa, the Middle East, and India before taking up his viceroyalty in 1943. While in France 1944–45, Janowitz perhaps took notice of the French West African troops in the Free French forces, or heard of the Indian soldiers and PoWs the Germans had, or was impressed with other foreign units of the Wehrmacht and *Waffen SS*.[3] However that may be, such colonial and foreign soldiers became more or less invisible in military sociology and have not informed its central ideas.

The distance between Janowitz's chair in Chicago and Wavell's office in Delhi helps identify two circular presuppositions at work in the debates over combat motivation. The first concerns Eurocentricm. It is mainly Western armies that are studied, and which have become the basis for general theories and concepts. The *Wehrmacht*, for example, is the site of paradigmatic studies.[4] The second is that armies and soldiers are seen as essentially *national* in character, serving in a world of *nation-states*.

With small exceptions, inquiry into and debate over combat motivation, cohesion, fighting spirit, and the like have concerned the national armed forces of Western states. Wherever one stands in the debates over combat motivation, the object of analysis is usually a national army, and most likely American, British, or German. The nature, character, and course of these debates, which began in earnest during and after the Second World War, were fundamentally shaped by the Western and national contexts in which both the researchers and the armed forces they studied were located.[5] When problems were identified with primary group theory, broadly the idea that soldiers fight 'for their buddies', scholars turned to national society, and to notions of citizenship, regime type, ideology, and culture or ethnicity for additional variables and new approaches.[6] The conceptual vocabulary and historical materials with which we study and think about soldiers and war are drawn nearly entirely from Western political modernity.

In Chapter 1, Anthony King shows how Janowitz and Shils contrasted an idealized image of the American citizen army and civil society with the Wehrmacht and its social context. This kind of problem is more widespread. There is a persistent essentialism evident in much military scholarship which attributes primary and enduring causal powers to national culture. The action of soldiers on the battlefield is to be explained by a deeply inscribed ethnic or national identity which determines what they do in various situations.

This chapter seeks to show how this Eurocentrism and national essentialism have shaped the course of debate and foreclosed analytic possibilities. This chapter seeks to develop King's 'practical' paradigm. On this account, distinctive national identities or policies are not dismissed as explanatory factors. Soldiers regularly appeal to them to engender solidarity; they often become self-referentially important. Appeals to apparently enduring concepts of national identity or culture are, ironically, responses to particular situations and specific contexts. Interestingly, irrespective of the cultural origins of soldiers, recurrent patterns of action are observable in war. In similar situations, soldiers, as human beings, react in commensurable ways. This chapter seeks to identify some of these dynamics and to encourage scholars to reject essentialist accounts of soldiers' identities, whether these are based on nation, ethnicity, or regiment.

In order to do so, it is helpful to think about foreign and subaltern soldiers, soldiers who fall outside the world of nation-states and national armies, such as those in the great colonial militaries of France and Britain. These played key roles in the world wars, could be found fighting into the 1960s, and became the armed forces of their own newly independent states.[7] The point in an essay such as this is not to recover the stories of brown and black soldiers. It is rather to use them to think about basic concepts, categories and approaches, to critique the usual ways of seeing an issue, and to open up new questions and perspectives that can in creative ways advance the combat motivation debates. What makes colonial soldiers theoretically significant is not that they are representatives of the 'East' or are somehow less modern than Western soldiers. It is rather that we are forced to confront the implication of categories like 'West' and modern in our understanding of who soldiers are and what they do, to rethink the presumed relations between terms like soldier, nation, and citizen, or between military discipline and modernity.

The discussion begins with the implication of the nation in the study of combat motivation. It then turns to the British Indian Army's Sikhs. The notion of a Sikh Martial Race was a colonial construct, not an essential property of those identified as Sikhs. This critique of national or ethnic difference in explanations of military behaviour is pushed further by a look at the idea of national differences among PoWs. The chapter concludes by returning to the ground zero of the combat motivation debates, the *Wehrmacht*, and to the relations between nation, modernity, and atrocity. Here again, attention to foreign troops highlights commonalities rather than national differences among soldiers. Military performance is explained more effectively by reference to the specific institutional and operational situations in which soldiers find themselves.

Military Sociology and the Combat Motivation Debates

The nation-state was right there at the origins of military sociology. During the Second World War, social scientists in the US were drafted into the national war effort, often serving in uniform and later using their wartime research as the basis for paradigmatic publications, as with Samuel Stouffer, and Janowitz and his colleague Edward Shils. The dominant role of the state and the military in commissioning and controlling research in military sociology continued after 1945, especially in the US, even as many researchers came to be based in universities.[8] Even for independent researchers, the widespread use of survey and interview methods required access to serving soldiers. Attention was directed towards contemporary problems occupying political attention, and towards national militaries. Key postwar publications with general titles, such as Janowitz's *The Professional Soldier* or Samuel Huntington's *The Soldier and the State*, were in fact concerned almost entirely with the US armed forces.

As military sociology became institutionalized in Western academies, the cross-national comparison became one standard form of inquiry and volumes with chapters or case studies on national militaries appeared.[9] Questions centred on similarity and difference between the armed forces of various countries. The military appeared always already as a national institution. Scholarly debate tended to track matters of national interest in the leading Western states, most particularly the US, concerning, for example, the integration of African Americans and of gays and lesbians into the armed forces.[10] In various ways, armies were reified in national terms, conceived within separate national containers.

In military sociology, and in much military history, an unhelpful framework governs thinking about 'military' and 'society'. They are conceived as separate but isomorphic domains, existing prior to one another and exercising independent causal force. Scholars can take positions on what better explains the variation in fighting spirit or combat behaviour: army *or* society. Is it what happens to people after they join the military, or is it the cultures, identities, and ideologies recruits carry with them from civilian society into the military that are decisive? Armed forces and society are isomorphic in that state, army, and society come in a nation-state package, coeval with sovereign territory, and with conjoined but analytically distinct histories over time. An explanatory compass is created which varies between national society and national military. Note that there is an 'ethnic' variant to this way of thinking. An armed force is seen as reflective of the people from which it is recruited, and one can look to the culture and society concerned to explain how this force behaves and fights.[11] Here the idea of a territorially based fighting people is retained even if such an entity is not formally speaking a sovereign and national state. The world is conceived as consisting of fighting polities and peoples, packages of more or less bounded societies and their armed forces. What armies and soldiers might *share* across national borders is analytically devalued in this set up.

The combat motivation debates began with a strong emphasis on the internal qualities of national military organizations, principally in terms of how they created and sustained 'primary groups'. Later, attention turned towards 'societal' phenomena such as nationalism, culture, and ideology. In the main, most military sociologists continued to emphasize the internal qualities of militaries, but usually in some combination with elements drawn from national society.[12] One reason for this is that 'primary group theory'—the principle basis of the organizational perspective—has a fatal ambivalence, an indeterminancy, when used as the main explanation for fighting spirit.

If soldiers 'fight for their buddies' in conditions of mortal danger, why would they vigorously follow orders placing comrades in harm's way in the first place? To be sure, once in combat, mutual protection is often best accomplished by fighting effectively together. But in many situations it is more directly achieved by avoiding danger. Moreover, disobedience and indiscipline often require the very cohesion provided by primary groups. An example is a patrol that finds a safe place to lie up and returns with false reports. Soldiers participate together in the deception and trust the secret will be kept. 'Fragging', the murder or attempted murder of officers considered dangerous by their own troops either for their offensive spirit or incompetence, also can require collusion among soldiers. Indeed, few things necessitate as much cohesion as mutiny, the collective willingness to defy orders. The group solidarity championed by primary group theory as the source of fighting spirit in fact cuts both ways. It can be the basis of discipline or indiscipline.[13] Nationalism, in motivating soldiers and pointing them towards the enemy, all too easily seemed the answer to this theoretic quandary.

Given the national policy science focus of military sociology, as well as the dominance of the American academy, it is not surprising that the Vietnam War became a significant occasion for the further critique and development of primary group theory. The morale problems in the US military and the contrasting performance of the opposing Vietnamese forces turned researchers' attentions to the social and political context of the war.[14] Ideology, nationalism, and cultural homogeneity seemed to play an important role in the success of Vietnamese communist forces.[15] At the same time, societal factors such as racial tension, drug use, and the political unpopularity of the war at home eventually impacted on discipline in the US military.[16] Military sociologists began to supplement primary group theory with qualities recruits brought with them from civilian society, especially beliefs and attitudes conducive, or not, to military service and obedience.[17]

A distinction was drawn between the 'coercive' discipline of pre-twentieth-century armies and the 'normative' discipline of twentieth-century armies, a version of Ardant du Picq's contrast between ancient and the modern battle.[18] As Janowitz put it, 'the integration of small groups into the larger military organization in the contemporary period required more explicit symbolic rationalization'.[19] His colleague Stephen Wesbrook spoke of the necessity of soldiers' 'moral involvement with a larger collectivity'. This depended on a soldier's 'sense of national identity, his belief that the sociopolitical system is meeting the basic needs of most members of society, and his acceptance of national ideology'.[20] Modern soldiers, apparently, are motivated by ideas, even amid the intense stress and fear of battle and the boredom of everyday service.

If military sociologists, social scientists concerned with general patterns, found themselves thinking about specific national ideologies and cultures, military historians were even more likely to do so. '[T]he predominance of studies focusing on only one army and society has sometimes led historians to overestimate the distinctiveness of their subjects.'[21] The use of memoirs, documents, and other first-hand accounts to write histories tends to privilege national interpretations. For example, in an important study, Craig Cameron writes of US marines that they drew on 'characteristically American perceptions and interpretations of the world'.[22] Given the kind of materials Cameron was working with, accounts of battle by US marines, this observation is not surprising. US marines inevitably experience and narrate their worlds in and through their particular historic contexts, in the idiom of their time and place. But what does this mean in terms of explanation, of why the marines did what they did, and came to think in the ways they did about their actions at the time and later? Since soldiers from different societies 'speak' in different languages, cultural and national difference will almost always be emphasized in studies of this kind.

Cameron raises the issue of parallels between his Second World War marines and other forces fighting in racially charged contexts, such as Wehrmacht and *SS* formations on the Eastern Front or the Japanese the marines themselves faced. In comparing the marines and the Japanese on Guadalcanal, Cameron observes that the 'marines did exhibit the same savage responses in battle as their enemy'. Indeed, 'the way in which the cultural values of the enemy, so often portrayed as alien to the West, were matched among the marines is striking'.[23] But he ultimately concludes that while '[p]arallels with other countries abound . . . for all their similarities, the marines were not the *same* as the German military elites on the Eastern Front or the Japanese on the other side of the lines'.[24] But is 'sameness' an appropriate or relevant criteria? German, Japanese, and American soldiers are necessarily shaped by different civilian and military contexts and will construct and describe their experiences accordingly. Given these differences, it is more surprising that there are striking parallels at all in their battlefield performance. Armies and battlefields pressure soldiers to think and act in similar ways despite their differing origins. Gwynne Dyer puts it this way: 'The government and society of the United States are greatly different from those of the Soviet Union or Egypt, but their armies are so close in structure and in spirit that their officers, when they come into contact, cannot help recognizing their common membership in a single, universal profession whose characteristics everywhere are shaped by the demands of battle.'[25]

Culture and Military Discipline: Sikhs, Chinese, and Humans

Dyer's point is not that soldiers from different armies are the same, or that American, Soviet, or Egyptian soldiers are not influenced by their cultural backgrounds. He can be read as suggesting that the really important cultural context is a *military* one shaped by the problem of battle and the regulative ideals of fighting in organized warfare. Moreover, this context is transnational as well as, on Dyer's rendering, historically recurrent in character (rather than specifically Western or modern). Any relations between the military and 'society' may be far more complex than a nation-state perspective allows. Soldiering is best understood not as the expression of essentialist national identity and culture but as something practised by humans in many different times and places, in response to commensurable situations and problems.

Certainly one army where both scholars and soldiers have seen 'culture' at work is the British Indian Army with its elaborate ideas about Martial Races. In his argument for the significance of local or national culture in shaping soldiers, John Lynn attributes the success of the Indian Army to the 'uniquely South Asian' values that underpinned service in it. European military organization, he argued, was 'fundamentally transformed' by 'indigenous culture' on the subcontinent.[26] 'European regimental culture only succeeded in South Asia by being profoundly altered so as to link the military unit with family, village, *jati*, and religion. Cohesion and motivation within native battalions depended on focusing the force of indigenous community identity and honor.'[27] Many British officers in the century after 1857 would have agreed with Lynn's views, and certainly there were important political-economic and disciplinary links between village, regiment, and army.[28] But questions of culture, and of relations between armed forces and society, were far more complex than Lynn allows.

The Martial Races did not exist in and of themselves, out there to be found in Indian society, as if Victorian racial theories were adequate accounts of real groups. They were constructed out of British and Indian imagination, memory, and identity. The codified, official versions of the Martial Races found in the Indian Army were *made real* through imperial power and military organization. The stereotypes of the Martial Races were realized through the disciplinary power of the military, shaping 'recruits in the manner of self-fulfilling prophecy'.[29] Sikhs who trimmed their hair or beards, for example, risked discharge from the army.[30] The various official component ethnicities of the Indian army were colonial constructs. Soldiers were organized and trained into the correct categories, bounded off from one another by unit and uniform, and by official interpretations of religious practice, diet, holidays, and so on. Caste, religious, regional, and other ethnic relations were sites of an evolving encounter between colonial rule and Indian society, in which Indian populations were disaggregated into 'convenient stereotypes'.[31] The British crystallized particular versions of Indian identities and instantiated them in Indian populations through governance and discipline in and beyond the armed forces. Derived variously from readings of Indian history, religious texts, colonial anthropology and sociology, and other 'investigative modalities', official identities were used to order populations and control access in politics, the economy, and in the army, shaping Indian society in multifarious ways.[32]

The Sikhs are paradigmatic.[33] Sikhism is a syncretic religion that developed out of Hindu and Muslim thought and practice in fifteenth century north India. There were ten Sikh Gurus, only the last of which, Gobind Singh, gave

Sikhism a primarily militant cast in the face of Mughal prosecution. Singh formed the Khalsa, a brotherhood of the pure, and promulgated rules of comportment and appearance. All were to take 'Singh' as their surname, which means 'lion'. The Khalsa was attractive to Jat peasants in the Punjab because it offered social advancement upon the adoption of Singh's rules, and many converted. Khalsa warriors provided the military basis for the rise of Sikh power as the Mughals declined, culminating in the powerful Sikh state established by Ranajit Singh and annexed by the British in 1849 after two wars. Like the Romans who often took in those they defeated, the East India Company began to recruit the conquered Sikhs, who had fielded a powerful and well-trained army that impressed the British. Recruitment was rapidly expanded during and after the 1857 revolt. By 1911, Sikhs accounted for 1 per cent of the Indian population but 20 per cent of the army.

For the British, the 'true' Sikhs were the Jat Sikhs whom they equated with the Khalsa, which in turn was equated with Sikhism. 'The British Indian army nurtured an orthodox, separatist, and martial Singh identity.'[34] Recruiting handbooks advised officers to recruit Sikhs only among the 'tribes' which had converted at the time of Gobind Singh: 'those tribes who, though they now supply converts to Sikhism, did not do so then, cannot be considered . . . as true Sikhs'.[35] Identifying these 'true Sikhs' was no easy matter because Sikhism was a variable amalgamation of Sikh and Hindu practices, something a recruiting handbook acknowledged by noting that 'the line between the strictest Singh and a Hindu is but vague'.[36] In his 1933 primer on the Martial Races for young officers and the public, Lieutenant General George MacMunn tells the story of the Sikhs 'from the beginning' based on the premise that they are a 'martial people'.[37] Such a reconstruction of the Sikhs as essentially martial enabled MacMunn to claim that by restricting recruitment to true Sikhs and disciplining them accordingly, it was 'the British officer who has kept Sikhism up to its old standard'.[38]

A Second World War British officer wrote home of his new experiences training Sikh troops: 'The Sikhs have many religious customs; we see that they keep them whether they like it or not.' When queried on whether the troops must wear their silver bangles, his battalion commander insisted '[o]f course they must wear it, it's part of their religion.'[39] Sikh company and platoon officers were convinced that religion was an important attribute of being a disciplined soldier and took a keen interest in ensuring their men lived up to the five visible signs of faith: hair, white underpants, the bangle, steel dagger, and comb.[40] A turban had not been one of the original items of dress in Singh's rules, and early representations of Sikhs show at least two different styles of headdress.[41] But by the late nineteenth century, along with a beard, the distinctive 'Sikh' turban had become an emblematic mark of Sikhs who served in the army, and as such of Sikhs generally. Army uniform regulations specified a turban for the Sikhs distinct from the one worn by Punjabi Muslims or by Hindu Dogras.

In the troubles that gripped Sikh units early in the Second World War, the turban became a rallying point for resistance as Sikh soldiers refused to exchange it for steel helmets. The cultural field shaped by British rule could be turned against military discipline. In maintaining ethnic distinctiveness, the British had created space for each class of troops to organize resistance on the basis of caste and religion and to draw on communal ties in doing so. Caste could be invoked to avoid distasteful jobs, whether through conviction or calculation. After the 1935 earthquake in Quetta, the 4/19th Hyderabad was sent to bury corpses and its high caste company of Kumaonis refused to touch the dead on grounds that it would break their caste. They were only convinced to do their share of the work when the British and Indian commissioned officers, among them a Brahmin, set to work first as an example.[42] Accompanied by a mortar unit portered by elephants in Burma in 1945, John Randle's Pathan company in the 7/10 Baluch refused, 'for cultural and religious reasons', to remove the tremendous amount of dung produced by the elephants, which was attracting flies and left a heavy stench around their positions.[43] These kinds of affairs were localized, but the communal organization of the army made it easy for trouble to spread, articulating men of the same class in different units.

This is the context against which to read the refusal in 1941 of the Sikh 20th Anti-Aircraft Battery, sent to defend Hong Kong from the Japanese, to don steel helmets.[44] The ostensible reason for this was religious, that the men were not being allowed to wear their turbans. Investigators thought that a communist or Ghadarite cell was active in the unit. In providing their Sikh troops with turbans, the British had also provided them with a totem around which to organize resistance. Ritual objects with powers of group formation were double-edged from a disciplinary point of view. Any Sikh who followed orders and wore his helmet could be seen to have violated his religion.[45] When the 2/14th Punjab arrived to reinforce the Hong Kong garrison, its Sikh company was jeered by the 20th AA for wearing their helmets. The 85 Sikhs in the 20th AA were told that if they did not relent and wear their helmets, they would be tried by court martial. Only two backed down. The next day Sikh quartermasters refused to handle boxes containing steel helmets. A factor underlying unrest in Sikh units in the early years of the Second World War, especially those about to embark for overseas service, were tensions in the Punjab arising from the Muslim League's goal of an independent Pakistan. Worried that these tensions would turn violent, as they eventually did, the Sikhs were concerned they would be unable to protect their families if sent overseas, a much more immediate reason for disobedience than religious sensibilities about headgear.

Putting the issue for the mutinous Sikhs in terms of safety of family moves analysis away from 'culture' per se, which becomes a field or medium for action in this instance. That is, it is not peculiarly Sikh cultural or religious practices, fears of pollution, and so on which are primarily at issue, but a concern which troops of *any* culture might share if they found themselves in similar circumstances, however differently they might choose to express it. Of course, interpreting the actions of the 20th AA and other Sikh soldiers requires attending to specificities, from the decline of British power in India to the particularities of local culture. But the forces driving the Sikhs' actions seem more immediate and general, whether they involved worries about their families, a disinclination to do dirty work, or some other disciplinary irritation. The Sikhs become comparable to other soldiers in similar situations.

Consider the further fate of the 20th AA. They and their guns fell into Japanese hands with the capture of Hong Kong in December 1941. At some point the Japanese recruited them and sent them to Burma, where they engaged Allied aircraft in the second Chindit operation in the spring of 1944. Some of the Sikh gunners were now captured again, this time by Mike Calvert's 77th Indian Infantry Brigade, involved in the battle for Mogaung. Calvert described them as 'an emaciated, miserable lot, completely resigned to whatever fate was due to them'.[46] They told him they would have been shot had they not joined the Japanese; that they were led astray by Japanese promises; and that they had only volunteered in order to get back to India and their families. A plausible account, but one intended to ingratiate themselves with their new captors.

Calvert was short of troops and decided to make use of the Sikhs. An experienced leader familiar with the Indian Army, Calvert knew how to make a pitch. He began by hinting at the possibility that they could recover their self-respect as soldiers. He told them that he had known their unit before the war and that they were efficient and accurate gunners. He went on: 'I had noticed that under the Japs [*sic*] they had been very inaccurate when firing at us. I considered that this must be because in their hearts of hearts they did not want to shoot at us.'[47] He then presented a stick and a carrot, a limited choice with harsh constraints. He told them he would temporarily give them the benefit of the doubt and not have them shot if they agreed to serve as stretcher bearers for his infantry before Mogaung. If they did so, although they would still have to be tried for their treachery on return to India, he would put in a good word for them. Of course Calvert had neither the jurisdiction nor the intention of shooting them. But with a threat on the one hand, and a ray of hope on the other, he secured their willingness to be put again under discipline and to function as soldiers in an arduous and dangerous task. Calvert's brigade suffered heavily at Mogaung.

The Sikhs 'did serve us well, but they never smiled. I did put in a good word for them. They were treated leniently.'[48]

These Sikhs were hardly elite shock troops, either for the Japanese or the British. But they provided both with elementary forms of military labour. Whatever views they may have had on the rights and wrongs of the war, their religion, their turbans, and so on could hardly explain a succession of moves from mutiny, to Japanese service, to British service again; survival, self-interest, and relative comfort were more likely motivations. While the political situation in the Punjab and their cultural background undoubtedly shaped their actions in various ways, to look for specifically 'Sikh' sources at every turn is to risk ascribing common human tendencies to a particular society. It is in this sense that the 20th Sikh Anti-Aircraft Battery helps identify some of the pitfalls involved in understanding soldiers in terms of nations and societies. Because the Sikhs are non-Europeans, Asians, the first move is to see them as driven by 'culture', and by irrational instincts associated with religious belief. But considering them in context, and without denying their cultural and historical specificity, they become comparable to other soldiers and people. In terms of the sources and motives for action, cultural difference serves more to obscure than to explain. The organizational and operational situation in which these soldiers found themselves becomes more relevant in understanding their actions. Their behaviour, in short, is traceable to the more immediate conditions that they faced.

This ultimately is the problem that Samuel Meyers and Albert Biderman run into in their classic military sociological account of the apparently '"unprecedented"' and '"strange"' behaviour of Chinese communist soldiers in the Korean War.[49] They were puzzled by the combination in Chinese soldiers of spirited combat behaviour and unwillingness to surrender, on the one hand, with a 'docile and cooperative' disposition as prisoners, on the other.[50] Chinese PoWs expressed discontent with the communist regime as well. 'What was there about the system of military-political control that could produce such excellent soldiers in the face of what seemed to be such widespread dissatisfaction?'[51] Their explanation, which combined organizational and social context in the manner of military sociology, was Orientalist. That is to say, it divided humans into Orientals and others, and took this division to be central in accounting for behaviour. For Meyers and Biderman, the People's Liberation Army produced strong primary groups which fought well and as directed, but the soldiers did not actually internalize communist values. Instead, turning to society, they exhibited 'the social behavior of the Oriental', conforming out of a desire for social harmony, seeking to avoid 'loss of face', and desiring social recognition through group membership.[52] When captured, this initially left them without guidance, hence 'docility' and willingness to cooperate, but they were liable to being organized into new primary groups and authority structures within the prison population. Janowitz, in order to account for this apparently perplexing behaviour, claims that Chinese soldiers were simply not 'modern' in his foreword to the Meyers and Biderman volume.[53]

Biderman had pursued a different strategy in his companion volume on US PoWs in Korea, *March to Calumny*. There he was concerned to show that the behaviour of US PoWs did not reflect defects in US society. Rather, as Janowitz put it commenting on this work, 'Americans in Communist prisoner-of-war compounds behaved very much like other human beings in terms of the amount of shock they experienced and the severity of the pressure applied.'[54] So, when captured and separated from their organization, American soldiers behaved like 'other human beings', while Chinese behaviour was to be accounted for by reference to the 'social-psychological character of the Chinese'.[55] The Orientalism of this construct, what with Westerners as humans and of Chinese as Orientals with 'traditional orientations to authority', is obvious.[56] What led researchers astray is not only a focus on cultural and racial difference, but the fact that PoWs, like national armies, are easy fodder for comparative study. Chinese PoWs were conveniently marked out for Meyers and Biderman's researchers by race, uniform, and the antagonisms of war. It is natural to assume that their behaviour derives from the specificities of the society they came from, as with the Sikhs. But, in so doing, behaviour that might reflect wider commonalities with 'other human beings' is instead reduced to an effect of particular societies, cultures, and politics, or to Orientalist envisionings of such societies. In fact, situational and collective dynamics were decisive.

This problem of reading human commonality as national and cultural difference is pervasive. German PoWs, for example, also apparently displayed 'traditional orientations to authority'. According to Henry Faulk, who studied German PoWs in the UK during and after the Second World War, German prisoners sought to conform to group norms, feared 'losing face', and were deeply desirous of group membership and acceptance.[57] Like Meyers and Biderman's Chinese, 'Their basic rule is Not to Lose Face, no matter what the cost' as one German officer wrote of his fellow prisoners.[58] Alternatively, consider the tendency of prisoners to bargain for comforts and perks. For Meyers and Biderman, bargaining among Chinese PoWs reflected not the behaviour of captives in general but a specifically Oriental propensity: 'The Oriental assumes that he can make bargains with the representatives of power, trading his active assistance for personal security and small favors.'[59] This is an excellent example of how, in inquiry into soldiers and war, a common human tendency can be ascribed to a particular society or culture. It is an inadequate way of understanding military practice.

Politics, Obedience, and Violence: Germans, Danes, and Luxembourgers

If one tendency in military sociology and history is to see soldiers and armies as differentiated by culture and society, another is to focus on political difference. We assume that soldiers, like the polities they serve, are fighting for *different* reasons. It seems obvious, too, that soldiers' actions, like those of PoWs, reflect distinct national political styles, on and off the battlefield. These notions are especially prevalent in studies of savage and intense fighting as well as of atrocity. Racial ideologies and national traditions of race hate are invoked in explanation. What is accomplished by the pervasive emphasis on national and political difference in accounts of battle and war? What is obscured and forgotten by the reduction of savage violence to the realm of politics and the perversities of certain fighting peoples?

The assumption that soldiers fight for their countries, and do so in distinctive national styles, makes common, or shared, sources of soldiers' actions difficult to see. It ignores situational dynamics in favour of cultural or political essentialism. By reducing wartime violence to politics, the idea is preserved that, ordinarily speaking, human beings are pacific. It is from the political world that their willingness to engage in organized violence arises, overcoming putative inhibitions and providing reasons for mutual slaughter. Political necessity or ideological fanaticism inspires the work of war and atrocity. The turn to national politics provides a meaningful frame for war, one which both accounts for and contains war's extraordinary eruptions of violence. Perhaps because few such eruptions were more cataclysmic than those set in train by Hitler's Germany, the work of nationalizing and politicizing the making of atrocity and combat is particularly evident in the debates over the motivations of German soldiers.

In Omer Bartov's work on German soldiers and Daniel Goldhagen's on 'Hitler's willing executioners', explanations for participation in savage combat and in genocide take similar form. On Bartov's account, German soldiers were 'devoted believers in a murderous ideology', which inspired them both to fight and to commit atrocity.[60] For Goldhagen, the members of the German reserve police battalions employed shooting Jews and others in Poland and the USSR 'wanted to be genocidal executioners' and 'considered the slaughter to be just', due to their 'demonological anti-semitism'.[61] A chain of cause and effect links political ideology to the practical making of wartime and genocidal violence. This chain creates a self-reinforcing cycle between soldier and polity. In respect of the Holocaust, this cycle is essential for connecting circuits of guilt and responsibility between the German people, the German armed forces, and genocidal crimes. The world of polities is presumed and naturalized; difference dominates accounts of war and its violence.

However, from the standpoint of the mass organization required for total war and industrial slaughter, the fervent belief of each individual is an unreliable source of obedience. In turning to organizational routine and bureaucratic dehumanization to account for the everyday making of the Holocaust, Christopher Browning, Zygmunt Bauman, and others relaxed the hold of national society on the perpetrators of atrocity.[62] For these scholars, the bureaucratic processes through which atrocity and genocide are made have a general character in which any ordinary person can be caught up. In Browning's hands, the feelings, beliefs, and actions of the German middle-aged police reservists he studied become understandable, given their situation and context, rather than exceptional. The policemen found the work of gunning down Jews and other victims physically revolting. They did not like shooting babies, required alcohol, and grew demoralized. Later as the killing process was institutionalized in camps and the policemen were mostly used in rounding up and deporting victims, they gave foreign auxiliaries the most distasteful jobs.[63]

At issue is the degree to which national state and society can be said to inform the beliefs and actions of soldiers and others, not just at the level of executive decision but at that of the everyday operation of military and security bureaucracies. While soldiers like other people have political and racial beliefs and attitudes, do these generally account for their regular obedience and any vigorous dedication to service? Goldhagen's avowedly 'monocausal' explanation makes a direct connection between specifically German anti-semitism, on the one hand, and obedience to orders to kill on the other. Politics and action, belief and behaviour, relate to one another as cause and effect with no need for any other 'intervening variable'.[64] State, nationality, and motivation are aligned isomorphically in a political and cultural geography composed of nation-states. 'The Holocaust was a *sui generis* event that has a historically specific explanation', which is German 'eliminationist' anti-semitism.[65] For Goldhagen, '[t]he notion that ordinary Danes or Italians would have acted as the ordinary Germans did strains credulity beyond the breaking point.'[66]

Browning does not in the first instance turn to 'national character' but to organizational routines and general categories of meaning and emotion, like fear of appearing cowardly or revulsion at dirty work. His work can be conceived as an effort to historicize the findings of Stanley Milgram.[67] Browning adds layers of context, such as immediate circumstances or career trajectories and family status. He notes that the sources of resistance in the police battalion he studied were not political or ethical in nature, but arose from the horrific nature of the work involved in shooting so many defenceless people at short range. Rather than invoke anti-semitism as their reason for killing, the policemen preferred to believe they had no choice in the matter. Their violence was not reserved for Jews but potentially directed at any one outside their unit.[68] Common and comparable social processes came to the fore for those who did the work of genocide, whatever the ideological designs that drove elites and 'high politics', and whatever political or racial views the policemen may have held.

From this standpoint, the specific content of ideology as a more or less coherent set of propositions is not so important, partly because many people lack knowledge of ideology in the required kind of way. Rather, the beliefs that drive social action are best conceived in terms of myth, rumour, and sentiment, with ideology influencing rather than determining popular attitudes. Mark Mazower, for example, shows how Nazi ideology shaped opinion about patriotism, women, law, and military honour, in ways which affected willingness to participate in atrocity.[69] What comes to attention from comparative work in atrocity and savage battle are similar patterns of belief and sentiment that serve psychic functions for groups of perpetrators and soldiers. Many different national myths and political and racial ideologies can supply the requisite cultural materials for these patterns of belief and attitude. One example is the stereotypical 'demonic other', which John Dower finds in broadly similar terms among both Japanese and Anglo-Americans in the Pacific War.[70] It can be moulded to specific circumstances by official propaganda and, in more informal ways, in the circuits of opinion which course through bodies of troops. Demonic others are put to work in the myths of betrayal and revenge that often accompany group participation in savagery. An example is the devious enemy who feigns death or surrender only to shoot you in the back, and who therefore must be killed outright. 'Surely it is in the coils of rumor, gossip, story, and chit chat where ideology and ideas become emotionally powerful and enter into active social circulation and meaningful existence.'[71]

In respect of the supposed unwillingness of Danes and Italians to participate in slaughter, Browning's riposte to Goldhagen involves some particularly willing Luxembourgers in Police Battalion 101.[72] It turns out that soldiering in Nazi Germany—a failed empire[73]—was often a more cosmopolitan affair than one might imagine. Nationality did not necessarily determine who was capable of what. Many of the paramilitary battalions involved in the Holocaust were composed of Ukrainians, Latvians, and Lithuanians.[74] By 1945, foreigners outnumbered Germans in the *Waffen SS*.[75] More Danes died fighting in SS Divisions 'Wiking' and 'Nordland' than against the Nazi invasion and occupation of their homeland.[76] West European SS formations were 'formidable fighters' and the 'backbone of the defense' in the Battle of Berlin in 1945, where a battlegroup of Danish, Norwegian, French, and Latvian SS mounted some of the last organized resistance around the Reich Chancellery.[77]

The ubiquitous presence of foreign troops in modern war poses a problem for the careful pairing of the ideological commitments of soldiers with their wartime objects of violence. Goldhagen addresses it by arguing that the principal 'national groups who aided the Germans in slaughtering Jews' were as anti-semitic as the Germans and that those who joined were 'animated by vehement hatred of Jews'.[78] He has a point, but one which opens the door to comparative anti-semitisms, making participation in the Holocaust less exceptional in character.[79] Fundamental premises are exposed by this turn to other national histories. If one conceives the world as populated only with potentially warring nation-states, each with their separate histories and societies, it is natural to conceive of soldiers and other agents of violence as necessarily rooted in the national. Organizational, psychological, or situational factors are at best an intermediate level of analysis, not a primary explanation. When confronted with the presence of foreigners in the Nazi war and genocide machine, this perspective asks why members of these other national groups might have sided with Hitler? What were their politics and national histories? Were they fascists or did they have their own national anti-semitisms? The world consists only of territorially separate, national societies each with their own histories and characters.

The debate over the Holocaust, with its charged politics, might be said to have stumbled across the presence of foreigners in its midst. Their presence is indicative of the transnational surfeit of barbarism and savagery in histories of warfare and mass violence, histories which did not begin or end with the Third Reich. At some point in the course of matching up specific national histories with specific instances of violence and atrocity, it becomes clear that there is a human potential for such actions across time and place. Here the situational and institutional dynamics are decisive in generating historically specific but also recurrent forms of social practice. Whatever the exceptional character of the Nazi Holocaust, organizing groups of humans to conduct the labour of mass killing is not historically exceptional.[80] Bureaucratically organized battle and atrocity are ubiquitous historical practices. Browning's approach, like primary group theory in military sociology, has the virtue of being a serviceable general explanation for a general phenomenon. He attends also to local circumstances and the trajectory of specific military formations and their personnel.

In not, in the first instance, turning to nation, race, and politics, Browning identifies dynamics that make his German policemen more human, and at the same time a threat to pacific conceptions of humanity. His approach challenges the idea that only *certain* nations are capable of such atrocity, which is perhaps why his work has generated controversy. For Goldhagen, the idea of the Holocaust as an exceptional crime is dependent on an account of the exceptional character of German society. The histories of different states and societies certainly lead to different forms and scales of mass killing, savagery, and atrocity. But such exceptionalism is not readily apparent in studies of why bureaucracies of death function, or in the elementary sources of obedience and discipline for militarized labour. Neither concentration camp guards nor Anglo-American bomber crews were prone to mutiny. It is not in the first instance a *German* matter why the soldiers of Police Battalion 101, or some other, followed orders, but a human and bureaucratic one.

Towards a Non-Eurocentric Military Sociology

In diverse and significant ways, soldiers and their 'fighting spirit' are conceived in terms of modern social and political phenomena. The national and territorial state, mass society and ideology, democracy and citizenship, bureaucracy and identity, all frame study and memory of soldiers and their wars. We imagine soldiering against the backdrop of a Eurocentric world—that of the national armies of the principal Western states.

The discussions above have mobilized 'subaltern soldiers' broadly understood—colonial soldiers, Chinese soldiers conceived through an Orientalists lens, Hitler's foreign auxiliaries—to trouble and disrupt the Eurocentric and national conceptual grid through which soldiers and armies are studied. Subaltern soldiers offer contrapuntal resources for thinking about army, society, and war. Forces like the British Indian Army fought in the same campaigns and against the same enemies as national armies, yet hailed from an imperial order. Colonial armed forces and society relations were used to draw comparisons with their oft taken-for-granted nation-state counterparts, where army, society, and state are conceived within a sovereign territorial package. The issue was not what the colonial situation *lacked* with respect to the nation-state, such as a putatively cohesive society.[81] But rather, what does the colonial situation enable us to say about the military and its relations with society *in general*? What happens when we take the service of imperial soldiers, and the multicultural formations of which they were a part, as the ordinary situation rather than an exception or derivative? How would we then need to understand discipline and the sources of obedience, or the relationship between military organizations and 'civilian' culture?

What the critique of Eurocentrism offers is the possibility of seeing questions afresh and answering them with creative combinations of traditional and new scholarly resources. Many will have noted above the turn to an older 'profession of arms' approach, one career soldiers themselves often invoke. It was also one that outlined the replacement of the division between 'military' and 'society' with a focus on the co-constitutive relations between armed forces and their social and cultural contexts. Military service shaped what Sikhism was, in and out of the army. Military practices recurrently arise in different historical contexts shaped by situational dynamics.

Armed forces are always articulated with some kind of social context, and draw upon some field of cultural resources in making soldiers, and both matter for outcomes, and for the character of bodies of troops. But the military is also a machine for group formation. Regular military discipline consists of a set of sturdy and robust ritual techniques, many of which double as training, easily adapted to diverse contexts and cultures. These techniques work more or less anywhere, but are always realized locally, in some specific context with its particular social, cultural, and political character. Soldiers always express themselves in a local idiom, but one organizationally transformed by military life. The soldier reappears as an anthropological and historical creature, of ancient and cosmopolitan origins. This suggests an alternate vocabulary to the modern and national grid of concepts that military sociology currently employs.

Notes

1. Morris Janowitz, *The Professional Soldier* (New York: Free Press, 1971), 82.
2. Field Marshall Lord Wavell, GHQ(I) Infantry Liaison Letters, No. 24, February, 1946, L/WS/1/778, Oriental and India Office Collection.
3. Myron Echenberg, *Colonial Conscripts* (Portsmouth: Heinemann, 1991); Roderick de Normann, 'Infantry Regiment 950—Germany's Indian Legion', *Journal of the Society for Army Historical Research* 75(303) (1997); George Stein, *The Waffen SS* (Ithaca, NY: Cornell University Press, 1966).
4. Omer Bartov, *Hitler's Army* (Oxford: Oxford University Press, 1992); Edward Shils and Morris Janowitz, 'Cohesion and Disintegration in the Wehrmacht in World War II', *Public Opinion Quarterly*, 12(2) (1948): 280–315.
5. See, for example, Lord Moran, *Anatomy of Courage* (Garden Park City, NY: Avery, 1987); Shils and Janowitz, 'Cohesion and Disintegration'; Samuel Stouffer, Arthur Lumsdaine, Marion Harper Lumsdaine, Robin Williams, M. Brewster Smith, Irving Janis, Shirley Star, and Leonard Cottrell, *The American Soldier*, 2 vols. (Princeton, NJ: Princeton University Press, 1949).
6. See, for example, Bartov, *Hitler's Army*; Elliot P. Chodoff, 'Ideology and Primary Groups', *Armed Forces and Society*, 9(4) 1983; Morris Janowitz and Stephen D. Wesbrook, eds., *The Political Education of Soldiers* (Beverly Hills, CA: Sage, 1983).
7. See, for example, V.G. Kiernan, *Colonial Empires and Armies 1815–1960* (Stroud: Sutton, 1998); David Killingray and David Omissi (eds.), *Guardians of Empire* (New York: St. Martins, 1999).
8. See Giuseppe Caforio and Marina Nuciari, 'Social Research and the Military: A Cross-National Expert Survey', in Caforio (ed.), *Handbook of the Sociology of the Military* (New York: Kluwer, 2003), 27–58. One American respondent to Caforio and Nuciari's survey wrote in: 'The United States' military does much more social science than any other country I can think of. I would like to see government social scientists, like me, get more freedom to determine what we will work on' (34). Caforio and Nuciari rank the US in their highest category of government control

over research on the military, along with Russia and Argentina among others (41). Overall, in their survey of military sociology in twenty countries, the main commissioner of research was the state (32). Notably, of the twenty countries, fifteen are European, the others being the US, Argentina, Israel, India, and South Africa (27). As these researchers mainly study their own militaries, not only does military sociology remain firmly in the grip of the nation-state, but it exhibits a heavy Western bias in its research topics.

9. See, for example, Martin van Creveld, *Fighting Power* (Westport, CT: Greenwood, 1982); William Hauser, *America's Army in Crisis* (Baltimore, MD: Johns Hopkins University Press, 1973); W.D. Henderson, *Cohesion* (Washington, DC: NDU Press, 1985); Charles Moskos, John Williams and David Segal (eds.), *The Postmodern Military* (New York: Oxford University Press, 2000). Tellingly, even reflections on military sociology itself take national form. See the special issue on 'Sociology at Military Academies around the Globe', *Armed Forces and Society*, 35(1) 2008.
10. See, for example, Sherie Mershon and Steven Schlossman, *Foxholes and Color Lines* (Baltimore, MD: Johns Hopkins University Press, 1998); Elisabeth Kier, 'Homosexuals in the U.S. Military', *International Security*, 23(2) 1998: 5–39.
11. See, for example, Rob Johnson, *The Afghan Way of War* (London: Hurst, 2011).
12. In 1981 among an interdisciplinary group of thirty-five social scientists who studied the military profession and military organization, 'there was widespread agreement that military morale and combat effectiveness rested on primary group and small group solidarity and cohesion'. Janowitz, 'Preface', in Janowitz and Wesbrook, *The Political Education of Soldiers*, 10. See also Moskos, *The American Enlisted Man* (New York: Sage, 1970), Ch. 6.
13. See in particular Tamotsu Shibutani, *The Derelicts of Company K* (Berkeley: University of California Press, 1978).
14. Jon Balkind, 'A Critique of Military Sociology: Lessons from Vietnam', *The Journal of Strategic Studies* 1(2) (1978): 235–59; John Helmer, *Bringing the War Home* (New York: Free Press, 1974); cf. Paul Savage and Richard Gabriel, 'Cohesion and Disintegration in the American Army', *Armed Forces & Society* 2(3) (1976): 340–76.
15. Henderson, *Why the Viet Cong Fought* (Westport, CT: Greenwood, 1979); *Cohesion.*
16. Hauser, *America's Army in Crisis.*
17. See, for example, Moskos, *The American Enlisted Man.*
18. Ardant du Picq, *Battle Studies* (New York: Macmillan, 1921).
19. Janowitz, 'Preface', in Janowitz and Wesbrook, *The Political Education of Soldiers*, 11. See also Chodoff, 'Ideology and Primary Groups'.
20. Wesbrook, 'Sociopolitical Training in the Military', in Janowitz and Wesbrook, *The Political Education of Soldiers*, 36, 44.
21. Alexander Watson, 'Culture and Combat in the Western World, 1900–1945', *Historical Journal* 51(2) (2008): 529–46, 545.
22. Craig Cameron, *American Samurai* (Cambridge: Cambridge University Press, 1994), 20.
23. Cameron, *American Samurai*, 104, 126. See also John Dower, *War without Mercy* (New York: Pantheon, 1986), 11–12.
24. Cameron, *American Samurai*, 20, emphasis added.

25. Gwynne Dyer, *War* (London: Bodley Head, 1986), 133.
26. John Lynn, *Battle* (Boulder, CO: Westview, 2003), 159, 174.
27. Lynn, *Battle,* 173. *Jatis* are the subcastes of the four major castes or *varnas.*
28. See, for example, Tan Tai Yong, *The Garrison State* (New Delhi: Sage, 2005).
29. Peter Fay, *The Forgotten Army* (Ann Arbor: University of Michigan Press, 1993), 21.
30. Charles Trench, *The Indian Army* (London: Thames and Hudson, 1998), 11.
31. David Arnold, 'Bureaucratic Recruitment and Subordination in Colonial India', in Ranajit Guha (ed.), *Subaltern Studies IV* (Delhi: Oxford University Press, 1985), 7.
32. Bernard Cohn, *Colonialism and Its Forms of Knowledge* (Princeton, NJ: Princeton University Press, 1996), 5.
33. See Cohn, *Colonialism,* 107–11; N. Dirks, *Castes of Mind* (Princeton, NJ: Princeton University Press, 2001), 177–80; R.G. Fox, *Lions of the Punjab* (Berkeley: University of California Press, 1985); Tan, *Garrison State,* Ch. 2.
34. Fox, *Lions of the Punjab,* 10.
35. R.W. Falcon, *Handbook on Sikhs for Regimental Officers* (Allahabad: Pioneer Press, 1896), 65.
36. Falcon, *Handbook on Sikhs,* 16.
37. Lt. Gen. Sir George MacMunn, *The Martial Races of India* (London: Sampson, Low, Marston & Co., n.d. [1933]), 118.
38. MacMunn, *The Armies of India* (London: Adam and Charles Black, 1911), 135.
39. Quoted in Hannah Cohen, *Let Stephen Speak* (London: Sylvan Press, 1943), 116–17.
40. Maj. Gen. Sir John Smyth, Mss. Eur. T 63, Oriental and India Office Collection, British Library.
41. Cohn, *Colonialism,* 108.
42. Humphrey Evans, *Thimayya of India,* (New York: Harcourt, Brace, and Co., 1960), 151.
43. Quoted in Jon Latimer, *Burma* (London: John Murray, 2004), 420.
44. See Disaffection of Sikh Troops, L/WS/1/303 and Note on Sikhs, L/WS/2/44, Oriental and India Office Collection, British Library; Disaffection in Indian Army, WO/208/763, Public Record Office, Kew.
45. A similar phenomenon was evident in the greased cartridges affair preceding the 1857 mutiny. Many Indian soldiers knew that their cartridges were not greased with pig or cow fat, in some cases because they had made up the cartridges themselves. But they still could not be seen to use them as long as it was popularly believed that the cartridges were greased with animal fat, for fear of social opprobrium. See John William Kaye, *A History of the Sepoy War in India 1857–1858,* Vol. I, (London: W.H. Allen, 1864), 553–9; J.A.B. Palmer, *The Mutiny Outbreak at Meerut* (Cambridge: Cambridge University Press, 1966), 6.
46. Michael Calvert, *Prisoners of Hope* (London: Leo Cooper, 1996), 210.
47. Calvert, *Prisoners of Hope.*
48. Calvert, *Prisoners of Hope.*
49. Samuel Meyers and Albert Biderman, 'Introduction', in Samuel Meyers and Albert Biderman (eds.), *Mass Behavior in Battle and Captivity* (Chicago: The University of Chicago Press, 1968), xvii.
50. Meyers and Biderman, 'Introduction', xxvii.

51. Meyers and Biderman, 'Introduction', xvii.
52. Meyers and Biderman, 'Introduction', xxviii–xxix; Samuel Meyers and William Bradbury, 'The Political Behavior of Korean and Chinese Prisoners of War in the Korean Conflict', in Meyers and Biderman, *Mass Behavior*, 243–4, 317.
53. Janowitz, 'Foreword', in Meyers and Biderman (eds.), *Mass Behavior*, vi.
54. Janowitz, 'Foreword', v.
55. Meyers and Biderman, 'Introduction', xxvii.
56. Meyers and Biderman, 'Introduction'; Cf. Edward Said, *Orientalism* (New York: Vintage, 1979).
57. Henry Faulk, *Group Captives* (London: Chatto and Windus, 1977), 67, 73, 175, 198, 205.
58. Quoted in Faulk, *Group Captives*, 73.
59. Meyers and Biderman, 'Introduction', xviii.
60. Bartov, *Hitler's Army*, viii.
61. Daniel Goldhagen, *Hitler's Willing Executioners* (New York: Knopf, 1996), 279, 392–3.
62. Christopher Browning, *Ordinary Men* (New York: HarperCollins, 1992); Zygmunt Bauman, *Modernity and the Holocaust* (Cambridge: Polity, 1989). See also Herbert Kelman and V. Lee Hamilton, *Crimes of Obedience* (New Haven, CT: Yale University Press, 1989).
63. Browning, *Ordinary Men*, 59, 61, 76, 77.
64. Goldhagen, *Hitler's Willing Executioners*, 416.
65. Goldhagen, *Hitler's Willing Executioners*, 419.
66. Goldhagen, *Hitler's Willing Executioners*, 408.
67. Stanley Milgram, *Obedience to Authority* (New York: Harper and Row, 1974).
68. Browning, *Ordinary Men*, 73, 76, 149–50.
69. Mark Mazower, 'Military Violence and Nationalist Socialist Values', *Past and Present* 134, February 1992.
70. Dower, *War Without Mercy*, 9–10.
71. Michael Taussig, 'Culture of Terror—Space of Death', *Comparative Studies in Society and History* 26(3) (July 1984): 494.
72. Browning, 'Ordinary Men or Ordinary Germans', in Robert R. Shandley, *Unwilling Germans? The Goldhagen Debate* (Minneapolis: University of Minnesota Press, 1998), 60–2.
73. Mazower, *Hitler's Empire* (London: Penguin, 2009).
74. Browning, *Ordinary Men*, 52; Stein, *The Waffen SS*, 112.
75. Stein, *The Waffen SS*, 137.
76. Approximately 2000 died of the 6000 Danes in the *Waffen SS*. Danes also served in the SS-sponsored Danish Legion. Around 900 Danes died in the resistance movement. See Claus Christensen et al., *Under Hagekors og Dannebrog: Danskere i Waffen SS 1940–1945* (Copenhagen: Aschehoug, 1998), 491–4; Ib Petersen (ed.) *Faldne i Danmarks Frihedskamp 1940–45* (Copenhagen: Gyldendal, 1970). When losses in the Danish merchant marine are included, however, the deaths in Allied service approach equivalence. Thanks to Lars Bangert Struwe for assistance on these points.
77. Antony Beevor, *Berlin* (London: Viking, 2002), 258; Stein, *The Waffen SS*, 164.

78. Goldhagen, *Hitler's Willing Executioners*, 409.
79. Timothy Snyder, *Bloodlands* (London: The Bodley Head, 2010), 191–4.
80. See, for example, Sven Lindqvist, *A History of Bombing* (London: Granta, 2001); *Exterminate All the Brutes* (London: Granta, 1997); Martin Shaw, *War and Genocide* (Cambridge: Polity, 2003).
81. See, for example, Stephen Rosen, *Societies and Military Power* (Ithaca, NY: Cornell University Press, 1996).

3

Killing and Cohesion in Close Combat

Contexts and Concepts from the First World War to the Present

Rob Johnson

Introduction

Close-quarter combat has always presented scholars with significant problems.[1] Given the great variety of individual reactions to fighting, it has been unclear whether responses to killing and the drivers which facilitate cohesiveness are universal human experiences or historically, culturally, or even regionally specific.[2] It is equally uncertain the extent to which killing at close quarters is embraced, is avoided, or enhances cohesion. Moreover, there have been significant problems in the interpretative frameworks adopted by scholars.[3] Historians and social scientists have struggled in their analyses of combat, not only because of the problems of archival work and field study, but also as a result of the contrasts and contradictions in the individual experience, the contextualized nature of the group's response, and the difficulty of extrapolating a general perspective on combat, cohesion, or killing. This seems paradoxical: the trial of close combat and the act of killing have been well documented, particularly in the twentieth century. So it seems appropriate, when the publication of this volume coincides with the centenary of the First World War, to make use of the archival record of that conflict and to compare it with more recent interpretations.

In this chapter, which compares methodology, representation, and the realities of combat, there are three observations.[4] One, given the confusion and fleeting character of close-quarter battle, the sources available to scholars through oral testimony and the archives present particular difficulties of interpretation. Two, the archival record indicates that representations of combat by participants are specific to their context. This indicates that while aspects of the combat experience may not necessarily change through time, any analysis must acknowledge the specificity of the context before suggesting any universal and unchanging nature in close combat or killing. Three, combat and killing are intensely individual experiences that, at the psychological level, elicit common emotional responses. However, the individual is profoundly influenced by the reactions of others in close proximity, situated within the identities of more remote organizations. Combat therefore seems to be an unchanging experience with common characteristics, but group responses can alter the individual's reactions, and these group responses are dependent on very specific contexts, such as levels of training, degree of cohesiveness, or leadership. Small, self-directed professional soldiers react very differently to untrained, uncertain, and mass formations of militia when under fire and subject to the intense stresses of close combat. Individuals are so profoundly affected by the group they belong to that it affects their endurance, their willingness to take risks, and their eagerness to kill. Individuals in cohesive groups can be inspired to behave more courageously in front of their comrades, although internal cohesion in wavering groups can seemingly induce mass flight. This chapter explores the methodological problems of studying combat in order to highlight these recurrent collective features of close combat. The chapter also touches upon the potentially distinctive experience of combat by today's professional soldiers as a result of their intense training and combat preparation. This distinction is not always easy to identity precisely because soldiers employ motifs and techniques to describe the often bewildering experience of combat in ways which seem to be compatible with accepted norms.

Interpreting Combat

Problems have arisen in our interpretive framework because of a tendency to conflate individual and group responses to combat and killing. Moreover, a variety of agendas have been asserted to contextualize killing, from national histories that exonerate or condemn particular behaviour, to sensationalist studies claiming to have uncovered a universal truth about killing. It is striking how many recent studies have sought to show that killing is rare and often avoided, when, in fact, what is remarkable is just how easy killing is, particularly since the advent of industrialized weapons. The First World War provides a rich documentary seam for scholars, but drawing general conclusions from only one conflict seems fraught with its own methodological problems, so, in this short survey, a sample of comparative observations from a variety of historiographical sources is offered in order to contextualize and conceptualize some nuanced conclusions on combat, cohesion, and killing at both the individual and group level.

Individual emotional reactions to combat, for professional and citizen soldiers, have been well documented. In anticipation of imminent violence, soldiers have recorded the sensation of involuntary clenching, chattering and grinding of teeth, jaw gape, or the loss of bowel or bladder control. Eyewitnesses speak of seeing their colleagues with pale and strained faces, smoking heavily, and sometimes sweating or shaking.[5] Others seem able to laugh off the tension, albeit briefly. Troops note the assault on mind and body of the smells, discomforts, and noise of battle: the ear-splitting concussion of artillery explosions, the sound of small-arms fire—like the 'tearing of paper, or a bundle of dry sticks in a fire'; the shocked and plaintive cries of the wounded, the proximity of engine noises, and the sickly stench of corpses.[6] Many have referred to the disorientation of war: the blinding dust, smoke, or flying debris.[7] Then there are the familiar environments suddenly rendered alien: the turfing of the ground by bullets, the cratering of fields and roads into 'an arid waste of tumbled earth', the torn buildings, and, of course, the shock of seeing the dead and wounded mutilated.[8] Veterans recall the unreality of the 'time-lapse' effect of combat: the speed of events does not seem to permit time to think. Many refer to exhilaration, adrenaline, and intense exertion, but also a mix of emotions, from fear and hate, to enjoyment and relief.[9]

In these environments, peacetime values and 'civilized' points of reference can quickly be discarded as irrelevant. The extreme conditions of trench warfare in the First World War increased the tendency for atrocities, leaving scholars with numerous examples in which prisoners were murdered and civilians were mistreated. This was a phenomenon which occurred in subsequent high-intensity conflicts, and it is tempting to conclude that modern war brutalizes its participants. Yet there are just as many examples of humanity in the First World War involving the rescue or succour of the wounded, regardless of their nationality, indicating the risks of selection in our sources. In fact, the instinct for survival was the more common denominator for many who found themselves in combat, and it is this survival instinct that could take priority over the imperatives to kill, or any remote political principles. Men might enlist for a 'cause' in the First World War, Captain Robert Graves noted, but while ideas motivated men, 'patriotism in the trenches was too remote a sentiment, and at once rejected as only fit for civilians or prisoners'.[10] Ideology was merely the basis or justification for going to fight, and while it may create, in part, the bonds of cohesion, it was not always the main motivation when men were faced with imminent destruction. Recent testimonies suggest that, despite the advent of information age warfare, the individual experience of war has remained as elemental, brutal, and callous as ever. In every conflict, the thresholds against atrocities, so quaintly delineated by wise counsels in peacetime, are crossed, and savagery escalates as war progresses, as Clausewitz observed, towards its absolute form and function.

Nothing, however, is as clear-cut as we assume. Oral testimony of combat is notoriously problematic for its analysts. Part of the problem is the compression of time, combined with the subjective reporting of the individual. Participants who have been through combat have often mentioned a confusion of chronology and events: it's hard to know what's really going on. Charles Carrington, in his work on the Third Battle of Ypres in the First World War, noted: 'What surprises me is that historians have elevated it into a tactical masterpiece like Messines. It was just all-in wrestling in the mud'.[11] Memory focuses on the visual stimuli that are often unsettling and disorientating, but sequences can become disjointed, especially when reactions at the time were so frenzied. Denis Winter noted that 'the upset in the body chemistry produced by a state of high fear long sustained gave strength to eliminate a calculating response for a limited time. This would explain... the difficulty in recalling battle details afterwards, even of comprehending the elation of a battle.'[12]

Nevertheless, the sheer emotion of close combat can make even more profound changes. William Manchester, a veteran of the Second World War, noted: 'some flickers of unreal recollection remain' and psychiatrists recognize that some traumatic moments are suppressed automatically, perhaps as a form of protection.[13] The exceptional and the unusual elements tend to linger, however, and that means extraordinarily vivid moments of face-to-face fighting have been recorded faithfully, albeit filtered through the interpretation of a single individual. Norman Lebrun, who fought on the Western Front in 1916, described his encounter in a way that seems familiar and perhaps typical:

> I didn't think of him as another human being. I just saw him as an enemy that had to be defeated and at the time that I sunk my bayonet into his body I didn't really give it too much thought. But when the time came to pull out my bayonet I found that it was quite difficult and so... I had to fire my rifle into his chest so that at the same time I could pull my bayonet out. I think I wasn't satisfied, I used the butt of my rifle and struck him somewhere on the head, I don't know exactly, to make sure it was all over.[14]

The situation and the sentiment appear to transcend time and place. The desire to assert oneself over an impersonal enemy was often powerful. Even during the unique Christmas Truce of 1914, Captain Bruce Bairnsfather of the 3rd Warwicks noted: 'There was not an atom of hate on either side that day; and yet, on our side, not for a moment was the will to war and the will to beat them relaxed.'[15] The will to assert oneself was manifest in other situations. The momentum of advancing appeared to enable men to overcome fear and develop a sense of confidence over their adversary. Professor Tawney, reflecting on his experience in the Battle of the Somme, concluded: 'I hadn't gone ten yards before I felt a load fall from me. I knew I was alright . . . I knew I was in no danger. I knew I wouldn't be hurt – knew it positively.'[16] H. Quigley, who advanced in the 3rd Battle of Ypres, explained the exhilaration: 'You know how excited one becomes in the midst of danger. I forgot absolutely that shells are meant to kill and not to provide elaborate lighting effects.'[17]

This was partly a phenomenon Randall Collins describes as 'forward panic' and which Ardant du Picq once identified as 'flight to the front'.[18] Nevertheless, analysing the emotions of combat is extraordinarily difficult because the range of responses is so wide. For the majority, the overriding recorded experience is fear, but for a few it is hate.[19] Some even find enjoyment, and try to convey the sheer flush of adrenaline that courses through them as a positive phenomenon, suggesting more than the body's physiological response to overcome fatigue. The intense experience of combat and the sheer exertion can, *in extremis*, prevent individuals being aware of their situation, their immediate comrades, or even of being wounded.[20]

Captain Wyn Griffith, Royal Welch Fusiliers, at Mametz Wood in 1916 recalled:

> It was life rather than death that faded into the distance, as I grew into a state of not thinking, not feeling, not seeing. I moved past trees, past other things; men passed by me, carrying other men, some crying, some cursing, some silent. They were all shadows, and I was no greater than they. Living or dead, all were unreal. Balanced uneasily on the knife-edge between utter oblivion and this temporary not-knowing, it seemed little matter whether I were destroyed, to go forward to death, or to come back to life.[21]

From this account, we might derive the observation that Griffith's experience was entirely individual, for he makes no mention of cohesiveness, while his descriptions of others suggest great variations in their responses: crying, cursing, or silence. Once again, we should exercise caution in our analysis. There is an attraction to the more profound and literary forms of interpretation when it comes to explaining the processes and mentality underpinning acts of violence. Captain J.E. Crombie, Gordon Highlanders, described marching through the ruins of Arras in March 1917 in a manner typical of the period, as: 'the long narrow ribbon of street was utterly silent, and the walls, with nothing but ruin behind them, aslant and tottering, [un]til it seemed a push with your hand would overset them . . . It is these ghastly, sightless, purposeless walls that catch you, and the silence.'[22] Yet men under his own command regarded the ruins of Arras as a place to shelter, protection against the enemy and the elements, and a place to fight from. In our analysis care must therefore be taken not to focus on favoured interpretations: the dilemma is to ascertain the relative value of typicality and the exceptional. The solution, perhaps, lies in being able to embrace the diversity of experience of the individual, including the perspectives outside of the Western canon, to acknowledge the cultural, educational, and social expectations of interpreting and experiencing combat, but to set all these within a context of ever-extending group identities, and, ultimately, a national or even transnational one. Not all experiences are universal, although there are patterns, and it is particularly important to situate killing, the ultimate act of combat, in its cultural and historical setting, as well as in the immediate situation facing the participants.

Cohesion and Killing

The interpretive difficulties involved with studying combat are clear. The evidence is potentially misleading, inaccurate, and unrepresentative. Moreover, individuals have multiple and often incompatible experiences from another: they interpret killing in a number of different ways. Nevertheless, as Randall Collins has demonstrated in his work on violence, patterns of social action (and the emotional responses they invoke) do recur in similar situations. Notwithstanding the great difficulties of investigating combat, therefore, much of the scholarship on warfare has sought to identify these patterns, and, specifically, the way in which the situational dynamics of the fighting and especially the cohesion of military units affects individual responses.

In the 1970s, John Keegan wanted to refocus the study of military history on the 'face of battle', that is, the actual individual experience of combat. The 'face of battle' genre was really introduced by Barrie Pitt in 1967 in his study of the two world wars, but Keegan popularized his own interests in combat at the lowest level, or, put another way, on what 'ordinary people' went through. Keegan was profoundly absorbed in the personalized perspective that would be appealing to non-academic and non-professional audiences, but he was also eager to highlight the demanding conditions of war and his contribution to the scholarship was to encourage a certain memorialization of veterans' personal experiences. The *Face of Battle* (1976) unpicked accounts of Agincourt, Waterloo, and the Somme and used eyewitnesses from the lower ranks. To fill in the lacunae, Keegan used imagined reconstructions to give a more 'honest' appraisal of what took place, and he suggested that there was far more confusion and error than the 'official histories' had admitted. His objectives were made clear in the preface, where he wrote:

> Since we appear to know a great deal more about generalship than we do about how and why ordinary soldiers fight, a diversion of historical effort from the rear to the front of the battlefield would seem considerably overdue. All the more does it seem desirable in the light of what little reliable information we do have about what goes on at the place soldiers call "the sharp end".[23]

Nevertheless, as critics pointed out, there were significant problems with reconstruction and attempts to recover soldiers' testimonies. Nostalgia could alter the record to make it more benign, while exposure to anti-war genres of film and literature created a form of historicism amongst veterans themselves. Having originally viewed their experience as one of contributing to victory in 1918, some First World War veterans gradually tailored their accounts to ones that satisfied the narrative of futility, and, in light of the Second World War, a conflict that no one won.[24]

Professional work on how men fight in combat has a longer historical precedent than the academic scholarship. For obvious reasons, all armies have wanted to know how to motivate men in combat and how to maximize their efficiency, particularly to overcome the paralysing effects of fear. Ardant du Piq exemplified such studies in his analyses of why the British infantry were able to defeat the French in the battles of Spain and the Low Countries in the Napoleonic Wars.[25] US Army Historical Teams during the Second World War were eager to discover how small groups fought once it was realized that men tend to fight for their immediate comrades rather than any higher cause. Edward A. Shils and Morris Janowitz, in examining German troops, famously concluded:

> It appears that a soldier's ability to resist is a function of the capacity of his immediate primary group (his squad or section) to avoid social disintegration. When the individual's immediate group, and its supporting formations, met his basic needs, offered him affection and esteem from both officers and comrades, supplied him with a sense of power and adequately regulated his relations with authority, the element of self-concern in battle, which would lead to disruption of the effective functioning of his primary group, was minimized.[26]

Sam Stouffer, who conducted thousands of interviews and questionnaires, suggested that individuals' actions defined and were defined by their primary groups and the context of their society, noting that status could be enhanced or diminished by their combat experience in a phenomenon of 'relative deprivation'.[27] The very combat experiences that encourage bonding also bring casualties, followed by a stream of replacements that lack primary group ties, and thus the cohesion of the group is eroded.[28] Nevertheless, even replacements can be rapidly assimilated. Filtering into existing units with whom they have yet to establish the close bonds, they will share some identities (language, regiment, region, and training experience) as the men they had joined.

S.L.A. Marshall's findings, published in *Men Against Fire* (1947), were controversial. Famously, his most polarizing conclusion was that most soldiers in close combat had not fired their weapons, and therefore, by extension, they had been reluctant to kill. He wrote:

> In an average experienced infantry company in an average day's action, the number engaging with any and all weapons was approximately 15 per cent of the total strength. In the most aggressive companies, under the most intense local pressure, the figure rarely rose above 25 percent of the total strength from the opening to the close of the action.[29]

His results have been challenged ever since.[30] Critics pointed out that he attributed non-firing to psychology (reluctance to kill) and firing to sociology (small-group cohesion), and therefore Marshall was inconsistent in his methods.[31] More practical objections have also been raised.[32] The reason why not all men discharge their weapons is because they don't get a clear target. US military training attempted to address the problem Marshall claimed to have identified and some 80 per cent of infantrymen in combat in Vietnam fired their weapon, but, as in previous wars, it was the hit rates that were the problem. There was greater ammunition expenditure because of the higher rates of fire, but hit rates were proportionally smaller because targets were concealed or protected. The results were clearly not a case of men being reluctant to fight but being constrained by other factors.[33] Comparison across periods was also difficult: hit rates had always been a problem historically because of the limited accuracy of earlier technologies.

Nevertheless, the primacy of small group cohesion, which Marshall, Shils, Janowitz, and Stouffer championed, has also been misrepresented. Stephen Ambrose, in his studies of GIs in 1997 argued that democratic soldiers may enjoy a stronger cohesion through shared values than soldiers of totalitarian regimes. Yet studies of German forces in Russia appear not to support this, nor studies of soldiers in less developed countries since the 1960s. Cohesion and motivation in groups are based on training, discipline, the ethos of the forces they belong to, exemplary junior officers, their time in combat, and the conditions they face. These differences are most easily discerned between 'professional' and 'citizen' soldiers, or 'seasoned' and 'raw' fighters.[34]

There may also be a correlation between groups and their legitimacy which determine the intensity of their support for killing the enemy. The anonymity of the enemy is referred to with some frequency—the emotional distance, in moral, cultural, and social terms makes it easier to engage and destroy others, but while these may be affected by attitudes held over long periods of time, there may be sentiments about the humanity of the adversary which are short-lived, temporary, conditional on the enemy's conduct, or only manifest after the event. Moreover, group attitudes appear to shape individuals' reactions to the foe. Stephane Audoin-Rouzeau and Annette Becker concluded the brutality soldiers described was 'always anonymous and blind', suggesting that this was an attempt to escape the guilt, imposed by social taboos, of having taken a life.[35] In contrast, Alexander Watson argued that portrayals of the enemy were universally anonymised because that was the common experience of the war: he posited that the number of wounds from hand-to-hand fighting and even from small-arms fire was negligible compared with casualties caused by artillery.[36] Nevertheless, the impersonal description of the enemy was formed long before combat was experienced at first hand, and could be reinforced by constant social interaction. Robert Graves explained that, between themselves, British soldiers never referred to 'Germans', only to the cyphers of 'Boches' and 'Huns'.[37] The German psychologist J.H. Schultz wrote in 1915 that it was far more difficult to kill an individual one regarded as a fellow human being.[38]

Richard Holmes conveyed the great range of individual experiences and reactions to combat in *The Firing Line* (1985), covering the anticipation (discipline, anxiety, performance, morale, cohesion), the first impressions (fear and how it is overcome), closing with the enemy (sights, sounds, senses, confusion, killing), and the post-combat experience (loss, grief, anger, fatigue, guilt, relief, pleasure, anti-climax), while dealing with the fundamental question of why men fight—which he attributed to an equally complex array of factors, from comrades, the cause, leader, religion, ideology, revenge, and the desire to protect or survive. Holmes also examined factors that lay behind the incidence of atrocities, and acts of compassion such as succour to the wounded. He examined exemplary courage and leadership, and he acknowledged the problems with the way each of these are recorded.

Holmes typically related vivid examples to illustrate each issue. He provided a case, where, immediately after an attack at Arras in 1917, which had resulted in the capture of a strong position, two Germans, one of whom was wounded, were taken prisoner, and raised their hands to ask for water, but an Australian officer had shot them dead. Holmes suggested that this may have been the result of the emotions of the attack still running high.[39] Another Australian eyewitness recorded he had received orders to take no prisoners, including the wounded, who were accordingly summarily bayoneted. Reports of this order were not unusual in the First World War, and the soldiers' reactions were either to accept this instruction from above grimly or to enjoy the licence to kill. Graves concluded the common motives were: 'revenge for the death of friends or relatives, jealousy of the prisoner's trip to a comfortable prison camp in England, military enthusiasm, fear of being suddenly overpowered, or, more simply, impatience with the escorting job'.[40] Holmes provided dozens of these events where context was crucial to our understanding.

New interdisciplinary studies emerged in the 1990s to develop the field. David Grossman's *On Killing* (1995) set out to examine the reality of combat killing, based on interviews and testimonies from Vietnam veterans. Reviewing his findings, he was deeply disappointed by the contrasting popular portrayal of combat he found in the United States.[41] He likened the appearance of killing in film and media to a form of pornography, far divorced from reality. His motivation was to warn of the psychological cost of learning to kill, and actual killing, for individuals, groups, and the societies they belonged to. Yet, as a soldier, he felt it necessary to applaud the nobility, courage, and honour of fellow servicemen, which jarred with his purpose of condemning the act of killing. He made a direct connection between the incidence of killing and Post-Traumatic Stress Disorder and contrasted the willing choice to *fight* with the common reluctance to *kill* in combat, even in situations where an individual's life was in imminent peril.[42] He examined killing at a distance and at close quarters, compared the effect of groups, authority, and the processes that sanctioned the dehumanization of the enemy. He concluded that only 2 per cent of combatants enjoyed killing, and that the majority of humans have always been reluctant to kill. Nevertheless, he acknowledged that, under the right circumstances, everyone is *capable* of killing. For Grossman, war and violence were more forms of ritual display to establish prowess and status in front of others, but, he argued, the act of killing another human is rare and even antithetical.

Grossman did not place his emphasis on morale, but instead interpreted a failure to engage an enemy as a general reluctance to kill. Nevertheless, despite Grossman's agenda, high motivation has often enabled units to willingly kill their enemies, take high casualties, and yet still remain cohesive and continue to function. Australian troops going ashore at Gallipoli under fire, according to one eyewitness: 'Crouched, sat nonchalantly', laughed and joked, or 'cussed with a ferocious delight' since 'fear was not at home'.[43] In the close-quarter fighting that followed, these units did not exhibit any obvious reluctance to kill and were self-consciously cohesive.

We should nevertheless guard against the temptation to attribute universal characteristics in small unit dynamics. Units in defensive positions tend to cling to their bunkers and trenches as the means for their survival, while some formations in the attacking role are animated by a sense of momentum. Some groups suffer light casualties in close combat and fall apart, panic, and run, while others suffer heavy losses and fight on, even against insurmountable odds.[44] Poorly trained men, in attack, frequently herd together and present a bigger target, or hit the ground at the first shot and refuse to move forward.[45]

If Grossman is correct, fear and social cohesiveness have been more important on the battlefield than the desire to kill. On the other hand, Grossman's thesis, based on accounts by Vietnam War veterans and a selection of secondary works, ought to be challenged. Other accounts across a range of wars indicate that killing is all too easy, and has been facilitated steadily and incrementally by more lethal, rapid firing, longer range industrial weapons since the beginning of the twentieth century.

In her *Intimate History of Killing* (2000), Joanna Bourke argued that military history needs to address the question: Are soldiers 'Natural Born Killers'?[46] Like Grossman, she wanted to know if killing was part of human nature or whether killing depended on environmental factors. She enquired whether soldiers start off as killers, or become killers because of the sights and experiences they are exposed to. Bourke posited that killing was not accepted reluctantly as necessary but was embraced; there was even elation and pleasure in killing related to a sexual experience.[47] This suggests that feelings of power, conquest, and superiority matter a great deal in framing analyses of combat, but one would also have to acknowledge 'elation' is also a reaction to survival, coupled with the feeling of being able to outwit and defeat an opponent. For some, a moment of frenzy and revenge, especially after the loss of popular friends and colleagues, created a sense of satisfaction that justice had been done.[48]

Robert Jay Lifton, with his background in psychiatry, believed that there is an automaton-like quality to killing, which has been enhanced by the long range weapon systems of artillery, missile batteries, and aircraft.[49] Gwynne Dyer highlighted the metaphoric distance of the term 'target': 'Gunners fire at grid references... submarine crews fire at "ships" (and not somehow at the people on the ships); [and] pilots launch their missiles at "targets".'[50] Bourke concluded, by contrast, that there is an *intimacy* in killing in the close-quarter battle. It is 'personal' in the sense that infantrymen see the people they kill. She found that when soldiers had shot someone and killed them, some would still bayonet their victims not only to 'make sure' of the task, but also to connect with their opponent. Many took souvenirs from their victims as a marker or a talisman. She examined motivations that lay behind this phenomenon and found, unsurprisingly, that military training deliberately enhances aggression.

For individuals and groups, the structure of training and motivation before battle does seem to have an impact on emotional reactions during the fighting, but it seems unlikely that views about killing the enemy are either entirely 'personal' or merely 'automatic'.[51] The difficulty is once again the sources we use. In the face-to-face encounter in most accounts of combat, fighters had expressed elation at momentary victory, but this exuberance can be mistaken for the displacing of fear.[52] One common method to expel anxiety which also bonded groups together was the use of war cries. In the First World War Lieutenant O'Sullivan said his Irish Rangers astounded him as they went 'over the top' with a 'concerted yell and cheer that could be heard even above the bombardment' and when they closed with the Germans there was 'a vortex of shrieking; of yells and brutish grunting'.[53] In the hand-to-hand struggle between combatants, there is a great deal of shouting and swearing, reminding us that the human voice has always been used to steady wavering troops, exercise command, threaten, cajole, rally, and encourage. The Turks at Gallipoli invoked their god and together shouted 'Allah u'Akbar' as they fought, but assumed the British and Australians worshipped a deity called 'Bastards' because of the frequency with which their adversaries used this term.

Generalizations have been a recurrent problem in several of the studies in the field, and, as well as identifying patterns in the past, it was surely the case that culturally specific examples of killing and combat were needed. The idea of the sacred in combat is well established, but there have been significant variations across time and place. Religious motifs and ideas were invoked in such a way as to compel individuals and groups to kill without the sort of consequences that Grossman had identified in his work. For many premodern societies, individual acts of courage in close combat were the means to achieve adulthood, transition into a form of citizenship by proving their worth, and be eligible for marriage and leadership. The transformative process was marked by ritual or the collection of a talisman, trophies, or even body parts of the enemy. Throughout history, rituals and beliefs in a society, or a section of a society, may affect behaviours towards violence and killing, and therefore individual and group responses to combat.[54]

Barbara Ehrenreich suggested that the attraction to violence and our warlike passions stem from our ancient fear of being the prey of the larger and stronger predators that roamed the plains alongside prehistoric man.[55] It was only much later that man as predator evolved. She argues that following the decline of hunting, gender roles developed because of the need to find a role for men. Warrior rites enabled men to make the transition to manhood, and, she posits, also terminate a young male's dependence on women. The rites themselves emphasized group cohesion and solidarity rather than bloodlust, but her assertion that it was war itself that was the predator on human history, which mankind has merely served, has drawn many critics. She insists that the future of combat will be characterized by the autonomy of war itself, that is, combat without humans:

> Once set in place, the cyber-automation of war is hard to stop. Humans will cling to their place 'in the loop' as long as they can, ... [but] we have to face the possibility that the institution of war might no longer need us for its perpetuation. Human desires, especially for the Earth's diminishing supply of resources, will still instigate wars for some time to come, but neither human courage nor human bloodlust will carry the day on the battlefield.[56]

There is widespread disquiet with the idea that combat and killing are moving out of the hands of humans towards computers. Drone strikes seem to raise inordinate amounts of passion even though discrimination in targeting through greater precision is, in principle, preferred over area bombardment. There seems to be a semantic aversion to the vocabulary of precision warfare, with its emphasis on 'targets' rather than terms like the 'enemy' which at least referred to a human being; there is equal disquiet about the use of 'collateral damage' to describe inadvertent civilian deaths in the combat zone, an opprobrium that began in the Vietnam War. Professional soldiers maintain their faith in neutral and corporate terms to describe their roles. Critics argue it is a dangerous self-delusion of misplaced confidence in both business ideas and technology. In *Virtuous War*, James der Derian argues there is a Western obsession with war technology but an increasing disconnection with the reality of combat.[57] Where Eisenhower coined the phrase 'the military-industrial complex', der Derian identifies 'the military-industrial-media-entertainment network', suggesting that the West is beginning to believe, again, its own aspirations and ideals to the exclusion of all others. 'Full spectrum dominance' did not prevent the 'Black Hawk Down' incident in Somalia in the 1990s, and, if we are to believe de Derian, in encouraging hubris, it almost certainly led to it.

While visions of future war focus invariably on technology, the trend of the present is of more close-quarters combat not less. It is suggested that the very advanced nature of the West drives rivals to seek military advantage by other means, and that implies getting in close to Western troops in order to avoid devastating air power. Akhunzada, a Taliban commander, recalled the bombing of 2001 on his position: 'When the bombing started, I was commanding some 400 fighters on the front lines near Mazar-i Sharif. The bombs cut down our men like a reaper harvesting wheat. Bodies were dismembered. Dazed fighters were bleeding from the ears and nose from the bombs' concussions. We couldn't bury the dead. Our reinforcements died in their trenches.'[58] The result was for the Taliban to become insurgents and to make assaults in the brief window before Western air power could come on station, or make greater use of close range assaults, single round sharpshooting attacks, or the improvised explosive device (IED). Shawn Brimley of the US Army War College concludes:

> America's continued strength in major force-on-force conflict will incentivize future adversaries toward distributed cellular forms of insurgency characterized by the improvised explosive devices and ambushes seen in Iraq and Afghanistan, or towards the use of organized small-unit kinetic operations buttressed with employment of advanced technology such as the anti-tank and anti-ship munitions successfully employed by Hezbollah in the 2006 Lebanon War.[59]

Anxiety about asymmetry, so prominent after the Iraq and Afghan insurgencies, nevertheless elicited volumes of work that reinforced traditions of combat. Thus, while the 1998 report from National Defense University had defined asymmetry simply as 'not fighting fair', by 2006, analysts and military practitioners were reaching for old doctrines that could provide solutions to such ancient concepts as ambush along the line of march, assassination, sniping, raiding, and insurrection. These were then republished and presented as 'new', with reassuring paragraphs about 'courageous restraint', the ability to conduct achievable operations during 'wars amongst the people', and the importance of legal approval for target selection in air-to-ground operations. Far from the choice about generalized or culturally specific reactions to combat, doctrinally we are left with a set of very functional responses.

Historian Jeremy Black not only criticized the overconfidence of existing assumptions in military circles about the character of future warfare and the place of close-quarter combat in it, but also the approaches of many scholars. In *Rethinking Military History* (2005), Black observed that too much scholarship is Eurocentric, Western-focused, and driven by an emphasis on technology, which he calls the 'machinization of war studies'. He favours an appreciation of the cultures of armies and politics that influenced the manner in which combat and killing was viewed and conducted in the past. Black has often warned against ignoring cultural difference but also the risks of assuming consistency and continuities with the benefit of retrospection.[60] He disapproves of the oversimplification of war studies in the armed forces. This manifests itself, he suggests, in training that gives the impression of set phases of combat, which don't exist in reality at all, or the absence of killing and mutilation except in 'petite mash' medical exercises. At least, in this regard, there is realism: the OPTAG (Operational and Tactical Group) exercises in Stanford Training Area in eastern England have an impressive regime which includes paraplegics enacting the loss of limbs and serious wounds caused by combat.

Despite the agonies of the analysts, for professional and seasoned soldiers in the heat of action, when under fire, killing does not always elicit much of a response. The enemy is identified in the sights, 'taken out' or 'down', and then the search is on for the next target.[61] The cold rationality of killing is most pronounced among snipers, but many professional soldiers express themselves with dispassionate candour.[62] In contrast to Grossman's thesis, killing is relatively straightforward. In intense combat around Hill 60 in 1915, CSM Shepherd of the Dorset Regiment recorded many details in his diary but remarked typically and rather prosaically: 'we slung up a barricade near the left end of [trench] 39 where the Germans were running towards us, and picked them off nicely there, some we bayoneted... we shot scores of Germans on the hill'.[63] Shepherd was particularly satisfied by the ability to inflict casualties after his battalion suffered a major gas attack in May 1915, and he survived dozens of near misses until he was killed the following year.

Nevertheless, even when surrounded by comrades, after the elation of survival, individuals sometimes had physical and emotional reactions to killing an enemy. Some men vomited or found their legs shaking; some expressed guilt or remorse at taking the life of another human; others were depressed, or seemed more conscious of their own mortality as they reflected on the similarity of their victims and themselves. Nevertheless, group cohesion appeared to help veterans through personal and individual reflexes to killing.[64]

Grossman suggested there was a universal reaction to the taboo of killing, and stressed in binary terms the formula that, in combat, while an empathetic distance is created as the physical distance to the enemy decreases, this breaks down when the perpetrator is confronted with the corpse of the adversary they have just killed.[65] Ethnicity, nationality, and other criteria for stereotyping are the means by which the enemy one confronts in close-quarter combat can be regarded as the Other; that is, as a cypher and not a human being. Among professional soldiers, though, hate is surprisingly rare.[66] Denigration of the enemy combatant is less common than disparaging remarks about non-combat troops, 'REMFs' (Rear Echelon Mother-F**kers), and civilians who are safe, suggesting that cohesion by troops who face killing, or are regularly exposed to danger, is exclusive.[67] Disgust for enemies seems to be based on known atrocities, long-term antagonism or conditioning, and, most common of all, death of immediate friends.

In contrast to the more lurid stories of reprisals, one British soldier at Arras in 1917 recalled how he crossed No Man's Land under fire and attacked a defended German trench. In the close-quarter battle, he bayoneted a German in the shoulder, but, the immediate peril now being over, he promptly bound the wound of his adversary, then gave him a cigarette and water. Tony Ashworth's research suggested that there was a thriving 'Live and Let Live' system which created quiet sectors, respected by both sides, in the First World War, but the principle of the idea, it seems, can also exist within episodes of more intense close combat.[68]

The protracted character of particular wars seems to lower the threshold of tolerance of an enemy, affecting the way they are treated in combat. In insurgencies, government or foreign security forces are confronted by enemies wearing civilian clothes, concealed amongst the population. It is hard to define and understand an enemy that one cannot encounter, know, or even see. Hugh Gusterson assesses the tendency of the West to stereotype its enemy as fundamentally 'Orientalist', but it is a common problem for all security forces on the modern 'empty battlefield'.[69] The guerrilla fighter, unable to confront the stronger conventional or state power militarily with any hope of success, kills at night, through ambushes, raiding, and assassination. The popular image is that Afghan insurgents are religious warriors and that ideology characterizes the Afghan national style throughout history. Nevertheless,

Tarak Barkawi and Keith Stanski have shown how war drives stereotyping to particular extremes, especially when it comes to interpretations of non-Western belligerents such as the Afghans.[70]

For the Afghans, as for many other insurgencies historically and geographically, religion offers a sense of righteousness in killing, bonds combatants together, and, conveniently embraces death in combat as sacred. The opposition to the Afghan government and its Soviet military support in the 1980s was described frequently as a struggle for liberation from foreign ideologues in defence of Islam. In fact, the resistance believed their interests were not so narrowly defined: they were defending the people and the very fabric of their life in a people's war. One fighter remarked to a journalist: 'Why do you differentiate between fighters and old men, women and children? If you have courage and treasure freedom, you are mujahed. We are all mujahedeen.'[71] Despite the alleged racist denigration of the Taliban by American troops in the conflict since 2001, many express admiration of the insurgents' courage, cohesion, and acceptance of casualties. One American officer, after a sharp fire fight in the Korengal Valley, concluded: 'I wish I was made as strong as "Haj" [nickname for insurgents]...They were balls[y] to do what they did.'[72] In contrast, some Taliban fighters refer to the Americans as 'monkeys', 'bastards', 'pigs', and 'kids'.[73] If it is easier to come to terms with killing Americans on the basis that they are *kufrs*, disciplining the local population by killing them can have a deeply detrimental effect on the morale of the Taliban. One fighter, Yahya, claimed: 'I'd like to delete my past from my memory...I'm worried about how Allah will treat me for what I have done.'[74] Another stated: 'We were playing with the lives of people...We killed and harmed innocents, just as the infidel Americans do...The commanders didn't care.'

Overcoming qualms about killing appears to be easier for thoroughly trained, professional soldiers compared with conscript citizen soldiers, a point made by Anthony King in his book, *The Combat Soldier*.[75] A form of professionalism, exhibiting indifference to danger, was manifest in the testimony of Trumpeter Jimmy Naylor, RFA at Mons in 1914. As German infantry drew closer to his battery firing over open sights, Naylor expressed rising anxiety, but he felt suddenly steadied by the calm voice of a British infantry officer nearby calling out the ranges of the German troops as if on musketry training: Naylor exulted the *sangfroid* of the officer, noting the enemy: 'fell down like logs. I've never seen anything like it, the discipline, the fire discipline of those troops...I thought: "What a marvellous army we are!"'[76]

Naylor's account points to the effect of inspirational leaders and courageous individuals on the cohesion and motivation of troops in close combat. A combination of loyalty, identity, and an individual's spirited response to the prospect of danger sometimes rallied entire units. William Slim (later Field Marshal) was in an attack that stalled in the First World War:

> As we wavered, a private soldier beside me, whom one would have thought untouched by imagination, ran forward. In a voice of brass he roared 'Heads up the Warwicks! Show the Blighters your cap badges!' Above the din, half a dozen men each side of him heard. Their heads came up. They had no cap badges – they were wearing steel helmets – but they had remembered their regiment.[77]

This may be dismissed as romantic and sentimental, even fanciful, but whatever its veracity it reminds us that bonds of loyalty and allegiance may overlap from the individual, to small groups, and up to larger associations. The cohesive effect of identity is not limited to small groups, although in the supreme crisis of close combat, the smallest group identity matters most.

On the other hand, as King points out, discipline, and the constant communication of it, has a part to play too.[78] Flight out of the combat zone carries penalties, either 'official' ones imposed by the authority of the armed force which might include punishments for cowardice, or 'social' ones that involve fear of peer judgement. Stephen Wesbrook argued there are three primary methods of obtaining the compliance of troops necessary in preserving unit cohesion and preventing military disintegration: physically coercing alienated individuals, remunerating those who engage in battle in order to achieve tangible rewards, and the use of normative power to bind morally involved individuals.[79] Combat motivation may be reinforced by the presence of battlefield police.[80] Provosts behind the firing line were permitted to shoot stragglers or deserters to discourage flight.[81] Nevertheless, battlefield conditions tend to eliminate coercion as the principal means of control at the frontline. It is by normative control of morally involved individuals that armies ensure compliance.

At the other end of the spectrum, troops that had become cut off by an enemy might not automatically give in. Their isolated and desperate position could, in some cases, generate a bloody defiance. When the Royal Lancasters were overrun by the German Army at Bethune in 1918, one eyewitness recalled seeing his Company Sergeant Major fighting with his fists and boots, surrounded by Germans, until he was shot at point-blank range by a German officer.[82] The entire 2nd Devons, twenty-nine officers and 550 soldiers, in a similar plight at Bois de Buttes on 27 May 1918, also fought to the death, even though flight had been a perfectly viable option for them. An honour code had asserted itself over the rational choice of self-preservation.

According to John Lynn, cultural conditioning has an important part to play in understanding combat.[83] He calls for an acknowledgement that politics and cultural values need more prominence. The difficulty, as Lynn points out, is establishing the parameters of a 'culture': does it rest on the imagined communities of nation-states to produce, for example, an 'American Way of War', or is there a regional dimension, a 'Western Way of War'?[84] This presupposes shared values, but they are rarely uniform enough for such a confident assertion. It is unlikely that there could be a fixed method of fighting, a way of war, that is unchanging through time, given the political, social, and military transformations that have taken place in human history. That said, there are characteristics of conflict for some societies that do endure, sometimes dictated by human and physical geography, and those with deeply engrained cultural markers have often interpreted events through a lens that is distinct, leavened, of course, by functionalism and experience. Hence, impoverished desert communities living a marginal existence were forced to fight in a manner which made use of distance, the environment, and concealment, evolving tactics of raiding as a result. Their emphasis would be on possession of resources rather than necessarily killing large numbers. On the other hand, groups that feared reprisals from more organized enemies or for whom it was impossible to carry away all the spoils of war, including slaves, might conduct a form of warfare that emphasized annihilation. The problem we have is that so many combat histories are no more than partisan national or cultural histories designed to fit a set of expectations and cultural norms within a particular society: experiences are reshaped and defined according to accepted terms, narratives, and standards.[85]

Chris Hedges, journalist of the *New York Times*, has argued that, culturally, we believe 'war is a force that gives us meaning'.[86] Despite the pervasiveness of war in literature, film, and in celebrations of nobility, sacrifice, and security, Hedges believes it is a grand delusion. War destroys culture, perverts desire, and asserts a preference for death over life. He cites those who enjoy war for its sense of purpose, and the resolve it generates behind a cause.

The value of combat for society appears to be a fascination with one of human society's great taboos: the violence of taking another's life. This contrasts with some historic views that combat was much more than a fascinating observation. Helmuth von Moltke, who influenced the generation of German officers that went to war in 1914, had written in 1880: 'War is part of God's world-order. Within it unfold the noblest virtues of men, courage and renunciation, loyalty to duty and readiness for sacrifice – at the hazard of one's life. Without war the world would sink into a swamp of materialism.'[87] The regulation of war, training individuals to kill, and overcoming an inherent reluctance to kill was, however, less about a particular philosophy, regardless of how persuasive Moltke appeared to be, than a pragmatic inculcation of discipline, training, and sanctions. Ideas might motivate men to war and even establish the infrastructure to make war likely, but at the sharp end of killing, the context was much smaller, practical, and clearly defined.

Despite state and sub-state organization for combat and for killing, close-quarter fighting has become *less* common in the last hundred years compared with previous eras. Despite the recent experiences of counter-insurgency operations and other 'wars amongst the people', most deaths in war in the recent past have been either through disease, massacres of civilians, indirect fire, or area weapons (including air bombardment).[88] Relatively few deaths have been a result of face-to-face fighting. In Afghanistan, the highest proportion of US deaths since 2001 was IED detonations (40 per cent) and the combined proportion of deaths *not* involving small-arms fire was 70 per cent while deaths from hostile small-arms fire made up the remaining 30 per cent.[89] Despite the overall decline, the face-to-face encounter still occurs with remarkable frequency, and highly technological armies regard training for the close-quarter battle as essential to soldiering.

Conclusion

This chapter has addressed a selection of the issues that arise from evaluating the individual and group experience of combat, killing, and the role of cohesion since the First World War. It has also highlighted problems in the methodology and analysis of these phenomena. For the individual combatant, what emerged were great variations in reactions, with dichotomies of courage and cowardice, hatred and comradeship across the lines, and intense and vivid experiences that contrast with those that spoke of a curious time-lapse 'unreality'. While some enjoyed the baptism of fire and the ability to kill, others were clearly appalled. Beyond the focus on individual experience, some scholars have concentrated on group motivations, and these are conceived of as ideological (religion, patriotism, and identity), social (comrades, leaders, and small groups), and thematic (revenge, assertion of power, acquisition, and protection). Scholars are left with the difficulty of how to solve this problem of diversity and apparent contradiction. Taking only individual emotional responses to combat, killing, and cohesion, one could be forgiven for assuming that the 'Face of Battle' is entirely randomized and unstructured. Combat appears to have functional, cultural, and historical specificities, but we are left with some serious challenges in conceptualizing group reactions to killing and the ways in which one interprets participants' accounts. The distance in time and expectations about an individual's encounters can make profound changes to the record of combat *wie es eigentlich gewesen* (how it essentially happened) and thus our record of what happened in any given battle, and, moreover, what happened to those engaged in combat.[90]

The conclusion we might draw is that individual emotional reactions possess similarities in psychology but can be shaped by, as well as shaping, group responses. In turn, group cohesion, while exhibiting some universal characteristics, is context-specific. Small, self-directed, highly trained, and well-motivated groups appear to manage the stresses of combat and killing with greater composure than a mass of inadequately trained, badly led, and poorly motivated groups. Well-trained, seasoned, and professional troops exhibit different reactions to taking or inflicting casualties from raw citizen soldiers. The boundary though is blurred and dynamic. Individuals can take on the collective conscience of a crowd, but single soldiers can sometimes behave more bravely in front of groups and, in turn, inspire others to act collectively. In the past, armies have expressed such reactions under the universal theme of morale, but scholars identify a greater complexity than this term would imply. The challenge remains not only to reveal the complex nature of morale, but to situate the experience of close combat and the response to killing within it. The track record of scholarship so far has been mixed, but the following chapters seek to clarify and qualify existing generalizations with direct experience and empirical research.

Notes

1. Randall Collins, *Violence* (Princeton, NJ: Princeton University Press, 2008), 10–11; Pieter Speirenburg, 'Faces of Violence: Homicide Trends and Cultural Meanings', *Journal of Social History* 17 (1994): 701–16.
2. See, for example, Brian M. Linn and Russell F. Weigley, 'The "American Way of War" Revisited', *Journal of Military History* 66(2) (April 2002): 501–33 and Lawrence Sondhaus, *Strategic Culture and Ways of War* (London: Routledge, 2006). For the idea that an unchanging 'Western' approach exists see Victor David Hanson, *The Soul of Battle: from Ancient Times to the Present Day* (New York: Free Press, 1999).
3. Jeremy Black, 'Determinisms and Other Issues', *Journal of Military History* 68 (2004): 122–7.
4. I am deeply grateful to Professor Anthony King for his comments and advice on an earlier draft of this paper, and the opportunity to present this work to the *Frontline* Conference, 18–19 March 2013, at All Souls, Oxford.
5. Captain Lionel Ferguson, 13th Cheshire Regiment, cited in Malcolm Brown, *Tommy Goes to War* (London and Melbourne, J. Dent and Sons, 1978), 176.
6. Cited in Lynn Macdonald, *1914* (London: Macmillan, 1987).
7. Second Lieutenant Blake O'Sullivan, 6th Battalion, The Connaught Rangers, in Brown, *Tommy*, 180.
8. Captain B.H. Liddell-Hart, *Memoirs*, I, (London: Cassell and Co., 1967), 23.
9. Richard Holmes and John Keegan, *Soldiers: A History of Men in Battle* (London: Konecky and Konecky, 1997 edn.), 262–6.
10. Robert Graves, *Goodbye to All That* (London: Jonathan Cape, 1929), 157.
11. Richard Holmes, *Firing Line* (London: Penguin, 1985), 155.
12. Denis Winter, *Death's Men* (London: Allen Lane, 1978), 181.
13. Holmes and Keegan, *Soldiers*, 276.
14. Cited in Holmes and Keegan, *Soldiers*, 267.
15. Tonie and Valmai Holt, *In Search of the Better 'Ole: The Life, The Works and the Collectibles of Bruce Bairnsfather* (Portsmouth: Milestone, 1985), 34.

16. Richard Tawney, *The Attack and Other Essays* (London: Allen and Unwin, 1953) cited in Winter, *Death's Men*, 180.
17. Hugh Quigley, *Passchendaele and the Somme* (London: Methuen, 1928) in Winter, *Death's Men*, 180.
18. Collins, *Violence*, 83ff; Ardant du Picq (1903), 88–9, cited in Collins, *Violence*.
19. Collins, *Violence*, 19–20; 43–6.
20. Collins, *Violence*, 404; Russell Glen, 'Introduction' to S.L.A. Marshall, *Men Against Fire* (Norman: University of Oklahoma Press, 2000).
21. Holmes and Keegan, *Soldiers*, 263.
22. Holmes and Keegan, *Soldiers*, 274.
23. John Keegan, *The Face of Battle* (London: Penguin, 1976), Preface.
24. Interview with Harry Patch, Wells, 2003.
25. Ardant du Picq, *Battle Studies* (reprintd. 2012).
26. Edward A. Shils and Morris Janowitz, 'Cohesion & Disintegration in the Wehrmacht in WWII', *The Public Opinion Quarterly* 12(2) (Summer 1948): 280–315.
27. Samuel A. Stouffer, *Studies in Social Psychology in The Second World War* (Princeton, NJ: Princeton University Press, 1949).
28. Hew Strachan, 'Training, Morale & Modern War', *The Journal of Contemporary History* 41(2) (April 2006): 211–27, <http://www.jstor.org/stable/30036383>, 212. Accessed 16 February 2011.
29. S.L.A. Marshall, *Men Against Fire: The Problem of Battle Command in Future War* (New York: The Free Press, 1947).
30. Anthony King, *The Combat Soldier: Infantry Tactics and Cohesion in the Twentieth and Twenty-First Centuries* (Oxford: Oxford University Press, 2013), 42–4.
31. Roger Spiller, 'SLA Marshal and the rate of Fire', *JRUSI* 133 (December 1988): 63–7; Frederic Smoler, 'The Secret of Soldiers Who Wouldn't Shoot', *American Heritage* 40 (March 1989), 36–45.
32. Peter Kindsvatter, *American Soldiers: Ground Combat in the World Wars, Korea and Vietnam* (Lawrence: University of Kansas Press, 1995), 222–3.
33. Holmes, *Firing Line*, 167–72, 325; Collins, *Violence*, 57–8.
34. For a full discussion of this issue, and its redefinition, see King, *The Combat Soldier*, 27ff.
35. Stephane Audoin-Rouzeau and Annette Becker, *1914–1918: Understanding the Great War* (London: Profile Books, 2002), 39.
36. Alexander Watson, *Enduring the Great War: Combat, Morale and Collapse in the German and British Armies, 1914–1918* (Cambridge: Cambridge University Press, 2008), 32.
37. Graves, *Goodbye to All That*.
38. J.H. Schultz, 'Einige Bemerkungen uber Feindschaftgefuhle im Kriege', *Neurologisches Centralblatt* 34(11) (1 June 1915): 376.
39. Holmes, *Firing Line*, 381–2; Keegan, *Face of Battle*, 322; on 'overkill' see Collins, *Violence*, 95.
40. Graves, *Goodbye to All That*, 153.
41. David Grossman, *On Killing* (New York and Boston, MA: Back Bay Books, Little Brown and Co., 1995).

42. Collins, *Violence*, 77, relating Grossman's view that most belligerents were afraid to kill.
43. Holmes, *Firing Line*.
44. Rob Johnson, *Outnumbered, Outgunned and Undeterred* (London: Thames and Hudson, 2011).
45. See Anthony King, *The Combat Soldier*.
46. Joanna Bourke, *An Intimate History of Killing: Face-to-face Killing in Twentieth-century Warfare* (London: Granta, 2000).
47. Bourke, *An Intimate History of Killing*, 1 and 6.
48. Collins, *Violence*, 89.
49. See for example, his *Home from the War: Learning from Vietnam Veterans* (Other Press LLC, 2005).
50. Gwynne Dyer, *War* (London: The Bodley Head, 1996), 119.
51. Collins, *Violence*, 67.
52. Collins, *Violence*, 85–7.
53. Blake O'Sullivan cited in Brown, *Tommy*, 182.
54. See, for example, Craig Cameron, *American Samurai: Myth, Imagination and the Conduct of Battle in the First Marine Division, 1941–1951* (Cambridge: Cambridge University Press, 1994); Gautam Chakravarty, *The Indian Mutiny in the British Imagination* (Cambridge: Cambridge University Press, 2005).
55. Barbara Ehrenreich, *Blood Rites: Origins and History of the Passions of War* (London: Virago, 1997).
56. Barbara Ehrenreich, 'War Without Humans: Modern Blood Rites Revisited', in <http://ehrenreich.blogs.com/barbaras_blog/2011/07/war-without-humans-modern-blood-rites-revisited.html>. Accessed February 2013.
57. James der Derian, *Virtuous War* (London: Basic Books, 2001).
58. Sami Yousafzai and Ron Moreau, 'The Taliban in Their Own Words', *Newsweek*, 5 October 2009, 1.
59. Shawn Brimley, 'Crafting Strategy in an Age of Transition', *Parameters* 38(4) (2009): 27–42, 27ff.
60. Jeremy Black, 'Determinisms and Other Issues', *Journal of Military History*, 68 (2004), 122–7; Jeremy Black, *Rethinking Military History* (London: Routledge, 2004), 142.
61. WOII Dave Falconer, 1st Battalion, The Princess of Wales Royal Regiment, Iraq, 2004 cited in Darren Moore, *The Soldier* (London: Icon Books, 2009), 224.
62. Collins, *Violence*, 381, 384–5, 405, and 430.
63. Ernest Shepherd, *A Sergeant-Major's War: From Hill 60 to the Somme*, ed. Bruce Rossor (Marlborough: Crowood Press, 1987), 42.
64. Vejeune D. Gota, 'The Aftermath of War: PTSD, social support and alcohol consumption in Lithuania', paper delivered at the 'Afghanistan, the Cold War and the End of the Soviet Union' conference, Hamburg, 14–16 March 2013.
65. Grossman, *On Killing*, 97–8.
66. Collins, *Violence*, 69. He estimates that only 6 per cent exhibit such sentiments.
67. Sam A. Stouffler, Arthur Lumsdaile, M.K. Lumsdaile, Robin Williams, M. Brewster Smith, Irving Janis, Shirkey Star, and Leonard Cottrell, *The American Soldier, Vol II: Combat and Its Aftermath*. (Princeton, NJ: Princeton University Press 1949), 158–65.

68. Tony Ashworth, *Trench Warfare 1914–1918, The Live & Let Live System* (London, 1965), 194.
69. Hugh Gusterson, 'Can the Insurgent Speak?', in Tarak Barkawi and Keith Stanski (eds.), *Orientalism at War* (London: Hurst & Co., 2012), 84; see also Patrick Porter, *Military Orientalism* (London and New York: Hurst & Co., 2009); Max Boot, *Invisible Armies: An Epic History of Guerrilla Warfare from Ancient Times to the Present* (New York: Liveright Publishing, 2013); Richard Holmes, *Firing Line* (London: Penguin, 1985).
70. See also David Welch and Jo Fox, (eds.), *Justifying War: Propaganda, Politics and the Modern Age* (Basingstoke and New York: Palgrave, 2012).
71. Cited in Edward Giradet, *Afghanistan: The Soviet War* (London: Croom Helm, 1985), 163–4.
72. Elizabeth Rubin, 'Battle Company is Out There', *New York Times Magazine*, 24 February 2008. <http://www.nytimes.com/2008/02/24/magazine/24afghanistan-t.html?pagewanted=all&_r=0>. Accessed 20 February 2013.
73. Elizabeth Rubin, 'Battle Company is Out There'.
74. Sami Yousafzai and Ron Moreau, 'How the Taliban Lost its Swagger', *Newsweek International*, 7 March 2011; see also Sami Yousafzai and Ron Moreau, 'The Taliban in Their Own Words', *Newsweek*, 5 October 2009.
75. Anthony King, *The Combat Soldier: Infantry Tactics and Cohesion in the Twentieth and Twenty-First Centuries* (Oxford: Oxford University Press, 2013).
76. Lyn MacDonald, *1914* (London, Macmillan, 1987).
77. Keegan and Holmes, *Soldiers*, 46.
78. Anthony King, 'The Word of Command: Communication and Cohesion in the Military', *Armed Forces and Society* 32 (2005): 1–20.
79. Stephen D. Wesbrook, *The Potential for Military Disintegration*, in Sam Sarkesian (ed.) *Combat Effectiveness: Cohesion, Stress and the Volunteer Military* (Beverly Hills, CA & London: Sage, 1980), 247–52.
80. William Ian Miller, *The Mystery of Courage* (Cambridge, MA: Harvard University Press, 2000), 131.
81. Keegan and Holmes, *Soldiers*, 56.
82. Winter, *Death's Men*, 182.
83. John A. Lynn, *Battle: A History of Combat and Culture from Ancient Greece to Modern America* (Boulder, CO: Westview Press, 2003).
84. See, for example, Russell F. Weigley, 'The "American Way of War" Revisited', *Journal of Military History* 66(2) (April 2002): 501–33; Victor David Hanson, *The Soul of Battle: from Ancient Times to the Present Day* (New York: Free Press, 1999); Lynn, *Battle*.
85. *Dead Birds*, directed by Robert Garner, Peabody Museum of Cambridge, CRM Films, 1962, cited in Collins, *Violence*.
86. Chris Hedges, *War is a Force That Gives us Meaning* (New York: Public Affairs, 2002).

87. Helmuth von Moltke to Johann Kaspar Bluntschli, 11 December 1880, cited in Harry Pross (ed.), *Die Zerstörung der Deutschen Politik: Dokumente 1871–1933*, trans. Richard S. Levy (Frankfurt am Main: Fischer Bücherei, 1959), 29–31.
88. Rupert Smith, *The Utility of Force: The Art of War in the Modern World* (London: Penguin, 2006).
89. <http://icasualties.org/OEF/Fatalities.aspx>. Accessed February 2013.
90. Leopold von Ranke, *Geschichte der romanischen und germanischen Völker von 1494 bis 1514* (1824), preface.

Part II
Current Debates

4

From a Sociology of Units to a Sociology of Combat Formations

Militaries and Cohesion in Urban Combat

Eyal Ben-Ari

This chapter represents an analysis of the social and organizational dynamics of military units in urban warfare. Specifically, I develop and expand ideas about how cohesion takes shape in urban circumstances and how it forms a necessary basis for the emergence of more complex social structures in these conditions. Extending previous work, I centre my analysis on combat in citified environments rather than the more general state of urban war because it is at this hitherto understudied level where warfare actually takes place. In other words, I go beyond claims that many current military activities take place in urban contexts to ask about how this very context influences sociological aspects of combat. This kind of analysis is important because armed conflicts are increasingly waged in various conurbations: the twentieth century is littered with city battles, and it is highly likely that the sprawling cities of the developing world will become one of the key battlegrounds of the twenty-first century.[1]

Analytically, I argue that urban combat provides us with a special insight into the question of cohesion because cohesion is both more difficult and more necessary than battling in open terrain. It is more difficult because the decline in force numbers (the end of conscription and downsizing) combined with the ways cities absorb, disperse, and fragment huge numbers of forces, means that small units and individual soldiers actually find themselves more isolated than in the past (and thus have to rely even more on their comrades). Cohesion is more necessary since combat in highly complex and multidimensional urban environments demands ever more of soldiers in terms of their ability to carry out closely coordinated collective practices. Accordingly, my focus in this chapter is specifically on urban combat operations rather than urban stabilization operations that are governed by much less intense dynamics. Theoretically, an analysis of urban combat necessitates the complementation of the conventional arsenal of social scientific concepts centred on cohesion with new frames. I build on new thinking in organizational studies, cognitive sciences, and anthropology to provide a theoretical account that complements the dominant sociological paradigm—essentially a functionalist one owing much to Durkheim's sociology.

Methodologically, I utilize an array of previous studies as well as some journalistic and autobiographical literature pertaining to the armed forces of the industrial democracies in Iraq and Afghanistan to illustrate my contentions. My special focus however is on the Israel Defense Forces (IDF) since its experience or urban combat is no less instructive than the experience of forces in other areas. Despite being a citizens army of conscripts militia and focused on territorial defence, the IDF is very useful as an example: in effect it is highly selective in the troops it chooses for the elite infantry and armour units I refer to. Moreover, these troops spend a full three years in service with at least another fifteen to twenty years in the active reserves. This situation means that apart from armed patrols and ambushes along the country's boundaries, troops almost always participate in at least two major armed conflicts during their careers. In addition, while selective service has brought into the ranks of NCOs and junior officers significant number of religious members, previous research has shown that, because of the professionalization of the elite forces, external features such as religious affiliation or political leanings usually do not have an impact on the actual performance of soldiers in units.[2] For these reasons the case of the IDF is roughly similar to that of the armed forces of the other industrial democracies and may be instructive for my purposes.

Urban Environments: Three Sociological Questions

Professionals in the industrial militaries have long had and continue to have an aversion to urban warfare because, like fighting in jungles, forests, or mountains, in citified environments the advantages of a mobile force with significant firepower are negated.[3] Because of the impairment of vision, intensification of sounds, and potential for surprise associated with close city quarters, urban combat is often perceived by soldiers to be more threatening than battles fought 'out in the open'. Take Evans' characterization of urban operations during the Second World War that

> demonstrated that fighting in towns and cities was extremely tiring both physically and mentally. Both the Axis and Allied forces found that urban combat combined military danger with industrial hazards. Fighting in cities came to represent a form of warfare that was multidimensional in character—demanding simultaneous tactical attention at horizontal, vertical and interior and exterior levels—a situation that necessitated constant alertness and frequent regrouping of forces.[4]

Accordingly, movement within cities and towns may take place on rooftops, roads, inside buildings, or through cellars, tunnels, or sewage systems. And

> Soldiers forced to deploy into sprawling industrial cities find that their military activity is immediately hostage to the spatial restrictions of built-up environments. Soldiers soon discovered that to operate in an urban area is to confront a succession of microenvironments ranging from narrow streets and alleys through corridors and courtyards to rooms and rooftops—all of which combine to effectively compress operational activity. Commanders on both sides in World War II soon discovered that the confined space of streets and buildings in cities did not permit the traditional deployment and manoeuvre of battalions and brigades led by senior officers. Rather, the prime responsibility for combat in an urban area rested with platoon and squad commanders and, as a result, most of the basic small unit tactics of modern urban combat familiar to today's soldiers were developed between 1939 and 1945.

Moreover, despite obvious differences—the city of Najaf is very different from Pristina, which, in turn, is very different from Gaza—urban complexes are marked by certain basic similarities. They contain densely situated multiple types of spaces and places within one environment that are used by regular troops and insurgents, tribal militias, warlords, and diverse non-state organizations. Accordingly, for soldiers, urban environments engender a curious mix combining claustrophobia, isolation, and separation along with feelings of being very exposed to multiple threats. At one and the same time as providing some cover and a sense of security inside buildings and rooms, soldiers moving along the corridors—avenues, streets, or alleyways—of cities and towns feel that they are uniquely exposed (sentiments often amplified by a cacophony of sounds and smells). Tactically, the problem is one of identifying enemies, yet the phenomenology is one of intense vulnerability brought about by the difficulties in spotting foes and once inside built structures a feeling that troops are enclosed and blind to what is going on outside. These emotions may be heightened for troops travelling in tanks of Armoured Personnel Carriers (APCs) where they are stuck inside moving containers with little awareness of what goes on the outside. For instance Williams talks about the 'vibrating darkness of the hull' of the armoured vehicle he rode into Iraq.[5] The social-psychological question here is thus connected directly to the core issues of this volume: the place of (small) groups of soldiers in providing a sense of security and support in such circumstances.

The characteristics of urban combat raise a set of issues, three of which I will deal with in this chapter. First, such environments entail a huge absorption of forces and their *fragmentation and dispersion.* In his study of the Battle of Stalingrad, the British military historian Beevor found such a pattern among German and Soviet forces.[6] He concluded that most of the combat in the city consisted of relentless, lethal little conflicts carried out by assault squads comprising between six and eight men. In urban battles, soldiers often find themselves in 'small packets' and where defiles are tightest may move only one man at a time.[7] This is the idea at the root of the US Marines' 'tactical bubble' that can manoeuvre abruptly when confronted by a new threat and maintain itself as it moves into another area of operations.[8] Being relatively autonomous in their actions, these tactical configurations can also be seen as *sociological bubbles* with some rather particular problems of coordinating with other units and maintaining internal and external leadership and communication. Sociologically, I ask about the effects of the creation of social bubbles for their internal dynamics. From the point of view of the sociology of combat the move is from a focus on individual units maintaining some kind of coherence with other detachments, to a situation in which units are more isolated than before and must now preserve strong distal relations with other units and with leaders.

Second, although the majority of engagements in urban settings take place at close range and infantry soldiers fight in their organic groupings, they rarely enter battles in one-arm formations. Thus, although urban combat has been described as being a peculiarly infantry skill, combined-arms teams are actually the norm for fighting in cities.[9] What one finds are *an array of temporary, modular assemblages* that are created for specific missions. While combined arms formations have been common in urban combat at least since the Second World War, today they comprise a greater mix that is evident down to the very lowest tactical level and that may include, apart from infantry and armour elements, snipers, intelligence operatives, bomb disposal squads, and even dog teams for identifying explosives.[10] The sociological question is what keeps these amalgams together, and involves an analytical move from a focus on individual units to the ways in which they combine and cohere into tactical formations. Accordingly, I suggest that we move from a sociology of combat units to a sociology of combat formations (in the dual sense of the word in English—as a noun denoting form and as a verb of forming). If the platoon, company, battalion nexus was the focus of the older sociology, the new one is on combined formations, on joint forces (using the word joint in the sociological and not doctrinal sense of assemblages designed for specific missions). In other words, this move opens up questions about the relations between internal unit cohesion and the capacity to perform together with troops of other units (often from other military arms and services). This point does not mean that cohesion is no longer important in impelling troops to fight or perform professionally but rather that we need to ask further questions about the way cohesion is a building block for (or an impediment to) these temporary assemblages.

The third feature is that of the *shift between multiple tasks* within one operation. The term 'Three Block Warfare' was coined by a former commander of the US Marines,[11] to characterize circumstances where troops are engaged in a spectrum of operations from humanitarian missions, policing-type operations, through peacekeeping and peace enforcement-type actions to full-blown combat—possibly within the space of three city blocks. In this view, civilian dimensions of actions in cities are not seen as somehow ancillary to military work but as part and parcel of the ways in which the armed forces act in contemporary conflicts. What is important here, however, is that there may be sudden shifts from one mode (say attacking armed foes) to another (such as dealing with civilians).[12] Indeed, the simultaneity of missions that militaries are tasked with in urban areas has become very much a permanent feature of contemporary conflicts. The organizational question here is not so much about the diversity of tasks but rather that of transitions between different situations: the cognitive, emotional, and corporeal move from one state—say heightened aggressiveness—to another—for instance, the restraint needed in policing civilians.

There are yet other peculiarities of contemporary urban combat—for instance, the exposure to the media and other actors—but the limitations of this chapter preclude an analysis of them.

A Sociology and Psychology of Contemporary Urban Warfare

What kind of interpretation would current sociological models offer in regard to urban combat? I suggest that such analyses would treat the dynamics of urban combat primarily in terms of an additive understanding based on the classic social-psychological model: more stress and additional anxieties, or less effective functioning and reduced control of units. Given that the classic sociology and social psychology of combat were developed in terms of the idea(l) of battles between regular armies in open terrains, when applied to urban contexts the idea is a simple one: urban circumstances represent added pressures and further friction leading to greater potential for individual and unit breakdown. Why is this so?

The classic social scientific models of combat are rooted respectively in clinical psychology and social psychology: while the first emphasized risk factors leading to stress, the second underlined cohesion and support as bases for small-group functioning; while the first saw stress as leading to individual breakdown and ultimately to combat stress reactions (CSR), the other saw the destruction of primary groups (peers, leadership, communication) as leading to lack of effectiveness. According to these models, because battles involve intense violence with one side trying to destroy the organizational structures and bodies and souls of the members of the other, military organizations have had to create and maintain abilities to withstand intense pressure through drills, discipline, leadership, and social support among members of small groups. This situation meant that the older sociology focused on the view that social structures and behaviours enhanced the resilience of units in the face of the potential breakdown of combat.

The clinical psychological model developed out of the work of physicians during the First World War and later developed closely in alliance with psychologists. This psychiatric literature sees adversity in combat as having inevitable and deleterious consequences, magnified in the setting of industrialized warfare.[13] Hence breakdown in battle in this model is a consequence of overwhelming fear and anxiety which because of either psychological conditioning and/or neurobiological changes may become fixed as a chronic anxiety disorder that is currently labelled as Post-Traumatic Stress Disorder (PTSD). The core understanding of the social psychology of combat, in turn, developed during and after the Second World War in terms of the dynamics of small groups, and was formulated by the efforts of what Wessely calls the 'big three':[14] Shils and Janowitz, Marshall, and Stouffer and his colleagues.[15] Accordingly, primary groups create cohesion, which in turn enables increased endurance during battles. Later Charles Moskos persuasively argued that it was task cohesion and trust in the professional ability of others and not personal intimacy or shared experience that was essential to cohesion.[16]

Through a host of organizational developments these essentially analytical insights were used as prescriptions for training combat troops:

> Conventional sociology and psychology of the military basically offer recipes for creating individuals and units that can perform military tasks within the stressful conditions of conventional combat.... In other words, this literature can be read as offering guidelines for countering group and individual disintegration brought about by 'typical' combat. The analyses of this scholarly literature are based on strong functionalist assumptions (common to more general strands in psychology and sociology) about the utility of certain behaviours. These assumptions center on the idea that military commanders can prepare units for combat by enabling them to withstand terribly stressful conditions for a certain period.[17]

And urban warfare? Helmus and Glenn offer the best example of this kind of functional view when they argue that urban combat operations can potentially lead to greater losses of manpower due to breakdown because the unique demands required by this operating environment may pose a threat to troops' psychological make-up.[18] Steed contended that urban warfare demands a great deal more of fighting soldiers and officers than combat in open terrain: it necessitates, according to him, higher levels of mental agility and psychological resilience, since troops fight in a degree of isolation far more psychologically demanding than in past wars.[19] The reasoning in such assertions involves the multiple stressors of urban operations derived from the unique character of citified environments outlined before: a three-dimensional setting with innumerable fields of fire, poor concealment for offensive forces, close-quarter fighting, diluted leadership, restrictive rules of engagement, and ambiguity as to the identity of hostiles. The advantage of these approaches lies in analysing the psychological aspects of urban combat.

To be sure, just as the older sociology of combat provided recipes for training and action, so does the new one specifically developed for urban operations. Thus military professionals are inundated by a prescriptive literature providing recipes and describing the ideal characteristics needed by military forces in urban combat. For instance, Glenn and Kingston argue that in citified contexts commanders and troops should be more flexible, able to join other forces, or decide quickly or that training needs to prepare commanders and troops to deal with the intermittent communications and perishable nature of situational awareness inherent in the high-density environment characteristic of built-up areas.[20] But these are prescriptions not social science.

Anthony King's recent tome represents the most sophisticated development of relevant assertions about urban or close-quarters combat.[21] King's thesis is that with the end of conscript militaries, there have evolved highly professionalized forces that emphasize training and discipline, and engender cohesion defined as the special solidarity which necessarily pertains in a platoon because all of its members are immediately known to each other and their actions are interdependent, mutually supporting and reciprocal. According to him, battle drills are the core of professionalism because they embody the central individual, communal, intellectual, and physical skills which are requisite for success at this level of combat. While there is a similarity with older drills, King continues, contemporary infantry doctrine is characterized by much greater detail and comprehensiveness than in the past. Furthermore, collective level teamwork and action are important because they have become impressed onto the mental and bodily responses of professional soldiers. Professionalism is so important because of the particular social and political environments within which contemporary militaries operate. One aspect is the requirement for accuracy because of the presence of civilians in current-day battlefields, especially in urban contexts. In terms of the psychology of combat (and building on the insights of Marshall), King says that ironically from such a psychological point of view the move has been from not shooting at all to shooting accurately.

Sociologically, King is aware of the fact that 'soldiers never operate alone (unless all their colleagues have been injured or killed). They operate as part of a platoon, squad, assault team or, in the urban environment, a "stack". A stack refers to the formation which an assault team makes as it enters a building.'[22] But while King brilliantly analyses the drills and manoeuvres that are internalized by soldiers, he does not go into the social dynamics of stacks, or what I refer to as the sociological bubbles, they find themselves in. Nor does he develop his thesis in regard to questions about multi-arm coordination and operation. Where he does develop such ideas is in regard to professionalism as a requirement for coordination between teams:

> Buildings necessarily channel assault teams down corridors or stairwells, separating them as sub-teams clear rooms individually. Unlike rural warfare, the sections and fire-teams can quickly become separated and dispersed, unable to see or hear one another. It is precisely because of the complexity of the urban environment and above all the bewildering collective action problem which any building can pose an assault team that collective drills have become so central to it.

But because King focuses on the actual assault and the (relatively) short time it takes, he does not go into questions of communication and leadership that may emerge after hours of fighting rather than the minutes of a specific assault. To conclude, the advantages of King's analysis lie in his historical contextualization of the impact on the tactical level of professionalization and an understanding of the functional equivalents of cohesion and local level leadership that have developed in order to allow fighting in cities. But his emphasis on professionalism seems to limit his view to the micro-tactical level of independent one-arm (infantry) units waging combat for a very limited period of time and which deal with no other missions except combat.

In what follows I lead my analysis to the three issues that complement King's analysis. Accordingly, my model should be seen as adding to previous analyses because many insights of the older paradigm (the importance of cohesion and leadership, the centrality of stressors and anti-stressors, and the significance of training) continue to be important for understanding urban combat. What is needed are some new questions.

Towards a Sociology of Combat Formations in Cities

Specifically I address three key issues.

Sociological Bubbles: Dispersion, Communication, and Autonomy

To be sure, sociologically speaking, cohesion within relatively isolated and independent forces—within the sociological 'bubbles'—implies that they do not disintegrate in combat. But under the circumstances of dispersion and isolation somewhat different social structures and dynamics than the ones that exist in conventional open war are created since units are often hidden from even occasional visual contact not only from enemies but from commanders and other units. Hence, in terms of communication, in urban contexts, troops clearing a building room by room require a denser signals system than is found in open-terrain combat and one that may reach down to cooperation between fire-teams since visual gestures are unavailable. One result is a heightened emphasis on go-betweens, soldiers who maintain eye contact between compartmentalized units and relay commands. Yet because of limited visibility there is a further heightened dependency on communications nets to understand what is going on. These experiences of isolation thus differ from the infantry's imagined battlefield of open spaces where control and coordination are often achieved primarily through sight, hand signals, and shouting. In other words, it is not enough that the soldiers be thoroughly drilled in urban combat, since their dependence on communications nets makes for a different quality for coordination, support, and leadership.

However, if the signals' net is devolved down to the level of fire-teams or even individual soldiers, it may actually create problems of transmission and interpretation overloads that may impede support and leadership.[23] Citing American evidence, Ackerman states that over the course of training, US Marines begin to economize their communication: they learn to listen to communications traffic among squad and fire-team leaders to acquire situational awareness.[24] Whether this is true of the context of combat as well, one implication is that individual soldiers become much more aware and dependent on the disembodied interaction on the airwaves. Moreover, at least since the Second World War, combat commanders have been highly aware that electronic communications carry huge implications for leadership: a calm clear voice inculcates a feeling of confidence among subordinates. Except for the immediate presence of junior commanders, within urban combat the transmission of messages by commanders is carried out almost completely via the airwaves, and social structures are hence often handled through disembodied voices. In this manner while new technological systems *enable* the movement of people, ideas, and information from place to place, person-to-person, or event to event,[25] they may also create new problems of dependence and overload. Interaction via electronic means thus, at once, allows the bridging of distance but comes at a price: there are much fewer non-verbal cues that are important for social support and guidance.

Next take leadership and decision-making. In urban circumstances the autonomy of the sociological bubbles implies that much more control and leeway for decision-making devolves down to lower level leaders than in open warfare. This point is borne out by the Israeli experience as a senior IDF commander in Gaza notes, 'Today the field is in practice the territory of the platoon commanders, they are the first to react to events, and in practice during the first moments of the event they must provide the solution, and on the whole, when we the [more senior] commanders, arrive, the whole thing is already over.'[26]

However, when a larger scale mission is initiated or when an emergency occurs, the autonomy of lower level commanders is often suddenly and radically restricted. This situation is exacerbated by technological developments. Current-day communications technologies enable self-organization and allow a much more fluid system of coordination—through a continuous stream of synchronization, negotiation, and movement with people who are present as well as those that are absent. But this kind of communication, in turn, is closely related to the overall control of tactical bubbles through aural or even visual control (via drone-mounted cameras) mediated through electronic means. It has been said that Mogadishu was 'warfare's first use of sophisticated imagery and real-time visual and communications links. Commanders and staff watched the units as each heard their requests'.[27] This situation implies that what appear to be autonomous units may actually be much more centralized ones than in the past and dependent on central decision-making and leadership. The problem in urban quarters then is constantly shifting modes of autonomy at local levels.

Next, increased autonomy, combined with the uncertainty of combat in general and urban warfare in particular, very often necessitates inventiveness on the part of lower level units. The unique circumstances of cities imply that forces do not always operate according to doctrinal directives but must constantly innovate. As Baum explains,[28] field artillerymen, tankers, and engineers have found themselves serving as infantrymen in Iraq, while infantrymen were building sewer systems and running town councils. According to him, platoon and company commanders exercised initiative in an impressive manner despite a lack of guidance from higher quarters. In 'Ramadi, the marines have rewritten their training manual for urban warfare. Having been taught to seize towns methodically, block by block—a mode more appropriate to Stalingrad than Baghdad—they have learned to patrol at high speed and on foot, sending snipers on the rooftops ahead, along streets littered with bomb debris and daubed with hostile slogans.'[29] In the Israeli case, fighting in urban Palestinian areas has meant the development of such practices as breaking walls in order to advance between dwellings. Thus in the Balata refugee camp the paratroopers brigade used sledge hammers or concrete saws for moving between houses so that they would not be shot by armed Palestinians (although this practice carried a price in terms of media reports about the destruction of houses). In Jenin, in order to lower risks to soldiers, one battalion developed a practice in which a bulldozer took out the corner of a house and then an APC would offload soldiers straight into the gap created so that they would be relatively protected.[30]

To refer back to the classic paradigm and the work of Anthony King, it is increased professionalism that actually allows improvisation that is the outcome of the interaction of the elementary forms of organization and action. Perhaps the best metaphor here is that of jazz musicians who improvise only once they have mastered the elementary moves of play: because they are masters they can improvise. Indeed, it appears that precisely because professional soldiers have mastered the complex drills involved in their repertoire of collective practices they can innovate and improvise.

Ad Hoc Units, Swift Trust, and Cohesion

The battles of Berlin and Stalingrad like many city clashes during the Second World War were marked by close cooperation between infantry and armour with occasional artillery using direct-fire techniques.[31] Today's urban encounters include other types of units working in close cooperation with tanks, infantry troops, and self-propelled artillery: for instance, bomb disposal squads and engineers, snipers and special forces, air control officers and (military or civilian) prisoner interrogators, or even Civil–Military Cooperation (CIMIC) officers. The idea behind these amalgams is to enhance the ability of a commander to use resources outside the formal boundaries of his (or very rarely her) force. For example, as King explains,[32] for the new European brigades: 'The role of the artillery officer and his signaller calling in indirect fire has become a small team, coordinating a diversity of fires. As a result of improved coordination, tactical capabilities are coagulating around these new concentrations of military competence in Europe.' What has happened during the past two or so decades is that ad hoc formations comprising heterogeneous components created for specific missions appear at a much lower tactical level: Tanks and infantry still work side-by-side, but instead of three companies of one supporting one of another, it is more likely that a single armoured vehicle works with a squad or platoon of foot soldiers.[33] Thus we are witness to the rise of new organizational forms: instant units—ad hoc aggregations—are often the preferred structures for urban combat.[34] The result is an increase in the number of separate elements that commanders must oversee, support, and lead (and often may also include indigenous forces or private contractors).

But it is only recently that key sociological questions about these assemblages have been asked: how do separate (often organic) units marked by internal cohesion work together? How do these configurations cohere under military leaders at the tactical level? An answer to these questions necessitates going beyond the level of the squad or platoon—those units that the older sociological paradigm focused on—to the links between such units and elements from other arms and military specializations. In other words, the emphasis on jointness as a sociological problem requires us to look outside the elementary forms of military organization—infantry platoons—as bounded units.

Here the crucial concepts are swift trust and professionalism as bases for ad hoc teams. In the formation of temporary units, it is the *process* which is of importance and not only structure. Meyerson and her colleagues developed the concept of swift trust to explain the dynamics of temporary teams whose existence is formed around a finite span of time and for a common task.[35] Such teams consist of members with diverse skills, with a limited common history of working together, and little prospect of working together in the future. Temporary or ad hoc teams include a large variety of groupings such as study missions, music performers for one-off performances, film crews, presidential commissions, construction gangs, cockpit crews, or medical teams in operating theatres. Such temporary groups constitute an organizational equivalent of a 'one-night stand': they have a finite life span, form around a shared and relatively clear goal or purpose, and their success depends on a tight and coordinated coupling of activity. But how do such configurations cohere?

In ordinary circumstances trust is created through minute processes of exchange: actors learn the relative value of things, establish new bonds, and convey the centrality of reciprocity. What happens in the regular formations of the military is that trust is built over a span of time and what is often formed is a cohesive group or unit. Often this trust, as King persuasively argues, is not necessarily based on intimacy but rather on professional conduct.[36] In the case of ad hoc units, however, the tight deadlines under which teams must work together leave little occasion for building professional relationships, and are marked by very little socialization, courtship, and other types of communication. The harsh schedules, moreover, are intensified by the deadly circumstances of combat. What we find in the case of instant units is an intensification of time through the constraints of assigned missions. The dynamic is one of amplification: relations and trust must be created much more quickly and swiftly. The mission becomes the stimulus that sets the dynamics of the instant unit and speedily coheres diverse team members.

> Within these circumstances, because time pressure hinders the ability of team members to develop expectations of others based on first hand information, members import expectations of trust from other settings with which they are familiar. In other words, individuals in temporary groups make initial use of category driven information processing to form stereotypical impression of others.

> One finds that troops... are aware of the reputations – some may say stereotypes – of different units. It is these professional reputations, in turn, that provide the initial knowledge of how to proceed with the collaboration necessitated in temporary combat formations.[37]

Yet the existence of reputations as a basis for action is not enough. Once units have met—very often for only a few hours or even minutes—a process of mutual testing ensues. During this stage a variety of temporary arrangements are used like a 'canary in a coal mine' to detect possibilities for further collaboration to more sensitive areas without suffering the consequences associated with plunging into untrammelled terrain.[38] Such tests may be trivial and include simulations over maps or observations of how a commander of a unit to be joined with one's own unit handles his troops.

Other factors contribute to the success of such instant units. First is the adoption by the 'host' unit of 'guest' elements in both logistical and professional terms: for instance by providing food and places to sleep or the sharing of knowledge and mutual learning. Such sharing, even in minuscule ways, involves a process of 'community-building'. Second, empirical studies of the Israel Defense Forces show that the most successful cases of cooperation were those in which small forces worked within larger ones (say, an armoured company commander under an infantry battalion commander or an engineering breaching squad within a company).[39] And third, in all of these cases a governing factor involved the ability of the smaller units to fully and freely express their professional knowledge and authority without direct relation to their military rank. To put this point by way of example, a lieutenant commanding an armour platoon is consulted by an infantry company commander with the rank of captain regarding the limits and potentials of using tanks in urban terrains.

For all of this, cohesion continues to be important. It seems that while cohesion may exist at the small-group level, other dynamics—those related to the creation and maintenance of swift trust—are important at the level of the ad hoc formations. Thus the elementary forms of military structure that King analyses are a *precondition* for the creation of larger formations even to the level of company or battalion-sized forces.[40] The very professionalism of current day forces allows the ad hoc amalgams to form quickly because cohesion is not the unconditional love or loyalty for others in a unit—popular cultural depictions to the contrary—but rather unity forms among members of a unit around training, drills, and the execution of collective drills. Potential differences are eliminated around the drills because these troops must perform concrete procedures to survive combat. In other words, unity in ad hoc units is dependent on the cohesion that emerges at the level of the forces comprising them. No less important, such professionals have trained, often hundreds of times, specific drills, but also have developed (over time) a meta-skill that can be called a capacity 'to join', to 'connect', and 'disconnect' with a variety of other kinds of units. These are the skills and processes that underlie the modular ability prescribed by military professionals and that complement the highly skilled performance of small-group tactics.

Transitions

Krulak's by now proverbial argument was that armed forces today have had to develop capacities for 'three block warfare', that is, the ability to carry out multiple tasks such as giving humanitarian aid, policing traffic, and engaging in high-intensity battles within, for instance, three city streets located near one another.[41] But from an organizational point of view, a key problem is not only developing these diverse capacities but the meta-ability to manage moves between the different missions. This notwithstanding, transitioning between military tasks is something that has not been hitherto studied systematically.[42] Previous studies have explored the cognitive difficulties in recognizing and defining situations in contemporary urban circumstances where enemies may appear to be friendly civilians or where peaceful demonstrations may suddenly turn violent.[43] Yet even if definitions of different situations are relatively clear or emerge quickly, there is still a difficulty in switching between tasks.

While transitions are not necessarily crises in and of themselves they do contain *potentials* for personal and organizational crises. Moreover, in situations where the actions of even small units may carry strategic implications an unsuccessful move between situations may be disastrous.[44] For example, the shift from routine barracks life to high-intensity engagements may be problematic because the character of everyday military life characterized by routine and boredom could impede the attentiveness and aggressiveness needed in combat. Probably the most glaring example of this difficulty was the entry of Israel's combat brigades into the Second Lebanon War. It took them quite some time to do what commanders referred to as 'change the diskette'. Because soldiers were expecting and operating according to the characteristics and modes of operation of the West Bank and Gaza they had problems in shifting to the high-intensity conflict of Southern Lebanon. Similarly, a switch from a highly intense set of circumstances to policing carries the potential of inappropriate violence that is out of control.[45] As a solution to the problems of shifts between modes of action, observers have suggested managerial prescriptions, that militaries be more ambidextrous,[46] or omni-competent.[47] The idea is that they acquire an ability to move physically, operationally, and psychologically between situations. But giving a label to this ability is not enough because what it does not delineate the dimensions of this capacity and a demarcation of what can be changed more or less easily within transitions.

My contention is that problems involved in transitions entail cognitive perceptions, emotional states, *and* bodily attitudes. Concretely, in transitions difficulties centre less on cognitive recognition than on the timelines or 'spillover' of emotions and body attitudes between circumstances: specifically, while the overflow in emotions is between specific situations (it takes time to calm down or to psych-up), bodily attitudes are often trans-situational and are embedded in the corporeal reality of troops. For example, fear, anger, or exhilaration experienced during high-intensity conflict may dissipate after an armed battle, but they take much longer to dissolve than the actual physical move away from the battlefield and the cognitive understanding that the firefight is over. In fact, sometimes after a battle forward panic emerges: tension and fear come out in an emotional rush that is often accompanied by the uncontrolled use of violence.[48] In other words, soldiers are caught up in emotional rhythms that carry them to actions they would not normally approve of in calm, reflective moments. Indeed at times transitions may even be accompanied by a visceral difficulty in shedding the previous identity—for example from the aggressiveness of an interrogator into a family member or from a soldier in battle to a trooper during a lull.

But the rhythms by which emotions wax and wane are very different from the slow changes bodies undergo. Bodily attitudes cultivated in combat soldiers during their professional socialization are sometimes so internalized that they do not leave the bodies of the troops even after military service;[49] they are much longer lasting than emotions. In other words, embodied stances have momentums that transcend specific circumstances and may contravene cognitive understandings and even emotional states. We are talking about things like posture, demeanour, and motion that are nurtured during formal training and informal exchanges between troops. One example involves the British troops during what has come to be known as 'Bloody Sunday' where the embodied aggressive attitudes of paratroopers came to dominate their actions.

Transitions may be especially difficult for combat soldiers since their socialization centres on high-intensity battles, and all the more so in today's highly professionalized armed forces. As Grossman and Collins show,[50] violence against humans does not come easily or automatically, and that is why so much effort is invested in military conditioning of soldiers to the use of weapons and the appropriate drills for using them. It is during the initial period in the military (and constantly reinforced along the military life course) that the corporeal necessities of waging violence in battle become internalized and embodied. As Hockey put it,[51] the somatic memory of drills becomes deeply embedded and practical mastery of them gradually penetrates the very being of individuals so that they become taken for granted in the soldier's flesh; they come to pervade corporeality so that knowledge, emotions, and action are meshed together. Hence, the increased professionalization of the armed forces of the industrial democracies may actually impede the ability to transition between situations. The closer a given situation is to that of high-intensity combat the easier it is for troops to switch into it, since they already have an embodied repertoire of appropriate actions. Conversely, shifts into missions located away from the violent battle are more difficult, especially when they are sudden. Accordingly, transitions from combat to more controlled apprehension of suspects are easier to manage than into policing of traffic regulation, let alone activities supporting local civilians involved in nation-building.

To be sure, there are individual, unit, and even force differences in this respect.[52] But, as Durch and England argue,[53] it is not clear 'that soldiers are collectively capable of doing much role-shifting as rapidly as doctrine now seems to require'. Such an ability to adapt could be expected only of a soldier with sufficient experience of training, for example, a special force sergeant aged thirty-four (with fifteen years in the forces) but less of a new nineteen-year-old infantry trooper. Their caution fits well with the overall preoccupation with cohesion. It seems plausible to suggest that the ability of professional soldiers to quickly cohere into functioning units with high performance levels is actually a contributory factor to the ability to transition. Because cohesion—again, the unity emerging out of highly coordinated collective practices—is such an inherent part of professionalism, just as it has become relatively easy for contemporary soldiers to be able to join rather rapidly with other units in new missions, then the moves necessitated by transitions between situations may also be rather easy. As such, this question is open to empirical research.

Conclusion: Special Forcification?

Let me conclude with a rather provocative summation. Could we argue that in urban contexts numerous 'regular' units take on many of the *organizational* characteristics of special forces: autonomy and independent actions of small groups, leadership devolved down to the squad and fire-team levels, dealing with civilians as an inherent part of their mission, abilities to move quickly and flexibly from one mission to another, and a capacity to join other kinds of forces and actors in temporary modules of coordinated action? To be clear, I am not proposing that all the combat units of the armed forces of the industrial democracies—or even their infantry components—have suddenly become special forces. Clearly they have not. The process of 'special forcification' I refer to implies that they take on certain organizational features that *resemble* those of special forces. Theoretically, the problem, of course, is that special forces have not been at the centre of the social scientific study of the military, if studied at all. Thus my more general assertion is that looking at such units as resembling special forces directs us to develop social scientific concepts to understand the new sociology of combat formations.[54]

Let me exemplify via two points. First, there is need for a much more dynamic view of combat formations than that provided by the classic sociological formulations, which have tended to be more structural. The move is from seeing the military not only as an entity that has to withstand the pressures of missions (basically battle) to an appreciation of the contingent often contradictory organizing processes that constantly take place within units and headquarters. Thus, an extremely cohesive unit may be characterized by troops who do not necessarily like each other yet participate flawlessly in collective actions. Structures and hierarchies, according to this view, are provisional and have to be constantly transformed, adapted, or indeed created to meet the multiple challenges that forces in the field encounter.

Second, there is a much greater appreciation of the combinatorial formations through which the armed forces now operate. This point entails understanding that militaries are much more open systems than was previously assumed and that the tasks of military formations today entails a constant move across organizational boundaries to create temporary frameworks that bring together the military and other actors. Hence, my argument is that the analysis of urban combat may aid us in *adding* to the older sociology of units a (new) sociology of combat formations. In terms of the analytical agenda of this volume, special forcification—in the sense of the increased professionalization of the militaries of the industrial democracies—is based on the cohesion of small groups. Indeed, this cohesion is the very basis for the ability to constantly form units in ever-increasing levels of complexity. Against the background of my chapter, I would go beyond King to suggest contemporary forces are characterized by three meta-abilities:[55] to cohere, to join (and disconnect), and to transition. As I have shown, each of these abilities—acquired and inculcated along the life course of combat troops—interacts with cohesion in various ways.

Notes

1. Paul Hirst, *Space and Power: Politics, War and Architecture* (London: Polity, 2005); Roger Spiller, 'Sharp Corners: Combat Operations in Urban Areas', in Michael Evans, Russell Parkin, and Alan Ryan (eds.) *Future Armies, Future Challenges* (London: Allen and Unwin, 2004), 82–95.

2. Yagil Levy, *Israel's Death Hierarchy: Casualty Aversion in a Militarized Democracy* (New York: New York University Press, 2012); Eyal Ben-Ari, Zev Lehrer, Uzi Ben-Shalom, and Ariel Vainer, *Rethinking the Sociology of Combat: Israel's Combat Units in the Al-Aqsa Intifada* (Albany: State University of New York Press, 2010).
3. Anthony Clayton, *Warfare in Woods and Forests* (Bloomington: Indiana University Press, 2012); Michael Evans, *City Without Joy* (Canberra: Australian Defence College, Occasional Papers No. 2, 2007).
4. Evans, *City*, 10–1.
5. Buzz Williams, *Spare Parts: A Marine Reservist's Journey from Campus to Combat in 38 Days* (New York: Gotham Books, 2004).
6. Anthony Beevor, *Berlin: The Downfall 1945* (London: Penguin, 2003).
7. S.L.A. Marshall, *Notes on Urban Warfare* (Aberdeen Proving Ground, MD: US Army Materiel Systems Analysis Agency, 1973).
8. Robert K. Ackerman, 'Training, Not Technology, Is the Key to Urban Warfare', *Signal Magazine* May (2001) <http://www.afcea.org/content/?q=training-not-technology-key-urban-warfare. Accessed 12 May 2012.
9. Evans, *City*, 10; Barry R. Posen, 'Urban Operations: Tactical Realities and Strategic Ambiguities', in Michael C. Desch (ed.), *Soldiers in Cities: Military Operations in Urban Terrain* (Carlisle, PA: Strategic Studies Institute, 2001), 149–66.
10. Ben-Ari et al., *Rethinking*; Dana Shen, 'The Psychology of Operational Joining: A Case-Study of a Dog Dispatch Unit', *Military Psychology* 2 (2003): 75–130 (Hebrew).
11. Charles C. Krulak, 'Forward: *Ne Cras*: Not Like Yesterday', in Richard H. Schultz Jr and Robert L. Pfaltsgraff Jr (eds.), *The Role of Naval Forces in 21st-Century Operations* (Washington, DC: Brassey's, 2000), xi–xvii; Mark Burgress, 'Marine's Three-Block War in Iraq', (Center for Defense Information, 2003) <http://www.au.af.mil/au/awc/awcgate/usmc/strategic_corporal.htm>. Accessed 5 June 2012.
12. For a US example from Iraq see Evan Wright, *Generation Kill: Devil Dogs, Iceman, Captain America and the New Face of American War* (New York: Berkeley Caliber, 2004).
13. Reuvan Gal and F.D. Jones, 'A Psychological Model of Combat Stress', in F.D. Jones, L.R. Sparacino, V.L. Wilcox, J.M. Rothberg, and J.W. Stokes (eds.) *War Psychiatry* (Washington, DC: TMM Publications, 1995), 133–48; Elmar Dinter, *Hero or Coward: Pressures Facing the Soldier in Battle* (London: Frank Cass, 1985); Simon Wessely, 'Twentieth-Century Theories of Combat Motivation and Breakdown', *Journal of Contemporary History* 4(2) (2006): 269–86.
14. Also W. Darryl Henderson, *Cohesion: The Human Element in Combat* (Washington, DC: National Defense University Press, 1985); Wessely, 'Twentieth-Century Theories of Combat Motivation and Breakdown'.
15. Edward Shils and Morris Janowitz, 'Cohesion and Disintegration in the Wehrmacht in World War II', *Public Opinion Quarterly* 12 (1948): 280–315; S.L. A. Marshall, *Men Against Fire: The Problem of Battle Command in Future War* (New York: The Free Press, 1947); Samuel Stouffer et al., *The American Soldier: Combat and its Aftermath* (Princeton, NJ: Princeton University Press, 1949).
16. Charles Moskos, 'Why Men Fight: American Combat Soldiers in Vietnam', *Trans-Action* 7(1) (1969): 13–23.

17. Ben-Ari et al., *Rethinking*.
18. Todd C. Helmus and Russell W. Glenn, *Steeling the Mind: Combat Stress Reactions and Their Implications for Urban Warfare* (Santa Monica, CA: Rand Corporation, 2004), 127–8.
19. Brian Steed, *Armed Conflict: The Lessons of Modern Warfare* (New York: Ballantine Books, 2003); also Paddy Griffith, 'Fighting Spirit: Leadership and Morale in the "Empty Battlefield" of the Future', in Michael Evans and Alan Ryan (eds.), *The Human Face of Warfare: Killing, Fear and Chaos in Battle* (London: Allen and Unwin, 2000), 109–23.
20. Russell W. Glenn and Gina Kingston, *Urban Battle Command in the 21st Century* (Santa Monica, CA: Rand Corporation, 2005).
21. Anthony King, *The Combat Soldier: Infantry Tactics and Cohesion in the Twentieth and Twenty-First Century* (Oxford: Oxford University Press, 2013).
22. King, *The Combat Soldier*, Ch. 10.
23. Griffith, 'Fighting Spirit', 114.
24. Robert K. Ackerman, 'Training, Not Technology, Is the Key to Urban Warfare', *Signal Magazine* (May 2001) <http://www.afcea.org/content/?q=training-not-technology-key-urban-warfare>.
25. John Urry, *Sociology Beyond Societies: Mobilities for the Twenty-first Century* (London: Routledge, 2007).
26. Vered Vinocker-Hai and Yotam Amitai, 'The War from the Perspective of Lower-level Commanders in the Continued Conflict in the Gaza Strip: An Interim Assessment', *Ma'arachot* 389 (2003): 30–5 (Hebrew).
27. Steed, *Armed Conflict*.
28. Dan Baum, 'Battle Lessons: What the Generals Don't Know', *The New Yorker*, 17 January 2005: 1–7. <http://www.newyorker.com/magazine/2005/01/17/battle-lessons>. Accessed 12 May 2012.
29. *The Economist*, 1 January 2005.
30. Amos Harel and Avi Isacharoff, *The Seventh War: How We Won and Why We Lost the War with the Palestinians* (Tel Aviv: Yediot Aharonot, 2004) (Hebrew).
31. Robert R. Leonhard, *The Principles of Warfare for the Information Age* (Novato, CA: Presidio, 2000); Gerhard L. Weinberg, 'Stalingrad and Berlin: Fighting in Urban Terrain', in Michael C. Desch (ed.), *Soldiers in Cities: Military Operations in Urban Terrain* (Carlisle, PA: Strategic Studies Institute, 2001), 17–29.
32. Anthony King, *The Transformation of Europe's Armed Forces: From the Rhine to Afghanistan* (Cambridge: Cambridge University Press, 2011), 255.
33. Glenn and Kingston, *Urban Battles*, xi.
34. Ben-Ari et al., *Rethinking*.
35. Debra Meyerson, Karl E.Weick, and Roderick M. Kramer, 'Swift Trust in Temporary Groups', in Roderick M. Kramer and Tom R. Tyler (eds.), *Trust in Organizations: Frontiers of Theory and Research* (Thousand Oaks, CA: Sage, 1996), 166–95.
36. King, *The Combat Soldier*.
37. Uzi Ben-Shalom, Zev Lehrer, and Eyal Ben-Ari, 'Cohesion During Military Operations? A Field Study on Combat units in the Al-Aqsa Intifada', *Armed Forces & Society* 32(1) (2005): 63–79.

38. Michael Barnett and Emanuel Adler, 'Studying Security Communities in Theory, Comparison and History', in Emanuel Adler and Michael Barnett (eds.), *Security Communities* (Cambridge: Cambridge University Press, 1999).
39. Ben-Ari et al., *Rethinking*.
40. King, *The Combat Soldier*.
41. Charles C. Krulak, 'Forward'.
42. Eyal Ben-Ari and Meytal Eran-Jona, 'The Israeli Military in Palestinian Civilian Contexts: Hybrid Conflicts, Multiple Logics and Organizational Transitions', unpublished manuscript.
43. Fabirizio Battistelli, 'Peacekeeping and the Postmodern Soldier', *Armed Forces & Society* 23 (1997): 467–84.
44. J.C. Azari, Christopher Dandeker, and N. Greenberg, 'Cultural Stress: How Interactions With and Among Foreign Populations Affect Military Personnel', *Armed Forces & Society* 36(4) (2010): 585–603.
45. E.H. Kramer, B. van Bezooijen, and R. Delhaij, 'Sense-making During Operations and Incidents', in J. Soeters, P.C. van Femema, and R. Beers (eds.), *Managing Military Organizations: Theory and Practice* (London: Routledge, 2010).
46. Joseph L. Soeters, 'Ambidextrous Military: Coping with Contradictions of New Security Policies', in Monica den Boer and Jaap de Wilde (eds.), *The Viability of Human Security* (Amsterdam: Amsterdam University Press, 2008), 109–24.
47. William J. Durch and Madeline L. England, *The Purpose of Peace Operations* (New York University: Center on International Cooperation, 2009).
48. Randall Collins, *Violence: A Micro-Sociological Theory* (Princeton, NJ: Princeton University Press, 2008).
49. Limor Samimian-Darash, *Violence, Control and Enjoyment—The Creation of a New Body: Studies in the Training of the Assassination Squads of the Israeli Military* (Masters Thesis, Hebrew University of Jerusalem, Department of Sociology and Anthropology, 2003).
50. Dave Grossman, *On Killing: The Psychological Cost of Learning to Kill in War and Society* (Boston, MA: Little Brown, 2005); Collins, *Violence*.
51. John Hockey, 'Switch On: Sensory Work in the Infantry', *Work Employment and Society* 23 (2009): 477–93.
52. Anders McD. Sookermany, 'The Embodied Soldier: Towards a New Epistemological Foundation of Soldiering in the (Post) Modernized Norwegian Armed Forces', *Armed Forces & Society* 37(3) (2011): 469–93.
53. Durch and England, *The Purpose of Peace Operations*.
54. Eyal Ben-Ari, 'What is Worthy of Study? The Sociology of the Armed Forces in Current-Day Conflicts', paper presented at a Conference on 'The Production of Knowledge on Peace and Conflict', Exeter University, June 2012.
55. King, *The Combat Soldier*.

5

Discipline and Punish

Encouraging Combat Performance in the Citizen and Professional Army

Anthony King

Introduction

In his 2006 article on the cohesion debate, Hew Strachan identifies three factors which have been recurrently recognized by scholars as central ways of explaining cohesion and combat performance: the 'small group theory' advocated by Janowitz and Shils, Marshall and Stouffer after the Second World War; ideological indoctrination and patriotism propounded by Omer Bartov, Stephen Wesbrook and Charles Moskos and the fear of punishment. To these three, Strachan adds a fourth: namely training. For Strachan, training is particularly interesting since not only does it enable soldiers to perform together on the battlefield but it provides a self-referential form of motivation. Strachan's article considers in detail the centrality of training to combat performance especially in the First and Second World Wars and records the great difficulty which the mass citizen armies of these conflicts had in adequately training their troops. Yet Strachan explicitly excludes any form of explanatory imperialism: 'None of these three [we could add four to include training] customary explanations is exclusive of the other. Moreover, however valuable comparative perspectives are, different factors operate in different armies at different times in different theatres.'[1]

Strachan's injunction to avoid explanatory imperialism is well taken. The historian of war's job is to identify the specific dynamics which engender cohesion in each army and in each war. However, while great care needs to be taken, it might be claimed that the balance of influence which the four factors—primary group theory, ideology, punishment, and training—have exerted can be seen to have shifted significantly since the 1970s and especially since the end of the Cold War. Specifically, a divide might usefully be drawn between the relative role these factors have played in the mass citizen army of the twentieth century and in today's professional forces. Indeed, it has been widely noted that in the Western citizen army of the twentieth century, and especially those which fought the two world wars, Vietnam, and Korea, both the primary group and ideology (of even a latent type) were essential, as Bartov's criticisms of the Janowitz and Shils' thesis shows. Mass armies typically had little time and inadequate resources to train their troops, and they consequently relied upon appeals to masculine comradeship and patriotism or ideology to unite soldiers as comrades and motivate them to fight. The situation has now changed markedly. In the professional army of the twenty-first century, far smaller in size and relatively better resourced with more time to practise and prepare, training has displaced patriotism and the 'pure' cohesion of the primary group as a decisive factor in combat performance. Naturally, as Strachan suggests, it would be unwise to dismiss ideology or masculinity entirely, but it is important to be sensitive to institutional transformations in which impersonal ties of professional competence and vocational honour have increasingly superseded the bonds of friendship and the national commitments of the citizen in arms. In this context primary-group theory and ideological explanations have been displaced by training, drills, and preparation, as a range of scholarship shows.[2]

However, while there has been extensive analysis of the relative roles of masculinity, patriotism, and training in the armed forces, the question of punishment has attracted less interest. The aversion to studying punishment seems to reflect the laudably liberal perspectives of most scholars. Because it jars with contemporary liberal values, it is presumed that military discipline and the threat of punishment cannot have had an important role in motivating troops. Soldiers—especially citizen soldiers—should be motivated positively by a national mission or loyalty to their fellows, not coerced by the threat of punishment and, ultimately, execution for cowardice and desertion. For instance, in his important recent book on authority and obedience during the First World War, Emmanuel Saint-Fuscien argues that French soldiers were primarily motivated by republican and nationalist morality rather than subdued by despotic military authority. Authority was dependent upon the civic consent of soldiers that officers actively inspired through education and example.[3] Similarly Charles Glass's recent work on desertion in the Second World War,[4] and the now classic literature on executions in the First World War,[5] are animated by a sense of injustice at the treatment of soldiers wrongfully accused of cowardice; the subtext of these works is that not only was punishment unfair but it was also ineffective. In his outstanding work on the

Italian army in the First World War, Mark Thompson discusses punishment at some length and also presumes that it cannot have had an effect—still less a positive effect—on the performance of Italian troops on the Isonzo.[6] Indeed, Stephen Wesbrook presumes, like many scholars, that not only are they undesirable but threats of sanction cannot have an effect on the terrifying and dispersed modern battlefield.[7] There, the soldier could shirk with impunity in the knowledge that his failings would never be reported. In this chapter, I want to explore the unfashionable and, indeed, uncomfortable question of the relationship between discipline—and especially capital punishment—and combat performance. This chapter addresses the third factor that Strachan identifies. Moreover, it tries to understand this third factor by reference to the historical divide between the citizen and professional army. Just as the primary group, ideology, and training play a differential role in the mass citizen and professional army, I want to suggest that discipline and punishment too have a different character and alternate dynamic in the eras of the mass army of the twentieth century and today's all-volunteer forces.

Discipline in the Citizen Army

The citizen armies of the twentieth century all employed capital punishment to encourage combat performance. Cowardice in the face of the enemy or desertion were normally punishable by death. In the First World War, the British Army tried several thousand for cowardice and eventually executed 340 soldiers.[8] The French Army shot approximately 600. After the mutinies of 1917, which involved some 30,000 troops, 554 soldiers were sentenced to death, but only forty-nine were actually executed; there seem to have been three summary executions.[9] The US Army shot a few soldiers, but their combat involvement in that conflict was limited to six months between March and November 1918. Distinctively as one of the major combatants, the German army shot only forty-six soldiers in the First World War. The situation was reversed in the Second World War when the Western Allies generally eschewed corporal and capital punishment. The British executed no one for cowardice during the conflict, although there was at least as much desertion as in the First World War, as Jonathan Fennell's chapter in this volume on the 21st Army Group confirms. He also shows, while they shot no one, the British Army court martialled many troops for desertion and cowardice. The United States executed only a very small number of soldiers, including the notorious case of Private Eddie Slovik.[10]

Historians and social scientists have sometimes tried to determine the effect of discipline—and especially the threat of capital punishment—on the citizen soldier. It is very difficult to establish the degree to which sanctions contributed to combat performance. Some scholars, noting that the immediate terror of combat must overwhelm all other considerations but the drive for immediate self-preservation, have suggested that discipline can have played little role. It is not an implausible argument and, certainly, among those forces where the sanction of death for non-combat performance was weak, it seems likely that formal discipline probably had little motivational effect on combat soldiers. For all the recent controversies about the executions conducted by the British and French military authorities in the First World War, the numbers executed were very small in comparison with the size of the armies and, especially in the French case, to the scale of mutiny which they faced. Moreover, although the military justice which prevailed in these cases might fall some way short of the standards of civilian justice, soldiers were not shot summarily. Consequently, it might be suggested that for all the combatants in the First World War on the Western Front, discipline was not a major motivating factor in enjoining cohesion in combat: it did not contribute markedly to performance.

This was quite different for the Italians fighting in the Dolomites and on the Isonzo, however. Perhaps because of the stereotype which coagulated around the Italian soldier in the Second World War, the extreme discipline exercised by the Italian army in the First World War is now often forgotten. Mark Thompson has usefully remembered the draconian discipline which was enforced by General Cadorno throughout the war until his removal after Caporetto. General Cadorno and his subordinates employed three disciplinary systems. Firstly, they used official forms of justice as a means of deterrent. As Thompson clarifies, official military justice in Italy during the First World War was independent of political control and, consequently, was so extreme as to border on summary justice. Army commanders regarded desertion as the greatest danger because in a mass army not only was desertion easy but if enough individuals decided to withdraw the army would collapse very quickly. Accordingly, the definition of desertion was extraordinarily wide; a soldier who returned to his unit at the front more than three days late after leave was considered a deserter and liable to execution.[11] In May 1917 alone, fifty-four men were executed and, in July 1917, some soldiers of the Catanzaro Brigade, exhausted from months of fighting, revolted at the prospect of being returned the front after only a few days' rest. Twenty-eight men were charged with rebellion and executed on the spot.[12] In all 729 executions for desertion alone were carried out in an army half the size of the French army and about the same size as the British army.[13] In addition to formal justice, summary justice was used liberally by the Italian army. Three hundred cases have been corroborated, but it is likely that several thousand soldiers were executed in this way.[14] Most disturbingly, the method of 'decimation' was introduced and employed on a number of occasions to discipline units collectively.

Thompson identifies the experience of the Ravenna Brigade in 1917 as one of the worst cases of the use of summary decimation in the entire war. Following the cancellation of leave, there was a minor protest by some members of the Brigade who were, according to their own commander, 'a bit annoyed, tired and almost all in a dreadful physical condition'. The men were 'amenable to reason, but a few shots were fired in the air'. At this point, the incident escalated and the corps ordered twenty soldier to be picked from the most rebellious company. Five of these men were selected to be executed by a firing squad made up of their comrades. Two weeks later, nine further soldiers were tried, four of whom were executed, followed by a further eighteen.[15] Finally, the Italian army employed carabinieri behind the frontline to ensure participation in assaults. Italian soldiers were reputed to know that a major assault was imminent when the military police mounted machine guns behind their trenches, shooting any soldiers who were reluctant to join the attack.[16] After one minor attack in the Dolomites, a doctor recorded twenty-five buttock wounds inflicted by the carabinieri.

Italian generals were clearly convinced of the necessity of harsh and summary justice in controlling a massive and generally very poorly trained army, many of whose soldiers were illiterate peasants with potentially little commitment to Italy or to the war. They were also concerned about the infiltration of socialism into the ranks. Some Italian scholars have argued that this disciplinary regime suppressed potential rebellion and, without it, 'mutinies on the French scale would have erupted'.[17] Giorgio Rochat suggests that so extreme was the discipline that 'disobedience lay beyond the conception of men'.[18] Certainly, it would seem to be difficult for troops to refuse to take part in an attack with the immediate presence of armed carabinieri behind them. Thompson perceptively observes that summary justice had 'disastrous side-effects'; he claims that 'evidence that soldiers' morale was harmed by the almost arbitrary killing of their comrades is strong and ample'.[19] Yet Thompson also admits that Cadorna's methods 'reinforced the obedience which almost all soldiers already displayed'.[20] Certainly, Italian soldiers cannot have enjoyed the regime, and their 'morale', by which Thompson seems to mean their individual and collective happiness (rather than their determination to fight), was doubtless undermined by it. Yet, as Thompson himself acknowledges, harsh justice does seem to have engendered discipline and ultimately enforced a fighting spirit on potentially reluctant Italian soldiers. For all its barbarity, summary justice does seem to have contributed to the combat performance of the Italian army during the First World War when it confronted uniquely appalling conditions. It was widely agreed that the Italian army, despite its poor equipment, training, and lack of educated troops, fought very hard and ultimately successfully on this front, albeit with assistance from the Allies following the disaster at Caporetto.[21]

It is instructive to compare the performance of the Italian Army in the Second World War with that of the First here. In the Second World War, the Italian army, fighting under quite different political and operational conditions, instituted far less draconian discipline, which was, in fact, more compatible with its democratic allies in the First World War; ninety-two soldiers were condemned to death between 1940 and 1943.[22] It does not seem implausible to suggest that the Italians' generally weak performance in that conflict, the object of much satire, may have been partly a product of this lax disciplinary regime. The Italian army's experience would seem to support the contention that severe discipline is potentially very important to the citizen army. Unpalatable as it is, the Italian example seems to suggest that the system of harsh military justice was not irrelevant to combat performance.

For the Allied forces in the Second World War, sanctions for cowardice became even weaker, and it would be very difficult to argue that the threat of discipline itself could have had any serious effect on Allied combat soldiers then. Indeed, discipline in general was not always enforced. Lieutenant General Lucian Truscott recorded, for instance, that he had issued two standing orders to the 3rd Army: helmets were to be worn in combat and only the specified number of individuals were to be carried in vehicles.[23] He noted that in each case his standing order was 'flagrantly violated'.[24] Indeed, the problem became so bad that he personally fined individuals he encountered without a helmet. He did not tell those soldiers that the fines would never be collected.[25] If a general was unable to enforce a minor but important dress regulation on his troops (for their own benefit), it is unlikely that the sanction of a distant and theoretical punishment had significant motivating effect upon US soldiers in combat. In the Second World War, Allied armies seemed to be genuinely imbued with a democratic and liberal spirit that reflected the polities which they defended. Armies may have been strict by civilian standards, and Truscott noted the discipline and tidiness of the British Army, but they were not authoritarian. In fact, the British Army treated not only its psychological casualties but even the underperforming and (before D-Day) near mutinous 7th and 51st Divisions with great leniency. So bad did the problem become that senior British generals in Normandy, including Lieutenant General John Crocker, commander of 1 Corps, advocated the reintroduction of the death penalty for cowardice and lobbied Churchill for its reintroduction. It is of course possible that the mere threat of a court martial for cowardice did have some effect on Allied soldiers. Social humiliation was used by the army and seemed to be mobilized by soldiers themselves to motivate themselves and make sense of their situation. However, given the complaints of senior officers about the lack of sanctions available to them, this threat does not seem to have been particularly serious. Discipline was not severe in the Allied armies, therefore, and, probably as a consequence, not very effective in encouraging combat performance among the democratic powers. It is perhaps because of this disciplinary latitude that the Allied infantry performance was so poor in a challenging campaign like Normandy where the defender held many advantages at the level of close combat.

The Wehrmacht and SS were, of course, quite different and, indeed, seem to illustrate precisely the importance of harsh discipline in the mass army. The German army executed 15,000 in the Second World War.[26] Under the authoritarian Nazi regime, discipline became almost fanatical, especially on the Eastern Front where thousands of soldiers were punished or executed for dereliction of duty and cowardice. Hein Severloh, who was stationed at Widerstand Nest 62 on Easy Red Sector of Omaha Beach on D-Day and, armed with an MG 42, personally played a significant role in disrupting the landing, might never have reached Normandy as a result of this draconian discipline. He had been deployed to Russia in 1942, where he contracted frostbite in his feet. He was going to present himself to the hospital but was advised against it by the regimental doctor who warned him that the condition would be considered a case of self-wounding, for which the punishment was death.[27] Clearly, the doctor may have been exaggerating in order to reduce his work or to avoid inquiries about his judgement from the rear, but the fact that Severloh believed him and returned to the front demonstrates that such punishments were regular. Indeed, the Wehrmacht actively publicized the sanctions for desertion and cowardice. Hans Kuss, an officer who initially served with the 94 Infantry Regiment on the Eastern Front but was reassigned throughout the war as his units were decimated, photographed a prominent warning sign on a road in East Poland which recorded the shameful execution of four soldiers, found guilty of desertion: 'Notice: On 21 February 1945, Grenadier G and Corporal de Lw. and, on 26 February 1945, Lance Corporal B and Engineer H were shot in Kahlberg following a court-martial.' The sign explicitly communicated the options for German soldiers by that stage in the war; 'The condemned came over the back-waters as deserters and draft-dodgers.[28] They shirked a hero's death before the enemy and died a death of disgrace and dishonour.'[29] It is not actually clear whether Kuss's sign referred to a genuine execution or whether it was merely a deterrent, but sentences of this kind were frequently carried out, many of them arbitrary and unjustified. Moreover, many thousands were also summarily assigned to penal battalions or to penal roles in which chances of survival were slim.[30]

Even in the apparently more civilized Western campaign, discipline was harsh, especially as the Wehrmacht began to be defeated. During the Normandy Campaign, after Operation Goodwood to the south of Caen, *Feldgendarmerie* detachments at bridges seized stragglers and deserters, executing many summarily, some of whom were hung from trees as a deterrent to others.[31] Of course, it is possible that faced with the prospect of immediate extermination by Allied and Russian firepower in combat, German soldiers forgot about the discipline which waited for them a few miles to the rear should they fail to perform. However, so severe were the penalties for even minor infractions and so common was recourse to summary justice that it seems at least likely that discipline did have some effect on the German soldier, not least because troops could report on each other. Indeed, it might be suggested that the further forward and the more extreme the discipline, the greater effect it was likely to have in motivating troops to fight. Clearly, the authoritarian culture of Nazism and the desperate strategic situation in which it found itself by late 1942 encouraged extreme discipline, but the fact that the German army had to resort to such extreme measures does seem to belie the notion that their troops were always willing combatants or that their units were always highly trained and competent. The German experiences seem to show that, in a citizen army, good performance in combat does not seem to be entirely dissociated from harsh discipline. Perhaps the willingness of German commanders to sanction their troops severely in Normandy might explain their generally superior performance over the Allies, rather than any greater military competence.

Discipline and the threat of execution can rarely, if ever, be invoked as a sole explanation of combat performance. Moreover, it is methodologically almost impossible to isolate the effects of discipline from factors of masculinity or primary-group loyalty, ideology, and training. In practice, they are typically fused. Moreover, although a comparison of German divisions in Normandy with Allied divisions, such as the 7th Armoured or 51st Highland Division, suggests that discipline may be effective, no genuinely scientific experiment is possible to isolate punishment definitively as an independent variable and to demonstrate its effect on combat performance. Such a comparison would require the analysis of the same unit, in the same tactical situation, one under the threat of punishment, the other without. Nevertheless, the evidence from the twentieth century suggests, even if it does not prove, a potentially distasteful truth. Discipline and, especially, capital punishment could be an effective if brutal response to the problem of combat inertia and straggling in the citizen army. Although it might not have had a major effect on individual deserters, especially when discipline was enforced through a slow judicial system as it was by Allied armies, the possibility of capital punishment for cowardice and desertion is likely to have provided some additional motivation to soldiers in combat. Where discipline was extreme and immediate, it seems probable that it had an effect on frontline troops. Minimally, discipline and especially capital punishment—and at least its threat—was important to the citizen army.

Discipline in the Professional Army

Historically, professional armies have often used savage discipline to maintain order and fighting spirit. The concept of decimation was, after all, developed by the Roman army. In the seventeenth and eighteenth centuries, discipline was harsh among professional standing armies. Frederick the Great was a notorious martinet, while the Duke of Wellington regarded harsh discipline, including corporal and capital punishment, as essential to maintaining good order in his army.[32] Corporal punishment became less brutal in the Victorian army. Nevertheless, it was still stern and it is very noticeable that in the First World War, capital punishment was carried out disproportionately on the professional troops of the original British Expeditionary Force, rather than on Kitchener's Armies. There seemed to have been an assumption that professional soldiers were expected to fight and to act as examples to citizen soldiers; they were therefore more deserving of punishment when they did not.[33] Discipline in professional armies has historically been severe.

Western armies are once again professional and with their emphasis on training and drill, they display many of the characteristics of Wellington's or Frederick the Great's infantry. Yet disciplinary regimes in the professional army of the twenty-first century are radically different from those of the early modern era. The death penalty and corporal punishment, so historically central to the professional army, have been abolished in most Western militaries today (only the US preserves the death penalty), while the sanctions available to the armed forces and the judicial process in the military has been brought into line with civilian standards. Moreover, it is very striking that the armed forces themselves have rarely applied military justice to cases of combat refusal. Specifically, and in contrast to professional forces in the past, the armed forces themselves have been very reluctant to court martial service personnel for desertion and cowardice. Thus, despite intense fighting in Iraq and Afghanistan and a number of serious operations in Africa, involving the deployment of hundreds of thousands of Western service personnel, the court martial remains almost entirely unused.

There are some exceptional cases when soldiers have been prosecuted for refusing to deploy. For instance, two British service personnel, Lance Corporal Joe Glenton and Flight Lieutenant Malcolm Kendall-Smith, refused, respectively, to return to Afghanistan or to deploy to Iraq on the grounds of conscience; they were charged and imprisoned for going absent without leave. In the United Kingdom, legal proceedings were activated against one British soldier, Guardsman Wayne Gorrod, in 2009 with a view to his being tried at court martial on charges of cowardice in the face of the enemy, but he was eventually discharged instead. It is possible, but unlikely, that Bowe Bergdahl, who disappeared from his patrol base in Afghanistan in June 2009 to be held hostage by the Taliban until March 2014, may be charged with desertion, should a current inquiry recommend it. However, in more than a decade of fighting, only one soldier has been charged with and tried by court martial for cowardice. In the United States, Sergeant Georg Pogany was charged and court martialled in October 2003: the first US soldier since the Vietnam War. Pogany had deployed to Iraq in early October as an interpreter with a Special Forces unit. However, following an incident in which an Iraqi insurgent was killed and his badly mutilated body brought back to the base, Pogany claimed to have had a panic attack and, therefore, to have been incapable of functioning properly. However, his pleas were rejected and he was court martialled for cowardice for refusing to perform his duties, which included going out on patrol.[34] Instructively, he was quickly acquitted on psychological grounds at his court martial in November 2003. In striking contrast to the citizen armies of the twentieth century—and even more stark contrast to professional armies of the past—no one has been found guilty and, indeed, only one Western soldier—Sergeant Pogany—has been formally charged with and court martialled for cowardice in the entire course of the Iraq and Afghanistan campaigns.

It is important to recognize that this is not because cowardice and desertion have not occurred; the issue of battlefield discipline remains very relevant to the military today. Professional forces in the twenty-first century are in no way immune to the problem of cowardice and desertion. On the contrary, there are certainly cases of 'combat refusals' in today's professional forces; cases which in the past might have been defined as 'cowardice'. They are often kept quiet, circulating only along informal networks, but there have been a number of refusals in both Iraq and Afghanistan by Western forces. Indeed, there was reputedly a near mutiny by the British Army regiment relieving the Princess of Wales Regiment in Al-Amara in 2005, so extreme were the conditions that they had been left to deal with after a—possibly unnecessarily—bloody tour by their predecessors. There have also been cases of individual 'cowardice', as the military would define it.

For instance, an unusually high number of examples were recorded in the British Army's Welsh Guards Battle Group's brutal tour of Helmand in the summer of 2009.[35] Following some traumatic events in which a number of his close colleagues were killed or injured, Guardsman Paul Carew broke down and had to be removed from combat. On one patrol, it was reported that 'he walked onto the road and fired on automatic, Rambo style... screaming his head off, changed magazines and used all his ammunition up'. Later he was found taking refuge in a mosque, sitting in the corner with his knees up to his chin, repeating: 'Get me out of here, get me out of here'.[36] He was evacuated and discharged from the Army. During another patrol in Nad-e-Ali, Private Wayne Gorrod ripped off his body armour and helmet, flung his arms in the air, and walked off towards Taliban lines. Fearing that he would be captured by the Taliban and, therefore, would become a major risk to security, he was eventually apprehended by an officer.[37] Finally, there were a number of self-inflicted injuries during the tour. One Guardsman shot himself in the foot, another had feigned a broken ankle, going absent without leave on his return to the UK. Towards the end of the tour, Guardsman Gavin Wynne deliberately shot himself in the leg while on guard duty.[38]

The Welsh Guards Battle Group was perhaps unusual in the number of cases which it experienced in 2009, reflecting the intensity of its tour and the scale of its operations, which ultimately involved the death of its own Commanding Officer, Rupert Thorneloe. Yet, while extreme, these incidents were not unique to that Battle Group. Moreover, previous conflicts suggest that occasional incidents of refusal have consistently occurred in professional armies. For instance, during the Falklands War, a Welsh Guardsman, Philip Williams, was widely regarded to have deserted on Sapper Hill and was found weeks later. In a lesser known but more striking case, a Parachute Regiment sergeant, apparently the 'victim of battleshock',[39] hid himself in a cave before the Battle of Mount Longdon only to be found after the ceasefire: 'the soldier reappeared and the realization grew amongst his comrades that he had panicked, taken a load of blankets from an incoming medevac chopper, and built himself a nest amongst the nearby crags in which to hide, curled up asleep in foetal security'.[40] It seems probable that the other battalions may also have had similar cases.

Professional soldiers demonstrate the same fears as citizen soldiers and, in some cases, they are prone to the same reactions; individuals have panicked, refused to fight, or deserted. Yet only one has been court martialled. This absence is all the more noteworthy and anomalous because, not only have combat refusals occurred, but there are numerous cases in which soldiers have been severely sanctioned for other offences. Courts martial have been most prominently prosecuted for abuse cases, of which the best known are Abu Ghraib and Haditha. However, there have been a number of other atrocities in both Iraq and Afghanistan which have been the subject of military justice. In September 2003, a local Iraqi, Baha Mousa, died as a result of mistreatment while being held in custody by the 1st Queens Lancashire Regiment in Basra; one soldier was eventually found guilty of inhumane treatment and imprisoned. On 12 March 2006, four members of First Platoon, Bravo Company, 1st Battalion, 502nd Regiment, 101st Airborne entered a house in Yusufihay, south of Baghdad, and murdered an entire family, whose daughter they also tortured and raped; four out of the five soldiers responsible were all found guilty and imprisoned by a court martial, the fifth, who had left the army, was charged in a civilian court.[41] In Maiwand District Kandahar in January and February 2010, five soldiers from 2nd Battalion, 1st Regiment, 5 Stryker Brigade murdered three Afghan civilians in separate incidents, allegedly collecting body parts as trophies; ten soldiers were eventually convicted with sentences up to life imprisonment. In March 2011, Sergeant Blackman, serving with 42 Commando Royal Marines, executed a wounded Taliban fighter in Helmand; he was found guilty of murder at his court martial and given a ten-year prison sentence. In March 2012 Sergeant Robert Bales left his patrol base alone and killed sixteen civilians in Kandahar Province.[42] He was found guilty and sentenced to life imprisonment in the United States.[43] The armed forces have prosecuted troops for criminal offences, then.

There have also been some very public cases when incompetent and unprofessional conduct has also been the subject of severe sanction. One of the most notorious Canadian cases involved Brigadier General Menard who was removed from his operational post as Commander Task Force Kandahar in May 2010 and court martialled following revelations of an affair with his female clerk, Master Corporal Bianka Langlois. Claiming that his career had been ruined by the case, Menard resigned from the military. Menard's affair was in contravention of rulings about fraternization and was certainly dimly viewed by Menard's superiors. Yet perhaps more damaging for Menard's reputation was the fact that he had had a serious negligent discharge on Kandahar Airfield while boarding a helicopter with the Canadian Chief of the Defence Staff, General Walt Natynczyk, in the same month.[44] Three rounds from his weapon passed between the Chief of the Defence Staff and his bodyguards. Although Menard was given a standard fine for the weapons handling misdemeanour, members of the Canadian Forces suggest that it is possible that his subsequent sacking may have been more to do with this nearly lethal negligent discharge than his affair. His credibility had already been irreparably damaged by the negligent discharge—not only within the Canadian forces but with NATO allies. The armed forces have been willing to punish unprofessional conduct, therefore.

Clearly, one of the factors which has discouraged professional forces from prosecuting their troops for cowardice is the reputational damage such a prosecution would inflict. However, the armed forces have been very sensitive to the public reputation and, indeed, in the case of some of the abuse cases, there have been some justified criticisms that the military judicial system was tardy in its response to the cases and lenient in its sentencing; in the British Army's Baha Mousa, although only one soldier was eventually found guilty of inhumane treatment, a public inquiry subsequently found that nineteen soldiers were directly involved in the abuses. Similarly, the acquittal of six US Marines involved in the Haditha massacre and the leniency shown towards a seventh have also engendered considerable disquiet.[45] As Pascal Vennesson reveals in Chapter 11 of this volume, the French Army attempted to conceal the murder of Firmin Mahé. Private Garrod does not seem to have been prosecuted, precisely because the British Army feared the negative repercussions of the case.[46] The armed forces have a significant interest in avoiding prosecutions in general, and courts martial for cowardice, desertions, and combat refusal, in particular. It would bring unwanted attention to their failures, calling their professional competence into question. If service personnel were found guilty, it might raise further unwanted attention into the military justice system and the appropriateness of its punishments.

Yet while the reputational benefits for avoiding prosecution seem clear, the armed forces have not always been lenient: the 5 Stryker, Blackburn, and Bales cases would all be examples where severe sentences were handed down. Indeed, in the case of Sergeant Blackburn, there were calls for leniency from the British public and press which the court martial ignored. The Canadian Army regarded the Brigadier Menard case as serious enough to warrant any negative publicity that it might have engendered. In these cases, the armed forces have quickly prioritized organizational needs about mere reputational considerations. Moreover, given the substantial risks to which the armed forces have been exposed, especially in Iraq and Afghanistan, where campaign defeat has been a very real prospect at various stages and the fighting at the lowest level has been intense, it might be presumed that the pursuit of soldiers who had demonstrably failed in combat—as some clearly have—would be rather more important than the protection of their domestic reputation. If the armed forces believed that combat effectiveness were being seriously jeopardized by individual cases of refusal, cowardice, or desertion or that combat motivation was being undermined by the non-prosecution of those who had failed in combat, it seems probable that they would have initiated courts martial against those individuals—whatever the reputational cost. Overwhelmingly, they have not. While the word 'cowardice' has been applied to those who have abused or murdered civilians, it has only very rarely been applied to soldiers who have failed in combat and the sanctions for it overwhelmingly not applied.

The lack of courts martial, even though it might have been legitimate for the armed forces to charge a small number of soldiers with cowardice, suggests that the military have not experienced the organizational or operational requirements to enforce discipline on the frontline in the same way as either their professional or citizen predecessors. There seems to be some institutional reasons for this. In *The Combat Soldier*, I sought to demonstrate that as a result of their intensified regimes of training, the combat performance of professional troops has improved substantially. Through training, professional forces have substantially overcome the inertia which S.L.A. Marshall observed on the twentieth-century battlefield. Overwhelmingly, professional soldiers have been able to follow their drills and to participate in combat even when enemy fire has been intense. The improved combat performance of the professional force and its higher participation rates seem to go some way to explaining why the armed forces, despite the very intense pressure under which their units have often been placed especially in Iraq and Afghanistan, have not felt the need to discipline soldiers or to court martial those deemed to be cowards. Firstly, as a result of their training and socialization for combat, combat refusal, and especially cowardice and desertion are much rarer than in the past. For instance, a British Army brigadier, who had commanded a battle-group at the height of the conflict in Basra, recorded that he had never once experienced nor heard of a single refusal in his entire career.[47] Clearly, as the evidence discussed above shows, some individual cases have occurred, but they are extremely infrequent. Secondly, precisely because all the other soldiers are highly trained, the cowardice of one soldier does not tend to have had the catastrophic effect that even small panics recurrently caused in the twentieth century. The cowardice or non-performance of a single soldier is simply that. The collapses and panics, which were very common in mass citizen armies, have not happened in either Iraq or Afghanistan, even though some of the fighting has been desperate at platoon and company level. The other soldiers have been so tightly committed to the execution of their drills by their training that they are far less prone to panic as a group.

Combat failure in today's professional force is simply not so prevalent nor as serious as it once was. Other organizational factors seem to have been at work, which have also allowed the armed forces to avoid court martialling their troops. In particular, informal compensatory mechanisms seem to have appeared, which have taken the place of formal disciplinary procedures. Specifically, just as training has reduced the problem and the risks of cowardice and desertion in battle, so professionalism has also generated informal mechanisms that have partially displaced the need for the court martial. Discipline has become less necessary precisely because soldiers are imbued with an accentuated sense of professional pride which commits them to heightened levels of performance. Professional soldiers mutually monitor each other and continually assess themselves to ensure that they are performing at expected standards. Chapter 9 includes an example of this when a British army corporal in Helmand consents to an almost suicidal order out of this moral sense of professional obligation. Increasingly, the prime motivating and disciplining factor in combat is that soldiers will be respected by their peers for upholding professional standards. This informal but powerful sanction seems to have substantially superseded the need for formal military discipline on the frontline.

Indeed, there is a very significant archive of evidence which suggests that formal discipline has been displaced by informal peer-group sanction in the professional force. A few examples drawn primarily from the British Royal Marines will illustrate the point. Clearly, the Royal Marines are an elite force whose professionalism is highly developed, but they are by no means unique or unusual among Western troops in either their levels of performance or in the importance that they place on professionalism. They usefully exemplify much more widely observed processes. In November 2007, a Royal Marine night patrol in Nad-e-Ali, Helmand, was informed that insurgents were operating on the roof of a compound near their location. One of the corporals led his section to the door of the compound, but before initiating an entry he turned to his section and asked them 'Does anyone not want to go through this door?' He explained his rationale subsequently: 'I would feel like shit if something happened to one of them and they hadn't wanted to have gone in.'[48] In the event, no one refused, and they entered to find that the fighters had disappeared, leaving only families inside the compound. Although the corporal was ostensibly concerned only with his own conscience, the incident illustrates a different point. Although his marines could theoretically have voluntarily withdrawn from the operation, to do so would have resulted in a catastrophic loss of face. Moreover, once they had voluntarily committed themselves to entering the compound, they were expected to perform whatever happened. Ultimately, in the explicit absence of any appeal to formal military discipline—no order was given—professional pride ensured that none of his marines voluntarily withdrew from this operation, even though they had every opportunity to do so.

There are other examples of precisely this process of appealing to the professional pride of individual soldiers in order to demand performance from them. At Khan Neshin, Helmand, in February 2009, a Royal Marines company commander with 42 Commando wanted to seize a fort which was known to be an enemy stronghold but knew he had to draw enemy fire in order to do it. Consequently, the company commander asked one of his troop (platoon) commanders to advance across more or less open desert in view of some compounds from which he would almost certainly be ambushed. The company commander subsequently described how he encouraged the troop commander to take on this unpalatable mission:

> He [the troop commander] was circumspect but quickly got on with it – he was engaged from the flank as he moved, and the game was on; just as I had wanted. So what? My take on this is that this acceptance of a 'tricky' task was enabled through mutual trust (the key pillar of mission command). I literally briefed the new troop commanders on arrival (they had turned up halfway through the tour) that it was likely I would have to ask them to do seemingly ridiculous things but that I would do so in the knowledge that we could discuss it, i.e. my hope was that by giving them the opportunity to voice a 'no', they would actually feel happy to always say 'yes'!.[49]

While individual soldiers might feel they have personally betrayed themselves if they refused these requests to perform, it should be noted that the main tribunal here is not psychological or internal but collective and public; it is the judgement of immediate colleagues. At the same time, professional soldiers are susceptible to these requests precisely because they actively want to sustain their reputation in the eyes of their comrades.

American troops seem to operate in a similar fashion. It is noticeable that in the case of Sergeant Pogany, where he pleaded trauma, the Special Forces claimed that they were not insensitive to legitimate reasons for combat avoidance: 'Special Forces soldiers are mature professionals, and they know that if someone is under stress and asking for help it is important to give it to them.'[50] Soldiers could, in particular circumstances, avoid combat duties without any repercussions for their reputation. There were valid professional reasons for excusing oneself from combat. The problem was that in the eyes of these Special Forces soldiers, Pogany, as a professional, had no legitimate reason to refuse to perform his duties in Iraq. Having volunteered for military service, he was obliged to fulfil his role. American soldiers seem to see military service as a vocation, which can only be performed properly by those who have voluntarily committed themselves to it. However, precisely because service is voluntary, professionals are expected to perform once they have committed themselves.

At the same time, the positive motivation of professional pride is supported by an unignorable sanction; ridicule, abuse, and ostracism by peers. While the corporal in Nad-e-Ali gave his marines the opportunity to withdraw without apparent loss of face, inadequate performance is, in fact, subjected to cruel and merciless criticism. The Parachute Regiment sergeant who panicked on Longdon in 1982 was threatened with execution by a fellow non-commissioned officer (NCO) on his discovery the following morning,[51] and his professional reputation was destroyed.[52] His treatment was typical and remains common to this day. During a tour of Helmand in 2006, one Parachute Regiment soldier was tied up by his corporal and members of his section and indecently assaulted as punishment for conduct which was deemed cowardly; he missed his final exercise before deploying to Afghanistan and avoided an operation by claiming to be suffering from stomach cramps.[53] He was subsequently subjected to systematic abuse: 'I was treated with open hostility . . . I was told I was a liar, a useless soldier, a coward.'[54]

Professional soldiers are well aware of these mechanisms. Paratroopers are far from unique in meeting out harsh, informal punishment for those deemed to have failed. Indeed, the chain of command sometimes seems to collude in this disciplinary process. For instance, one Royal Marine lance corporal was publicly ridiculed in a report on Helmand in the Regimental journal in 2007 when, at the start of a mission, he cried for personal 'all round defence' because he was married with children. Bizarrely, he wanted his subordinates to form a cordon around him to ensure his safety throughout the operation. In the report, his rank, Lance Corporal, and name, Terry, was changed for satirical effect to 'Supreme Commander Allied Force (LCpl) Terry-fied'.[55] His professional reputation was destroyed publicly and permanently in the official regimental account. Officers are equally open to condemnation of this kind. In 2007, the Royal Marine's 3 Commando Brigade, organized a raid against a fort in Helmand, which was an important base for Taliban fighters in the area. The mission was badly conceived and ended disastrously in total failure and the death of one of the marines. The officer, who commanded the failed assault, was vilified by his subordinates. Significantly, he was regarded not only as incompetent but also, specifically, as a coward. He was removed from his command immediately and his reputation in the Royal Marines both among officers and non-commissioned officers was irredeemable.[56]

In his work on leadership, Bryan McCoy has perhaps surprisingly emphasized the importance of shame to maintaining and enforcing professional standards, which he calls 'the virtue of shame'.[57] McCoy outlines the general dynamics of shame and notes the great influence which the fear of shame exerts over soldiers:

> It is shame that prevents us from doing shameful things; it is shame that propels us forward despite fear, hunger, and sleeplessness. Shame is the knowledge that one's behavior is less than what is expected by the group. Shame in the eyes of our brothers is a powerful motivator. No one wants to be known or remembered for coming up short when most needed.[58]

As American citizens and US Marines, McCoy and his battalion were probably sensitive to the international reputation of the United States, of which they were presumably proud. Yet the shame which McCoy describes does not refer to patriotism or national pride. On the contrary, the shame is entirely professional: 'Our greatest fear in battle was not death or maiming. It was the fear of being a coward in the eyes of our comrades'.[59] Professional failure was for McCoy and his subordinates the ultimate shame. Critically, failure in combat would mean a betrayal of the Marine Corps and inevitable exclusion from the bonds of fraternity within it. Indeed, in his pre-invasion speech on 19 March 2003, McCoy did not refer to the strategic rationale of the operation the Marines were about to conduct: 'I did not give a rousing "Sands of Iwo Jima" speech, but I spoke firm and soberly.'[60] The content of the speech was illuminating. He tied the performance of his marines to the obligation to which the marines had committed themselves by volunteering:

> I reminded them that we, as a Corps, exist for one reason – to win on the battlefield. The day the Marine Corps fails to win in combat is the day we will cease to exist. I reminded them that our ancestors from Tripoli, Belleau Wood, Guadalcanal, Iwo Jima, Inchon, Chosin and Vietnam would be watching us. It was now our turn, our duty to take our place in the line of history and fight well.

> I appealed to the pride of being a Marine. Up till now they had reaped all the benefits of wearing the uniform, the prestige and honour of belonging to such a storied fighting unit.[61]

As members of the Corps, marines had reaped very significant advantages from the institutional goods which the Corps possessed. However, to remain as marines and to sustain the Corps as a credible military organization, it was essential that McCoy's marines contributed to its battlefield performance, even at the cost of their lives. Professional shame was, for McCoy, the central motivating factor which drove him and his marines into battle and sustained them in combat. This obligation was not merely theoretically. For instance, at Al Kut on 3 April 2003, one marine was killed and several others wounded when McCoy's unit was caught in a major ambush. They demonstrated the potent influence of 'mere' reputation for the professional soldier. They were struck as they conducted the drills they had been trained to perform and which both they and their peers expected them to perform, whatever the costs.

The US Marines have been far from alone in institutionalizing a concept of professional shame in order to encourage combat performance. Canada's Special Forces unit, Joint Task Force 2, have sought to harness the power of shame to encourage performance in a way that is closely compatible with Bryan McCoy's account. In the last few years, Canada's Joint Task Force 2 began (like the Navy Seals before it) to explore whether sports psychology, which had been successful in raising pass rates in the US Navy Seals from one quarter to two thirds, might alleviate the problem. To that end, and on an informal basis, the Training Squadron at Dwyer Hill hired the services of a well-known sports psychologist who had been part of Canadian's Olympic team for a number of years. One of the sports psychologist's innovations was the introduction of the 'Banquet Speech', the oration which is given to Canadian soldiers at the formal dinner for all their comrades at the end of their service, in training. The Banquet Speech required each Special Forces candidate (and badged members of Joint Task Force 2) to articulate, preferably in writing, how they would like to be remembered at the end of their careers to their fellow soldiers. By publicly announcing how they would like to be remembered, Special Forces soldiers could be called to account by their colleagues when they fell short of the standards which they had set for themselves. Soldiers could point out when a comrade was behaving in a way which was inconsistent with his own standards. The Banquet Speech, therefore, deliberately exposes every aspect and every act of each soldier to public accountability,[62] employing mechanisms of shame and honour in the most extreme way.

The disciplinary regime for professional forces in Iraq and Afghanistan has been distinctive. Courts martial have not been staged for cowardice, even though some individuals might be guilty of acts consistent with the military definition of it. Soldiers who failed professionally are subject to a perhaps worse sanction; public shaming. Their professional reputation is ruined and they are dismissed and ostracized by former colleagues. They have no prospects of promotion and ultimately they are no longer part of the military profession, even though they may continue to be paid by defence ministries. They are excluded from their professional group, having offended its central status honour: the competent execution of military duties. In this context, discipline is reserved either for relatively minor infractions or for illegal acts such as abuse. The sanction for combat failure is ostracism; it is professional rather than actual death. Yet, while roundly loathed, soldiers and officers who have been incompetent or cowardly on operations are not typically subject to formal discipline. They are an embarrassment rather than criminal. At worst, they are quickly removed from their unit and from command.

Conclusion

Strachan is correct to emphasize the four possible and interrelated explanations of combat performance: the primary group, patriotism, discipline, and training. Yet while extensive scholarly work has been carried out on three of these factors, discipline tends to be ignored, and even disparaged, as a possible factor in combat performance. By contrast, military discipline has historically been viewed as critical by armies themselves, both as a means of retaining control over their troops and of engendering combat motivation. It is all the more striking then, that cowardice, desertion, and punitive discipline have almost disappeared from military consciousness and the military vocabulary, although both cowardice and desertion still occur in the professional army. The apparent disappearance of cowardice, however, indicates that while formal punishment may indeed be irrelevant to the professional army in generating combat performance, this is only because an informal, self-generated, and constant process of mutual surveillance and disciplining have superseded formal processes. Formal tribunals and courts martial may, indeed, have fallen into desuetude for acts of cowardice, but only because professional soldiers have themselves internalized and enforced upon each other an intricate system of governmentality.

There is, of course, a final question that this exploration of discipline in the citizen and professional army raises. Although these mechanisms of shame and honour seem to have been deeply institutionalized into the armed forces today, it cannot be assumed that formal discipline will always be absent in the future. Punishment for desertion and cowardice has been notable by its absence in recent conflicts, but that cannot mean that severe discipline could never return. In particular, it is important to recognize that the campaigns which Western forces have fought in the last decade were generally conceived as wars of choice rather than as ones of existential national survival—even in the United States. Accordingly, they were not invested with the same political significance as the First or Second World Wars, for instance. In the future, Western forces might have to fight a more conventional war of national defence of the highest strategic importance in which defeat would be politically catastrophic. In this situation, the failure of individual soldiers might be looked on less generously than has been the case during the post-9/11 wars. Cowardice and desertion might become extremely serious issues for an army—and a state—facing potentially disastrous defeat. In such a situation, it is not inconceivable that severe sanctions and even capital punishment could be applied or even reintroduced in the professional forces and a regime of martial justice instituted that is closer to that of the nineteenth century—or the First World War. This is a possible scenario.

Yet, against this picture, it is also worth noting that, even during conflicts of the utmost political seriousness, such as the First and Second World Wars, only the most authoritarian regimes ever instituted a system of draconian military justice. The armies of the liberal democracies did not favour punishment and were always restrained in their use of it. There is no reason to think that professional forces and states in the twenty-first century would conduct themselves differently even in future wars of national survival. Moreover, it seems unlikely that it would be difficult to motivate professional soldiers in such a conflict of clear national interest. In the last decade, Western troops have engaged willingly in high-intensity and dangerous military operations in which significant casualties have sometimes been taken even though the strategic goals of these campaigns have often been obscure or even nugatory. Professional status honour has been enough to motivate soldiers on missions of questionable strategic value. It would seem probable that if these same soldiers were fighting a genuine war of national survival, the mechanisms of professional shame and honour would be vastly magnified by a heightened sense of political motivation and, even, patriotism. The performance of American troops in Iraq and Afghanistan provides probably the best insight into how a professional force fighting for important national causes would operate in the future. In this context, it seems not only unlikely that the armed forces would want to resort to corporal and capital punishment but also that, even more so than in Iraq and Afghanistan, such sanctions would be necessary. Clearly, great caution needs to be exercised here, but it seems plausible to claim that the internal sanctioning methods—the mechanism of shame and honour—evident among professional armed forces in the last decade are likely to remain the norm in the coming decades of the twenty-first century. A new paradigm of military discipline seems to have emerged.

Notes

1. Hew Strachan, 'Training, Morale and Modern War', *Journal of Contemporary History* 41(2) (April 2006): 214.
2. Strachan, 'Training, Morale and Modern War'; Uzi Ben-Shalom, Zeev Lehrer and Eyal Ben-Ari, 'Cohesion During Military Operations: A Field Study on Combat Units in the Al-Aqsa Intifada', *Armed Forces & Society* 32(1) (2005): 63–79; Anthony King, 'The Word of Command: Communication and Cohesion in the Military', *Armed Forces & Society* 32(1) (2006): 493–512.
3. Emmanuel Saint-Fuscien, *À Vos Ordres? La relations d'autorité dans l'armée française de la Grand Guerre* (Paris: Edition de l'Ecole des hautes etudes en sciences sociales, 2011).
4. Charles Glass, *Deserter: The Last Untold Story of the Second World War* (London: HarperCollins, 2013).
5. See Cathryn Corns and John Hughes-Wilson, *Blindfold and Alone* (London: Pheonix, 2005); Anthony Babington, *For the Sake of Example: Capital Courts Martial 1914–1918, The Truth* (London: Martin Secker and Warburg, 1983); Julian Putkowski and Julian Sykes, *Shot at Dawn* (Barnsley: Pen and Sword, 1988); André Bach, *Fusillé pour l'Example 1914–15* (Paris: Talladier, 2013).
6. Mark Thompson, *The White War: Life and Death on the Italian Front 1915–1919* (London: Faber and Faber, 2004).
7. Stephen Wesbrook, 'The Potential for Military Disintegration', in Sam Sarkesian (ed.), *Combat Effectiveness: Cohesion, Stress and the Volunteer Military* (Beverly Hills & London, 1980), 251–61.
8. Significantly, and in contrast to professional armies today, the British Army executed more professional soldiers than citizen troops so that, while the size of the Army increased dramatically, the rate of capital punishment declined. At this point, professionalism seemed to have involved a sense of duty (compatible with contemporary professional soldiers) but was distinctively enforced by physical sanctions including execution. A greater leniency was exercised towards citizens soldiers who had volunteered or been conscripted to fight.
9. Thompson, *The White War*, 266.
10. Steven Ambrose, *Citizen Soldiers* (New York: Touchstone, 1997), 343–4.
11. Thompson, *The White War*, 269–70.
12. Thompson, *The White War*, 261.
13. Thompson, *The White War*, 273–4.
14. Thompson, *The White War*, 273–4.
15. Thompson, *The White War*, 264–5.
16. Thompson, *The White War*, 227.
17. Thompson, *The White War*, 275.
18. Thompson, *The White War*, 276.
19. Thompson, *The White War*, 275.
20. Thompson, *The White War*, 275.
21. Brian Sullivan, 'The Italian Soldier in Combat, June 1940–September 1943: myths, realities and explanations', in P. Addison and A. Calder, *Time to Kill: The Soldier's Experience of the War in the West, 1939–45* (London: Pimlico, 1997), 179.

22. Sullivan, 'The Italian Soldier in Combat', 178.
23. Lucian Truscott, *Mission Command: A Personal Story* (Novato, CA: Presidio, 1990), 544.
24. Truscott, *Mission Command*, 544.
25. Truscott, *Mission Command*, 544.
26. Strachan, 'Training, Morale and Modern War', 215.
27. Hein Severloh, *WN 62: erinnerungen an Omaha Beach Normandie, 6. June 1944* (Garbsen: Hek Creativ, 2010), 26.
28. Back-waters (Haff) refers to the rivers and fens which predominate in that region of Poland.
29. Hans Kuss, *Schiksal einer Generation* (Selent: Pour la Mérite, 2008), 194.
30. Jason Mark, *Island of Fire* (Sydney: Leaping Horseman Books, 2006), 125.
31. Antony Beevor, *D-Day* (London: Penguin, 2010), 392.
32. Beevor, *D-Day*, 175, 163.
33. I am grateful to Hew Strachan for this point: See Corns and Hughes-Wilson, *Blindfold and Alone*; Babington, *For the Sake of Example*; Putkowski and Sykes, *Shot at Dawn*; Bach, *Fusillé pour l'Example 1914–15*.
34. Jeffrey Gettleman, 'Soldier Accused as Coward Says He Is Guilty Only of Panic Attack', *New York Times*, 6 November 2003 <http://www.nytimes.com/2003/11/06/us/soldier-accused-as-coward-says-he-is-guilty-only-of-panic-attack.html>. Accessed 22 January 2015.
35. Toby Harnden, *Dead Men Risen: The Reality of Britain's War in Afghanistan*. (London: Quercus, 2011).
36. Harnden, *Dead Men Risen*, 443–4.
37. Harnden, *Dead Men Risen*, 444–5.
38. Harnden, *Dead Men Risen*, 524–5.
39. Christopher Jennings and Adrian Weale, *Green-Eyed Boys* (London: HarperCollins, 1996), 165.
40. Jennings and Weale, *Green-Eyed Boys*, 167.
41. Jim Frederick, *Black Hearts* (London: Macmillan, 2010).
42. Reuters, 'US soldier apologises for cowardly massacre of Afghan civilians', The *Daily Telegraph*, 22 August 2013, <http://www.telegraph.co.uk/news/worldnews/northamerica/usa/10260874/US-soldier-apologises-for-cowardly-massacre-of-Afghan-civilians.html>. Accessed 22 January 2015.
43. BBC News, 'Afghan massacre soldier Robert Bales gets life', 23 August 2013, <http://www.bbc.co.uk/news/world-us-canada-23820790> Last accessed 10 February 2015.
44. CBC News Montreal, 'Ex-Canadian Forces commander Daniel Menard loses rank', 21 July 2011, <http://www.cbc.ca/news/canada/montreal/story/2011/07/21/montreal-menard-hearing-721.html>. Accessed 10 February 2015.
45. Charlie Savage and Elisabeth Bumiller, 'An Iraqi Massacre, a Light Sentence and a Question of Military Justice', *New York Times*, 27 January 2012, <http://www.nytimes.com/2012/01/28/us/an-iraqi-massacre-a-light-sentence-and-a-question-of-military-justice.html?pagewanted=all&_r=0>. Accessed 10 February 2015.
46. Harnden, *Dead Men Risen*, 444–5.
47. OF-7, British Army, email communication, 20 October 2014.

48. Sergeant (he was promoted after the tour), Royal Marines, personal interview, 6 September 2011.
49. Major, Royal marines, email communication, 30 September 2011.
50. Jeffrey Gettleman, 'Soldier Accused as Coward Says He Is Guilty Only of Panic Attack', *New York Times*, 6 November 2003, <http://www.nytimes.com/2003/11/06/us/soldier-accused-as-coward-says-he-is-guilty-only-of-panic-attack.html>. Accessed 23 April 2009.
51. Jennings and Weale, *Green-Eyed Boys*, 166.
52. Vince Bramley, *Excursion to Hell* (London: Pan, 1991).
53. The Daily Telegraph, 'Young soldier "humiliated" by paras in Afghanistan for being "coward"', 23 April 2009, <http://www.telegraph.co.uk/news/uknews/law-and-order/5209325/Young-soldier-humiliated-by-paras-in-Afghanistan-for-being-coward.html>. Accessed 10 February 2015.
54. The Daily Telegraph, 'Decorated paratrooper took part in indecent assault on Afghanistan "coward", court told', 24 April 2009, <http://www.telegraph.co.uk/news/uknews/law-and-order/5209000/Decorated-paratrooper-took-part-in-indecent-assault-on-Afghanistan-coward-court-told.html>. Accessed 15 February 2015.
55. David Gaunt, 'Vikings VM'ingS in the 'Gan', *Globe and Laurel* (March/April 2007): 107.
56. Sergeant, Royal Marines, personal interview, 22 July 2011.
57. Bryan McCoy, *The Passion of Command* (Quantico, VA: Marine Corps Association, 2007), 54.
58. McCoy, *The Passion of Command*, 54.
59. McCoy, *The Passion of Command*, 54.
60. McCoy, *The Passion of Command*, 55.
61. McCoy, *The Passion of Command*, 55.
62. Fieldnotes, 12 October 2011.

Part III

Historical Experience

6

Combat and Cohesion in the Early Twentieth Century

Strengthening the Inverted Pyramid

Antulio J. Echevarria II

Western militaries took great interest in the issue of unit cohesion in the early twentieth century, though not in the same way nor for the same reasons that most modern observers do. Cohesion was as much a means to an end for *fin-de-siècle* military thinkers as it is today, but both the means and the ends were different. Whereas modern observers study how cohesion is formed outside drill, military theorists a century ago saw cohesion as a product *of* drill. Cohesion was, in a sense, the sine qua non of a good unit. A military formation that could not carry out its collective orders was a failure as a unit. While today's analysts may divide unit cohesion into subcategories, such as task and social cohesion, theorists in the decades before the First World War saw cohesion simply as the intangible bond that kept units from breaking apart—from atomizing into so many individuals—under fire. As French Colonel Ardant du Picq declared in his nineteenth-century classic *Battle Studies*, 'Combat requires today, in order to give the best results, a moral [psychological] cohesion, a unity more binding than at any other time;' cohesion was, in his view, to be thought of as 'unity . . . teamwork . . . dependability'.[1]

Du Picq's concept of cohesion, which was largely representative of Western views before the First World War, consisted of unity of action and individual psychological resilience or steadfastness.[2] These ingredients were mutually reinforcing. Psychological resilience helped individuals withstand physical adversity and battlefield stress as they performed their collective duties. The presence of one's comrades, in turn, bolstered an individual's resilience. Such qualities were instilled chiefly through discipline and battle drills, which were repeated until soldiers performed their tasks well (until they were second nature) even under adverse conditions. The belief that social bonds could play a role in maintaining cohesion was largely accepted; but such bonds were also regarded suspiciously, as they sometimes worked against good order and discipline by encouraging subversive loyalties that could prove divisive in combat situations (and possibly lead to mutinies). It was preferable, therefore, to build cohesion around loyalty to the unit (usually the regiment), to the service, and to the monarch or the nation, and in roughly that order.

The acceleration of technological and social changes over the last half of the nineteenth century made cohesion both more important and more difficult to preserve. Improvements in firearms made it easier for the densely packed military formations of the early nineteenth century to be shot to pieces; and new conditions brought on by the forces of industrialization and urbanization called into question the psychological resilience of the very individuals who were to serve in the ranks of those formations. As du Picq had stressed in the 1860s, advancements in modern firearms meant that 'skill and morale remain factors whose importance ever increases'.[3] Western militaries responded to the techno-social crisis introduced by the forces of modernity with a combination of rhetorical and practical measures aimed at strengthening the ability of combat units to hold together under fire.

The Technological Crisis

Several inventions since the middle of the nineteenth century—breech-loading firearms, magazine-fed rifles, smokeless powder, machine guns, and quick-firing artillery—contributed to making the infantry attack, never the easiest of battlefield tasks, progressively more difficult. By the 1890s, crossing the 'deadly zone'—the distance at which the defender's fire was most effective—was commonly referred to as the tactical problem of the day. It was actually much more than that. Western paradigms of war resembled an inverted pyramid balanced on the necessity to achieve tactical victory. Without it, neither operational nor strategic successes were possible. If the final infantry assault could not be driven home, ground could not be taken, offensive campaigns would gain nothing, and war would truly become 'impossible', as Ivan Bloch would proclaim in 1898.

While the need for tactical reform was well documented as early as the 1850s by the Prussian general staff and its chief, Count Helmuth von Moltke, it reached a new level of urgency in the 1870s due to the high casualties suffered by German units in the Franco-Prussian War. At engagements such as Mars-la-Tour (16 August 1870) and St Privat (18 August 1870), attacking Prussian formations were demolished by French units armed with longer range, breech-loading *chassepots* and rapid-firing *mitrailleuses*. In the former case, the Prussians lost between 40 and 50 per cent of their combat power, and many units simply disintegrated under the fire. In one attack, the 2nd battalion 48th (Brandenburg) Infantry Regiment lost more than 70 per cent of its combat strength—all its officers and some 600 of its soldiers—in a matter of minutes.[4] Similarly, attacking Prussian units (mostly Guard infantry) suffered 50 per cent casualties at the Battle of St Privat.[5] And these were only the most studied cases. Similar results occurred elsewhere in the campaign theatre. Correspondence by the renowned Prussian historian, Hans Delbrück, who served in the 29th (Rhenish) Infantry Regiment during the war, provides a first-hand account of his unit's loss of cohesion during an attack:

> We quickly passed through a village and were in open fields. Here we were exposed to the most terrible shelling, *mitrailleuse* and *chassepot* fire; we ran as far as we could, then rested. During our run across the fields, our regiment became completely scattered and lost all order. Many tried to find cover by lying in a gully in the middle of the field. As it was relatively safe, they could hardly be motivated to go forward. In the confusion, two or three hundred men from different units became mixed together. Leadership either did not exist or was poor and the soldiers showed no courage: it was a pitiful sight.[6]

Discipline, order, command, and control had clearly broken down. Delbrück's unit was essentially *hors de combat*. Even if many of his comrades were not wounded and were thus physically capable of fighting, they were shattered psychologically and intermingled with troops from another unit. To be sure, it was not uncommon for a unit's cohesion to hang in the balance during a battle, or to be temporarily disrupted, until officers could rally it. As Napoleon famously said, victory and defeat were often 'separated by a thin line'.[7] However, the distinctive uniforms and markings of the officers, which made them recognizable to their own soldiers, also made them easier for an enemy to identify. The greater range and accuracy of modern firearms meant that officers were often among the first to become casualties; without its officers, a unit's chances of rallying were poor. Then, as now, a unit's rate of recovery was a function of the number and quality of its leaders, the perceived safety of its environment, the training and experience of its soldiers, and their knowledge of the overall situation.

Six years after the Franco-Prussian War, egregiously high casualties were also noted in the Russo-Turkish War (1877–78). At the first Battle of Plevna (30 July 1877), for example, a Russian division of 9000 troops suffered 50 per cent casualties in an assault against hastily prepared Turkish defences.[8] Two decades later, engagements in the Boer War (1899–1902) showed that advances in modern weaponry in the form of magazine-fed rifles, quick-firing artillery, and smokeless powder had made the problem even worse. At the Battle of

Magersfontein (11 December 1899), an assault by some 900 troops of the British Highlander brigade, the pride of Scotland, was broken up and driven to shameful panic by accurate fire from 'Dutch farmers' armed with Mauser rifles. The Highlanders suffered some 400 casualties, and a loss of collective honour.[9] What this event meant for military tacticians was that the deadly zone had grown from 400 metres or so, as in the Franco-Prussian War, to nearly 1000 metres at Magersfontein. By 1914, the deadly zone had grown to 2000 metres due to the steady and accurate fire possible with machine guns, and the artillery's ability to concentrate its fire in pre-registered zones.

It was not just the greater volume and accuracy of fire that caused concern for military tacticians. It was the fact that both could be achieved with fewer troops and in less time. The traditional attack ratio of 3:1 was, thus, obsolete, as small units were now capable of holding off far larger numbers of attacking troops. As Clausewitz duly noted, battles were won or lost through the 'steady accumulation' of physical and psychological events until the scales tipped in one direction or the other.[10] However, *fin-de-siècle* combat conditions meant the tipping point could be reached in far less time. What was not clear was the percentage of losses an average infantry unit could endure before it broke. While casualty figures of 50 per cent were commonly referred to, military analysts suspected (correctly) that the majority of these casualties occurred after a unit had disintegrated and the troops were attempting to flee. The actual breaking point was thus unknown.

The Social Crisis

If the advance of weapons' technologies were not enough, the rapid pace of social change added further complications to the tactical crisis confronting nineteenth-century military thinkers. The collective forces of industrialization and urbanization along with the rise of anti-militarist sentiments due to the growth of socialism and pacifism were thought to be sapping the psychological and physical strength of the human material that would make up the West's armed forces. It was not clear, therefore, whether new conscripts would have the psychological resilience necessary to endure the privations of modern war. As one esteemed Prussian officer explained:

> We should not forget that our modern personnel have become more susceptible to the impressions of battle. The steady improvement of standards of living tend to increase the instinct for self-preservation and to diminish the spirit of self-sacrifice. The spirit of the times looks upon war as an avoidable evil and militates directly against the kind of courage that despises death. The fast manner of living at the present undermines the nervous system; and the fanaticism, the religious and

national enthusiasm of yesterday are gone; finally, the physical powers of the human being are also on the wane.[11]

Many British officers agreed, ascribing a certain 'restlessness, nervousness, and excitability' to the growing pool of 'town-bred' masses.[12] Respected French officers were of like mind, arguing that families, schoolmasters, and professors must endeavour to give the youth 'a sound moral and physical education' in order to give the army 'young men who can calmly and fearlessly look danger in the face'.[13] In this view they reflected du Picq's belief that, 'The warrior is the ideal of the primitive man, of the savage, of the barbarian. The more people rise in moral civilization, the lower this ideal falls.'[14] American officers harboured similar concerns, noting that modern society 'has undoubtedly made men more nervous, more hysterical and less able to face danger, suffering and death'. The influx of immigrants since the mid-nineteenth century also raised anxieties about the 'mixed character' of the modern American population, and whether its apparent lack of 'independence and resourcefulness' and 'love of country' might make it less capable of enduring the hardships of war.[15] Even though the US army was more isolated than its European counterparts, scattered in small, relatively remote forts where the influences of urban life had difficulty penetrating, concern over modern 'nervousness' and the anti-militarist attitudes that characterized the 'spirit of the times' was still present.[16] Western militaries were, thus, united in fearing that civilized people had lost something natural and vital.

While it is easy to dismiss such concerns as mere handwringing by a martial culture clinging to outmoded values and romantic notions of war, such worries were actually quite widespread, having found their way into popular journals and magazines of the day. Moreover, the notions that underpinned these concerns were introduced and advanced by several accomplished scientists and academics. Military writers were not at all original in their worries; but actually derived much of logic and terminology regarding modernization's downside from the growing body of psychological and sociological literature. Literacy rates were rising rapidly in Western societies. However, with this measurable step forward in the name of Progress came the fear of another, perhaps greater, step backward in terms of the loss of emotional stability and resilience.

In 1881, for instance, George M. Beard, a fellow of the New York Academy of Medicine and a member of the New York Academy of Sciences, the New York Neurological Society, and of the American Neurological Association, published the most important of such studies, *American Nervousness: Its Causes and Cures*. Beard argued that nervousness—a deficiency of vital nerve- or life-force—had recently increased dramatically within the United States, particularly in its more urbanized northern and eastern regions. The primary cause of this nervousness, Beard claimed, was '*modern civilization*', which was characterized by the use of 'steam-power, the periodical press, the telegraph, the sciences, and the mental activity of women [*sic*]'. The outward signs of nervousness were 'certain forms of hysteria, hay-fever, sick-headache, inebriety, and some phases of insanity'.[17]

Beard also asserted that nervousness was becoming more frequent in Europe; and, indeed, his ideas found ready acceptance on the other side of the Atlantic Ocean. Sir James Crichton-Browne, a prominent neurologist and member of the British Medical Association, agreed with Beard and found much evidence of increasing nervousness in British society. Other men of science and medicine exclaimed that the 'nerves of the modern civilized man are less fit than they were in the past' to meet the trials of battle. The 'increasing rush and worry of our existence, the railway, the telegraph, the herded aggregations of human beings in cities, conduce to nervous complaints'.[18] In 1902, the social psychologist Willy Hellpach advanced his own version of the Beard thesis, declaring that 'nervousness' was the 'sickness' of the modern era. Nervousness resulted because human feelings were constantly pulled between opposing emotive poles, 'desire and reluctance'. The remedy was further economic and spiritual development, but in balance. Other German writers found symptoms of nervousness in schools, factories, and cramped residential areas.[19] Concern over the psychological and sociological effects of rapid modernization also appeared in France, where the popular scientist and prolific author Gustave LeBon published his classic, *The Psychology of the Crowd* (1895), arguing that the modern age had become a period of 'transition and anarchy'. The ideas of the past were only 'half destroyed' and those of the new era 'still in the process of formation'.[20] For scientists and pundits alike, the cure for the ills of modernity was often greater participation in 'dangerous sports and adventures', a belief that gave rise to an epidemic of sports clubs and back-to-nature activities. Given such diagnoses by the scientific community, it is little wonder that military leaders grew concerned with the fitness of the modern 'citizen' for combat.

The senior ranks of Western officers' corps were also changing, gradually losing their social homogeneity. In 1875 British officer ranks had 50 per cent commoners; by 1912, that percentage had risen to 59, though commoners accounted for only 36 per cent of officers above the rank of major general. Concern over the rising number of officers with 'base origins' was growing in Russia as well. By 1895, just over half of the officers were hereditary nobles and most of these were not of the landed nobility.[21] By 1914, commoners accounted for 36 per cent of the Kaiserheer's division commanders, 20 per cent of its corps commanders, and 52 per cent of the deputy chiefs of staff of the general staff. By 1899, in the French army, purportedly the most 'democratic' of all, 70 per cent of ranks above the grade of major general and almost

90 per cent of lower officers' ranks were filled by commoners. These percentages are suspect, however, as the French Revolution made aristocratic origins unpopular; and hence many officers disguised their noble roots by removing the *de* from their names. When aristocratic origins were back in vogue at the end of the century, some middle-class officers added the *de*, though they were not actually of noble blood. Accordingly, by 1914, change had clearly altered the social homogeneity—and with it perhaps the social cohesion—of Western officers' corps. The loss of such cohesion may well have added more friction to whatever prejudices and animosities already existed. Remarkably, this topic is understudied. While much is known about the changing composition of Europe's officers' corps, the same cannot be said for the extent to which the loss of social cohesion impaired military performance at the operational and strategic levels of command, that is, among and within corps and army headquarters respectively.

Unity of Action

To accommodate the greater destructive power of modern weapons, Western militaries revised their tactical doctrines in three important ways. First, they encouraged liberal use of open-order and extended order formations so attacking units would present less lucrative targets to defenders, and so the troops could more easily use terrain features for cover as they advanced. Second, they emphasized greater synchronization between the attacking infantry and its supporting artillery fire. The principal idea was to use the attacker's firepower to break or at least suppress the defender's firepower at the decisive point as the infantry made its approach and final assault. Third, they put more importance on the initiative and decision-making of subordinate leaders, junior officers, and noncommissioned officers, as well as soldiers. On the modern battlefield, soldiers would be expected to decide for themselves how to move forward under fire, especially as they were likely to be beyond the visible range of their commanders. These doctrinal changes occurred incrementally, after many years of debate and experimentation; but preserving cohesion or unity of action was at the core of them.

The most celebrated example of such debates occurred in the German army between the proponents of standardized tactics (*Normaltaktik*) and those of mission-oriented tactics (*Auftragstaktik*).[22] The former argued that the only way to maintain cohesion on the modern battlefield was to keep military units in open but orderly formations for as long as possible. According to this view, allowing the troops to become individual fighters or skirmishers would inevitably result in disintegration. Thus, an attack was to proceed according to prescribed stages in which units changed their formations from company columns to extended order as they approached the enemy. The proponents of mission-oriented tactics, in contrast, contended that the forward movement of units and the formations they used had to be dictated by the nature of the mission, the terrain, and the situation. Cohesion would be preserved by training units according to proven principles rather than tactical templates. Similar debates appeared in the professional literature of the British, American, French, and Russian armies. In the end, the doctrine Western militaries took into combat in 1914 was a blend of both views.[23]

However, the actual training that Europe's vast conscript armies undertook before taking the field in 1914 was carried out largely according to prescribed templates. Also, training exercises in the pre-war era could test the cohesion produced by small-unit battle drills, but not how exposure to physical and mental stress might break down that cohesion over prolonged periods of time. Nor could they reveal the genuine difficulty of achieving unity of action with artillery fire and infantry movements over vast distances, and over terrain partitioned by earthworks and trenches or torn to pieces by artillery fire. While military doctrine had generally kept pace with the demands of modern warfare, the same was not true of training procedures. Tension persisted between the need for dispersion and decentralization, on the one hand, and cohesion and centralized control, on the other. The tendency within each army's regulations, which explicitly avoided prescribing universal rules, was to give officers a set of guidelines, or principles, and to leave the choice of specifics, such as formations and control measures, to individual unit commanders as far as possible.

In opting for this sort of synthesis, or compromise, however, the armies of Europe and the United States had, paradoxically, decided not to decide. Warfare was in a period of profound transition, and strict decisions one way or the other might prove premature. Traditional weapons such as the bayonet were not obsolete, as the bloody fighting in the trenches showed; nor were machine guns and artillery always sufficient to win the day. The military regulations of the period thus acknowledged not only the uncertainty that attends warfare in general, but also the difficulty of forecasting the nature of the next war. While this acknowledgment was commendable in some respects, it assumed that the many young officers and soldiers who entered military service possessed the acumen to penetrate the fog and friction of war and make tactical decisions wisely.

The military also placed greater emphasis on the roles of noncommissioned officers. Not only were the numbers of sergeants and corporals increased, their roles were expanded. The Imperial German army seems to have led the way in this regard, as it had twelve long-term noncommissioned officers per company, while the French army had six, and the Russian army only one.[24] It is thus a tribute to the stoicism of the Russian soldier that, without a robust corps of noncommissioned officers, he still endured some two-and-a-half years of suffering at the front, before joining the revolt against the Tsar in 1917. Noncommissioned officers in all Western militaries were expected to stem the 'demoralizing' influence of pacifism and Social Democracy, two of the more contagious ideological plagues of the times. In Germany, instructional handbooks, such as *Der Feldwebel* (The Sergeant), stressed the critical nature of the role of the non-commissioned officer within the unit. The *Feldwebel* the 'Mother of the Company' and formed the essential link between officer and soldier.[25] Likewise, the German army's Infantry Drill Regulations and training pamphlets repeatedly underscored the point that the 'noncommissioned officer supported the officer and stood in for him whenever necessary. Upon his trustworthiness and dedication to duty rested the inner *cohesion* of the company.'[26]

Psychological Resilience

Contrary to expectations, the young British, French, Germans, Russians, and Americans of the early twentieth century—those who would become the fabled 'lost generation'—were more likely to err on the side of audacity than caution. Time and again, that inclination was to prove costly indeed. The literature on the 'ideology of the offensive' has grossly misunderstood the military thinking of this period.[27] The message military theorists were, in fact, trying to drive home by emphasizing the spirit of the offensive was not that the human element alone could overcome material deficiencies, but that it had to be strong enough to complement the new destructive power of modern weapons. Ironically, in the age of mass industrialized warfare, it was the psychological resilience of the individual—itself buoyed by a unit's collective resilience—that was the key to maintaining unit cohesion. The individual both contributed to the collective strength of the unit, and was in turn strengthened by the unit. For that reason, malcontents or weak-minded individuals, unwilling or unable to commit to the unit, posed a threat to its cohesion and had to be punished or removed. Put differently, the individual soldier had become the tip of an inverted pyramid upon which the success of the squad, the platoon, the company, and higher formations depended. Unfortunately, and quite incorrectly, the human material that was to make up the armies of the early twentieth century was thought to have been psychologically weakened by the strains of modernization, and the perceived influence of new attitudes, articulated most loudly in socialist and pacifist literature, which called into question the utility, if not the necessity, of war.

Military writers of the *fin de siècle* thus waged an ideological war against the core ideas of social democracy and pacifism, and renewed their efforts as 'schools of the nation' by instilling military values in their recruits. They also participated in the founding of youth groups, such as the Boy Scouts in the United Kingdom and the United States and the Young German League in the Kaiserreich, and served as their guides and mentors.[28] Drill remained the best guarantee that soldiers would do their duties when the moment arrived. But time of service was reduced in many militaries, and hence the battle for a nation's hearts and minds had to begin well before its youth were eligible for conscription. Still, for all their efforts, the effectiveness of the militaries' endeavours remains difficult to measure.

As mentioned earlier, military thinkers perceived an important relationship between the physical and psychological dimensions of warfare. Engagements were *fought* within a material dimension, with rifles, machine guns, and artillery. However, they were *won* in the psychological one, when one side or the other broke or simply lost hope. Psychological factors were, thus, considered more important than material ones, and in many respects are still considered so today. If all other factors were equal (they rarely are), superiority within the psychological dimension would prove decisive. Even in cases where nothing was equal, superiority in moral considerations was desirable. Such superiority translated into a will to win. An army thus armed would have a higher threshold for enduring casualties and physical hardships. It would continue to find 'work-arounds' in the face of adversity and struggle to prevail. It would also adapt itself to conditions faster than its opponent, and capitalize on such errors and opportunities as presented themselves.

Conclusion

As events in the First World War were to show, the problem of the infantry attack was essentially solved by the time the guns of August began their grim work. Time and again, combat units crossed no-man's land and took the enemy's first line of trenches, albeit sometimes with heavy losses. On 21 March 1918, during the opening assault of the German Michael Offensive, for instance, Lieutenant Edgar von LeSuire of the 1st Bavarian Infantry Regiment could well record in his diary how his unit left its trenches with good 'soldierly spirit' and swiftly took the first line of British defences. The advance benefitted from heavy fog, smoke, and clouds of gas, and was well covered by a creeping artillery barrage. LeSuire reported that the Tommies in the first line 'begged for quarter and showered us with cigarettes, food, and the like'. Nonetheless, as swift as the Bavarian advance was, the creeping barrage that covered it 'rushed ahead of us too fast', and thus left LeSuire's regiment exposed to British artillery and machine-gun fire as it moved toward the second line of defence. Cohesion was lost at this stage. However, the Bavarians regrouped and reconstituted, and pushed on to take the second line of British trenches by nightfall. Reorganization and reconstitution continued into the night as units and leaders were linked up once again, ammunition and food were brought forward, communication was re-established, and hasty defensive positions were established. LeSuire's unit resumed its advance the next day and took its objective, the Crozat Canal near the village of Jussy.[29]

In short, even amid the chaos of close-quarters combat, attacking units might lose cohesion temporarily. But the leadership of junior officers and noncommissioned officers, as well as good training and discipline, could restore it enough for units to continue fighting. However, exploiting a tactical formation's initial successes remained a problem throughout the war for virtually every army's higher headquarters. Sufficient supporting elements and supplies could not be pushed through the gaps to the forward units faster than the opponent could add depth to the defence. The process of crossing the deadly zone thus had to be repeated time and again, which wore down the material strength and morale of the attacking forces. By the end of 1917, General Erich Ludendorff calculated that he had enough seasoned officers to command the German army's battalions, but not its companies.[30] It was a war of attrition after all, and as such numbers would ultimately tell.

By the time the 'Guns of August' opened their historic barrages in 1914, maintaining unit cohesion was no longer quite the problem military commanders had anticipated. The statement of tactical innovator British General Sir Ivor Maxse about tactics and battle drills needing to be 'so ingrained on the training ground as to become second nature in battle' proved largely correct with respect to preserving unit cohesion.[31] Instead of failing at the tip of the inverted pyramid, therefore, the system broke down too often at higher command levels, where cohesion is arguably more important but seldom recognized. It was nightmares like that of the US 2nd Division, which were vastly more common than they should have been, that were ultimately more damaging to morale and to combat performance. Scholars would later refer to the division as the best American unit of the war; however, on the eve of the Battle of Soissons (18 July 1918) that was not yet the case.[32] As it moved into its attack positions, the US 2nd Division had managed only sporadic liaison with units on its flanks, had no experience in working with tanks (which were to support its attack), had conducted no reconnaissance of its attack zone, had insufficient ammunition, had been separated from its heavy weapons, and had been on the march non-stop for forty-eight hours owing to a series of confusing orders and counter-orders.[33] That its attack succeeded at all on the following day was due more to the German commander's decision to withdraw most of his depleted forces. Unfortunately, the story of the US 2nd Division is not unique; it had the opportunity to grow more cohesive over time, whereas other units did not. Regardless of their nation of origin, military staffs in brigade headquarters and above struggled to execute the myriad tasks involved in the large-scale coordination of fire and movement: reconnaissance, synchronization of artillery fires and infantry assaults, logistical planning and follow through, intelligence collection and dissemination, and preparation and revision of orders. Despite that, unity of action at lower levels, which has become a preoccupation if not an obsession among modern observers since the 'primary group' studies of the Second World War, continues to receive an overwhelming amount of attention. Perhaps the greater irony, therefore, is that the higher level task cohesion necessary for directing divisions, corps, armies, and army groups remains understudied. This is true despite the widespread contemporary belief that strategic activities, even at campaign-level headquarters, are far more important than tactical ones.

Notes

1. Charles Ardant du Picq, *Battle Studies: Ancient and Modern Battle*, trans. John N. Greely & Robert C. Cotton (Harrisburg, PA: Military Service Publishing, 1947), 102; from *Études sur le combat: Combat antique et combat modern* (Paris: 1868; reprinted Musée de l'Armée, 1978). This chapter draws on research conducted for Antulio J. Echevarria II, *After Clausewitz: German Military Thinkers before the Great War* (Lawrence: University Press of Kansas, 2000).
2. Carl von Clausewitz, *Vom Kriege*, 19th edn., Werner Hahlweg, ed., (Bonn: Fred. Dummlers Verlag, 1980), Book IV/Chaps. 4 and 9. Rory Muir, *Tactics and the Experience of Battle in the Age of Napoleon* (New Haven, CT: Yale University, 1998), 68–104, 193–216; also Brent Nosworthy, *Battle Tactics of Napoleon and His Adversaries* (London: Constable, 1995); Paddy Griffith, *Forward into Battle: Fighting Tactics from Waterloo to the Near Future* (Novato, CA: Presidio, 1991), 12–49.
3. Du Picq, *Battle Studies*, 216. He did not live to see the measures Western militaries would take to preserve moral factors; he died near Metz in 1870, a casualty of the Franco-Prussian War. Nonetheless, his *Battle Studies* articulated the relationship between unit cohesion and military success.
4. The German III Corps, which bore the brunt of the fighting, lost a total of 307 officers and 6300 soldiers, or 22 per cent of its strength; many of its subordinate units could no longer carry out their orders. David Ascoli, *A Day of Battle: Mars-la-Tour, August 16, 1870* (London: Harrap, 1987), 138. By the end of the day, the Germans had suffered 15,780 casualties to 13,761 for the French, Michael Howard, *Franco-Prussian War* (London: Methuen, 1981), 161. Wilhelm von Scherff, *Studien der neuen Infanterie-Taktik* (Berlin: A. Bath, 1872–74), 33–4. Fritz Hoenig, *Untersuchungen über die Taktik der Zukunft: Entwickelt aus der neueren Kriegsgeschichte*, 4th edn., (Berlin: R. Felix, 1894), 141–2; trans. as Fritz Hoenig, *Inquiries into the Tactics of the Future. Developed from Modern Military History*, 4th edn., trans. Carl Reichmann (Kansas City, MO: Hudson-Kimberly, 1898).

5. Theodor Fontane, *Der Krieg gegen Frankreich* 4 Vols. (Berlin: R. v. Decker, 1877–78), I, 380.
6. Arden Bucholz, *Hans Delbrück & The German Military Establishment: War Images in Conflict* (Iowa City: University of Iowa, 1985), 23.
7. Napoleon's quote is from Wilhelm Balck, *Taktik,* 6 vols., 4th edn. (Berlin: R. Eisenschmidt, 1908–12), vol. V, 329. *On War,* VI/25, p. 469; VII/4, 5, 22, pp. 527–8, 566, 570–2.
8. Bruce W. Menning, *Bayonets before Bullets: The Imperial Russian Army, 1861–1914* (Bloomington: Indiana University, 1992), 62.
9. Byron Farwell, *Queen Victoria's Little Wars* (New York: W.W. Norton, 1985), 344–5.
10. *On War,* I/1, 2, pp. 80, 90–9; and IV/3, 4, 11, pp. 227, 231, 233–4, 258.
11. William Balck, *Tactics,* 2 vols., 4th rev. edn., trans. Walter Krueger, (Fort Leavenworth, KA: 1911–12), vol. II, 194; Michael Howard, 'Men against Fire: The Doctrine of the Offensive in 1914', in Peter Paret (ed.), *Makers of Modern Strategy from Machiavelli to the Nuclear Age* (Princeton, NJ: Princeton University Press, 1986), 519; and Hoenig, *Untersuchungen,* 170–1: 'A certain intellectual and political apathy is evident in some nations, and it is doubtful whether their national sentiment can produce those advantages [necessary] for tactics'. Comparisons between the effete civilized man and his virile primordial ancestor began to appear more frequently. Goltz praised the 'half-tamed' nature of the Boer people with their 'warlike instinct' and 'pride of independence', characteristics which stood in marked contrast to those of the 'modern civilized man'. Gen. d. Inf. Frhr. v.d. Goltz, 'Moralisches Heimweh', *Deutsche Revue* (January 1903): 1–5. C. Freiherr v.d. Goltz, 'Luxus im Heere', *Deutsche Revue* (December 1903): 257–63; and Schlieffen, 'Krieg der Gegenwart', *Deutsche Revue* (January 1909): 13–23.
12. Timothy H. E. Travers, *The Killing Ground: The British Army, the Western Front and the Emergence of Modern Warfare 1900–1918* (London: Allen and Unwin, 1987), 4–5, 46. The quote on bravery is from 'The Human Animal in Battle', *Journal of the United States Cavalry Association* VIII (1895): 44–6; Field Marshal Wolseley, 'War and Civilization', *United Services Magazine* 14 (March 1897): 560–77. Maj. Stewart L. Murray, *The Future Peace of the Anglo-Saxons* (London: Hugh Rees, 1905), 56.
13. Colonel Hippolyte Langlois, *Lessons from Two Recent Wars,* Translated for the General Staff (London: His Majesty's Stationery Office, 1909), 89, 94.
14. Du Picq, *Battle Studies,* 216.
15. Maj. R.L. Bullard, 'A Moral Preparation of the Soldier for Service and Battle', *Journal of the Military Services Institute* XXXI, No. 120, (1902): 779–92, here 785. Theodore Roosevelt, 'Twisted Eugenics', *Outlook* 106 (1904): 32. Capt. Leroy Eltinge, 'Psychology of War', *Journal of the United States Cavalry Association* XXI, No. 84 (May 1911): 1015–52. For more on anti-modernism in American society see T.J. Jackson Lears, *No Place of Grace: Antimodernism and the Transformation of American Culture 1880–1920* (New York: Pantheon, 1981).
16. On the awareness of greater responsibilities, see Lieut. Joseph T. Griffin, 'Military Instruction for Schoolboys', *Journal of the Military Services Institute* 35 (1905): 61–7. On the influence of geographic isolation, see David I. MacLoed, 'Socializing American Youth to Be Citizen-Soldiers', in Manfred F. Boemeke, Roger Chickering, and Stig

Förster (eds.), *Anticipating Total War: The German and American Experiences, 1871–1914* (Cambridge: Cambridge University Press, 1999), 137–66. On the threat of socialism, see Lieut.-Col. James S. Pettit, 'How Far Does Democracy Affect the Organization and Discipline of Our Armies, and How Can its Influence Be Most Effectually Utilized?' (Seaman Prize Essay), *Journal of the Military Services Institute* 38 (1905): 1–38, who reported that the socialist vote had grown from 60,000 in 1900 to 600,000 in 1904.

17. George M. Beard, *American Nervousness: Its Causes and Consequences* (New York: Putnam's, 1881), vi–viii; emphasis original. Beard also held that nervousness was a mark of distinction: it appeared only in 'advanced' races, religions, and walks of life. His views were widely publicized during the 1880s in such journals as *North American Review, Atlantic Monthly, Appleton's Journal, Yale College Courant*, and the *Virginia Medical Journal*.
18. Beard, *American Nervousness* , xiii; see also Sir James Crichton-Browne, 'La Vieillesse', *Revue scientifique* 49 (1892): 168–78. Stephen Kern, *The Culture of Time and Space 1880–1918* (Cambridge, MA: Harvard University Press, 1983), 110–13. H.W. Wilson, 'The Human Animal in Battle', *Fortnightly Review* summarized and reviewed in *Journal of the United States Cavalry Association* VIII (1895): 44–6. See also Tom Lutz, *American Nervousness, 1903: An Anecdotal History* (Ithaca, NY: Cornell University Press, 1991), 3–7.
19. Willy Hellpach, *Nervosität und Kultur* (Berlin: J. Räde, 1902), 12, 221ff, emphasis original; also 'Soziale Ursachen und Wirkungen der Nervosität', *Politisch-anthropologische Revue* (1902): 43–53, 126–34; and 'Hysterie und Nervosität', *Psychische Studien* 30 (1903): 14–21, 89–95, 152–9. Walter Stallmeister and Helmuth E. Lück (eds.), *Willy Hellpach: Beiträge zu Werk und Biographie* (Fankfurt a.M.: P. Lang, 1991). Gerhard Heilig, 'Fabrikarbeit und Nervenleiden', *Medizinische Reform* 16(31) (1908): 369–97. Else Hasse, 'Nervosität und Geistesleben', *Der Türmer* 13 (1911): 577–87.
20. Charles Féré, 'Civilization et névropathie', *Revue politique* 41 (1896): 400–13; Gabriel Hanotaux, *L'Energie française* (Paris: E. Flammarion, 1902). Gustave LeBon, *The Crowd: A Study of the Popular Mind*, Intro. Robert A. Nye (New Brunswick, NJ: Transaction Pub., 1995), 34; emphasis original. See also Robert A. Nye, *The Origins of Crowd Psychology: Gustave LeBon and the Crisis of Mass Democracy in the Third Republic* (London & Beverly Hills: Sage, 1975). E.A. Ross, *Social Psychology* (New York: Macmillan, 1908); William McDougall, *The Group Mind* (Cambridge: Cambridge University Press, 1921); Sigmund Freud, *Modern Sexuality and Modern Nervousness* (New York: Critic, 1915).
21. Ulrich Trumpener, '*Junkers* and Others: The Rise of Commoners in the Prussian Army, 1871–1914', *Canadian Journal of History* 14 (1979): 29–47; Wilhelm Deist, *Militär, Staat und Gesellschaft: Studien zur preußisch-deutschen Militärgeschichte* (Munich: R. Oldenbourg, 1991), 43–56; Friedrich Forstmeier, et al., *Deutsche Militärgeschichte 1648–1939*, 6 vols. (Munich: Bernard & Graefe, 1983), Vol. 3, Part V, 85–91; and Hughes *King's Finest*. P.G. Razell, 'Social Origins of Army Officers', *British Journal of Sociology* 14(3) (1963): 253. Douglas Porch, *March to the Marne: The French Army 1871–1914* (Cambridge: Cambridge University Press, 1981); John Bushnell, 'The Tsarist Officer Corps, 1881–1914: Customs, Duties, Inefficiency',

American Historical Review 86(3), (June 1981): 753–80, 759; Peter Kenez, 'A Profile of the Prerevolutionary Officer Corps', *California Slavic Studies* 7 (1973): 121–58, 123; and Hans-Peter Stein, 'Der Offizier des russischen Heeres im Zeitabschnitt zwischen Reform und Revolution, 1861–1905', *Forschungen zur osteuropäischen Geschichte* 13 (1967): 409–18.

22. For more detail, see Echevarria, *After Clausewitz*, 32–42.
23. See, for example: Preussen Kriegsministerium, *Exerzier-Reglement für der Infanterie* (Berlin: E.S. Mittler, 1906), paras. 443–4; and 'Exerzier-Reglement für der Infanterie', in *Beiheft zur Militärwochenblatt* 4 (1908): 287; Preussen Kriegsministerium, *Felddienst-Ordnung* (Berlin: E. S. Mittler, 1908), paras. 29, 31; and 'Die neue Felddienst-Ordnung', in *Beiheft zur Militärwochenblatt* 4 (1908): 163–214, esp. 167.
24. Walter M. Pintner, 'Mobilization for War and Russian Society', John L.H. Keep, 'Soldiering in Tsarist Russia', and Bruce Menning, 'The Army and Frontier in Russia', in Col. Carl W. Reddel (ed.), *Transformation in Russian and Soviet Military History* (Washington, DC: Office of Air Force History, 1986), 39–50, 5–20, and 25–39, respectively. John L.H. Keep, *Soldiers of the Tsar: Army and Society in Russia, 1462–1874* (Oxford: Oxford University Press, 1985); John Bushnell, *Mutiny Amid Repression: Russian Soldiers in the Revolution of 1905–1906* (Bloomington, IN: Indiana University, 1985); 'Peasants in Uniform: The Tsarist Army as a Peasant Society', *Journal of Social History* 13 (1980): 565–76; Kenez, 'Prerevolutionary Officer Corps', 129–33.
25. *Der Kommandierende General des 18.Armeekorps, von Eichhorn, über den Dienstunterricht von Offizieren an Mannschaften*, Frankfurt am Main, 19.9.1905. Karl Demeter, *Das deutsche Offizierkorps in Gesellschaft und Staat 1650–1945*, 4th edn., (Frankfurt am Main, Bernard und Graefe, 1965), Appendix 20, 310–13; Kitchen, *German Officer Corps*, 100–3. Holleben, *Der Feldwebel: Aus der Praxis für die Praxis* (Berlin: Liebschen, 1911), 4, 6–7.
26. Oberleutnant Weberstedt, *Der Unteroffizier und Unteroffizierschüler als Lehrer im Gefechts- und Felddienst* (Leipzig: Friedrich Engelmann, n.d.), 8, emphasis added; Friedrich Immanuel, *Der Unterricht an die Unteroffiziere der Infanterie* (Berlin: Liebelschen, 1914), 17; Oberst, z.D, Spohn, *Der Unteroffizier als Lehrer und Erzieher. Unteroffizier-Bibliothek*, Vol. 2 (Oldenburg i. Gr.: Gerhard Stalling, 1909), 8, 10, 40–1, 55–6, 60–1; and *Berufs- und Standespflichten der Unteroffizier. Unteroffiziier-Bibliothek*, Vol. 6 (Oldenburg i. Gr.: Gerhard Stalling, 1909), 22, 53.
27. Jack Snyder, *The Ideology of the Offensive: Military Decision Making and the Disasters of 1914* (Ithaca, NY: Cornell University Press, 1984).
28. For more detail, see Echevarria, *After Clausewitz*, 105–20.
29. Cited in Holger H. Herwig, *The First World War: Germany and Austria-Hungary 1914–1918* (New York: St. Martin's, 1997), 403–4.
30. Herwig, *First World War*, 395.
31. Paddy Griffith, *Battle Tactics of the Western Front: The British Army's Art of Attack 1916–18* (New Haven, CT: Yale University, 1994), 99.
32. Mark Ethan Grotelueschen, *The AEF Way of War: The American Army and Combat in World War I* (Cambridge: Cambridge University Press, 2007), 200–5.
33. Douglas V. Johnson II and Rolfe L. Hillman, Jr., *Soissons 1918* (College Station: Texas A&M University, 1999), 59–62.

7

Re-evaluating Combat Cohesion

The British Second Army in the Northwest Europe Campaign of the Second World War

Jonathan Fennell

As the United Kingdom's involvement in combat operations in Afghanistan came to an end in 2014, it is appropriate that the contributors to this edited collection reflect upon the thirteen years of fighting. Such reflection, however, is best situated within a broader context than the pressing military and political imperatives of the early twenty-first century. Thus, this chapter aims to broaden the historical and conceptual perspective by building on the considerable body of research on the material and tactical underpinnings of successful military performance in the twentieth century and examining the role of morale. Indeed, Stephen Biddle has pointed to the need to give morale 'systematic theoretical attention'.[1] The chapter explores the conceptualization of military morale and argues that morale remains ill-defined, inconsistently used, and poorly understood. It proposes that in order for the concept of morale to have appropriate explanatory value in a critical analysis of the past, particularly in a combat environment, it should be clearly differentiated from definitions associated solely or primarily with mood or group dynamics. A functional conceptualization is, therefore, proposed, which focuses its meaning and relevance on motivation and the willingness to act in a manner required by an authority or institution. The chapter then takes a step towards developing a methodologically sound quantitative approach to assessing morale in an historical battle context. It makes use of a diverse range of new and underused sources, from archives across the Commonwealth, to quantitatively assess on a weekly basis the morale of the British Second Army in the Northwest Europe Campaign of the Second World War. The chapter demonstrates, for the first time, how, at least in the context of the British experience in Northwest Europe, morale can be accurately and robustly quantified over the course of a campaign. These conceptual and methodological innovations facilitate an in-depth analysis of the complex web of interrelationships that shape military performance. The findings point to the need to re-evaluate the dominant theories on combat cohesion and put strategy, understood as an iterative multilevel process rather than just a bridge between national policy and tactics, at the heart of any understanding of combat cohesion.

Combat Cohesion

Much has been written on and around the subject of combat cohesion in the twentieth century.[2] Anthony King's recent study *The Combat Soldier* has built upon this literature and offered a reconceptualization of cohesion and a reinterpretation of the factors that influence cohesive military performance. King defines cohesion as the 'successful coordination of actions on the battlefield'; it refers to the ability of soldiers 'to act together and to achieve their mission in the face of the enemy'.[3] Thus, King, like Sam C. Sarkesian,[4] points to the connection between cohesion and successful or victorious military performance. This understanding of cohesion is subtly different to that propounded in other studies where cohesion incorporates a greater notion of efficiency.[5] Allan R. Millet, Williamson Murray, and Kenneth H. Watman argue that 'judgments on effectiveness should retain some sense of proportional cost and organizational process'. In their view, the German army in the Second World War was more cohesive than its enemies because it converted national resources into fighting power in a more efficient manner. The fact that Germany lost the war was irrelevant.[6]

This chapter embraces the King definition of cohesion. While efficient military performance is certainly desirable, it is only one aspect of the far greater endeavour of using military means in the pursuit of national policy. There is a real danger that military practitioners and scholars, by lauding the efficient performance of military organizations such as the Wehrmacht, lose sight of the purpose of war and the fundamental art that governs its use, strategy.

In light of this understanding of cohesion, it is possible to identify three broad trends, or 'metanarratives', in the literature on cohesion in Western armies in the twentieth century.[7] The first, which can be referred to as the 'brute force narrative', stresses the importance of firepower and materiel to successful combat performance. In the First World War, these ideas were captured and expressed in the attritional warfare that developed on the

Western Front. Sir John French, the British Commander-in-Chief, wrote in January 1915 that

> Breaking through the enemy's lines is largely a question of expenditure of high explosive ammunition. If sufficient ammunition is forthcoming, a way can be blasted through the line. If the attempt fails, it shows . . . that insufficient ammunition has been expended, i.e. either more guns must be brought up, or the allowance of ammunition per gun increased.[8]

One of the more influential authors on the subject during and after the Second World War was S.L.A. Marshall. He contended that what was needed in battle was 'more and better fire'.[9] He argued that 'fundamentally fire wins wars and that every other aspect of operations is important only in the measure that it contributes to this grand object'.[10] His ideas became dogma in the US Army, and there was, as Hew Strachan has pointed out, a clear trajectory from the publication of *Men Against Fire* to the performance of the US Army in the wars that followed, such as in Vietnam.[11] In the decades following the Second World War, the 'brute force narrative' gained credence. Widely read historical studies by Max Hastings, John Ellis, and Carlo D'Este, among others, stressed the material and technological underpinnings of successful combat performance:[12]

> In the broadest sense . . . it was economic might and productive capacity that determined the outcome of the Second World War. In the final analysis, once America and Russia had been drawn into the war and once each had blocked its opponents' first mad rush, then there was absolutely no chance that the Axis powers could salvage even a negotiated peace . . . a Reich that would wage Blitzkrieg with only 47,000 tanks against its enemy's 227,000, 116,000 guns against 915,000 and 350,000 trucks against 3,000,000, or a would-be maritime empire that aims to maintain its grasp by producing only 13 aircraft carriers against its opponent's 137, has not much real chance of imposing its will.[13]

More recently, a vein of scholarship has stressed the importance of conceptual factors in combat cohesion.[14] These works have emphasized the influence of doctrine and tactics on effective combat performance. In fact, according to Stephen Biddle, the effects of technology and materiel are 'secondary to force employment'.[15] Biddle has presented evidence to suggest that one method in particular, the modern system (based on cover, concealment, dispersion, suppression, combined arms, and independent small-unit manoeuvre at the tactical level and depth, reserves and differential concentration at the operational level),[16] 'broke the trench stalemate in 1918 and defined the standard for successful military operations throughout the post-1918 era'.[17] The 'tactics narrative' does not reject the significance of material and technological preponderance, but argues that victory and defeat result from the way material and technology are employed on the battlefield. This approach to understanding cohesion places a considerable weight on the institutions and individuals whose responsibility it is to intellectualize, imagine, and resolve the problems of battle performance.

Anthony King has outlined a third approach to understanding combat cohesion, a 'morale narrative'. King argues, building on the work of authors such as Ardant du Picq, John Baynes, and John Keegan, that in the context of mass conscript armies in the twentieth century, tactics (usually enshrined in doctrine) were not central to successful combat performance. Although 'western powers had developed refined infantry doctrine', their platoons were 'normally unable to execute [it] in combat, despite expectations that they would do so'. This was due to the fact that 'the vast majority of infantry was poorly trained'. In the absence of adequate training, success in battle was the product of a 'mass-individual dialectic' aided by supporting fire and, above all, artillery. Either the highly motivated group charged the enemy or the highly motivated individual (usually an officer or noncommissioned officer) assaulted the enemy for the group. Success or failure was, therefore, contingent on the willingness of the soldier to close with and kill, wound, or capture the enemy. Success was dependent on morale.[18]

The 'morale narrative' also recognizes the significance of materiel in war. It emphasizes the role played by massed firepower in suppressing enemy resistance to allow attacking forces to cross the fire-swept zone. It constitutes, in many ways, an important counterpoint to the dominant approach of historians and social scientists interested in the study of war. It builds on a 'Tolstoyan' understanding of combat (or history for that matter), that battles (events) are not merely the execution of a general's or leader's concept, but rather the product of 'the grubby actions of small groups of men, often confused and frightened'.[19] However, unlike the 'brute force' and 'tactics' narratives, the 'morale narrative' has received little 'theoretical attention'.[20] This may be because any study of morale faces somewhat unique conceptual and methodological problems in that scholars and military practitioners struggle to agree on a workable definition of morale or a way to meaningfully measure it. These shortcomings have potentially profound implications for the study of combat cohesion. First, as Michael Handel has argued, there is a danger that historians, social scientists, and military practitioners 'focus on that which can be measured' and ignore 'the critical but more elusive factor of will'.[21] Second, assessments of morale based on too little evidence can misrepresent patterns of events and skew understanding of the past, with serious implications for present and future decision-making.

Defining Morale

A number of definitions of the term morale have been proposed in the literature. Robert Guion, Reuvan Gal and Frederick Manning, Stephan

Motowidlo et al., and Thomas Britt and James Dickinson have broadly claimed that morale refers to various affective states, such as contentment and happiness, and group dynamics.[22] Although not without foundation, there are major problems with approaches that identify affective states and/or group dynamics as central to morale. Specifically, there is much evidence to suggest that troops can experience positive affective states while also behaving in manners that are completely contrary to the interests of the military establishment. For instance, a combatant might feel 'happy', 'satisfied', or 'optimistic' due to the fact that he has run away and is now safe from harm. Similarly, as Chiara Ruffa's chapter illustrates, strong group bonds and 'high morale' can undermine positive military performance, where for example a soldier might stop to aid a wounded comrade in spite of orders to press the attack.[23] At the same time, group desertions and mutiny can evidence small group bonds, yet they are clearly actions contrary to the needs of the military institution.[24] For instance, in Vietnam, the importance of group survival often outweighed the need to complete assigned tasks.[25] Definitions of morale based on affective states or group dynamics can lead, therefore, to the paradox, demonstrated in all these cases, of high morale leading directly to poor military performance.

Motivation is a further aspect of morale that requires consideration, but it is important to stress that motivation does not require the individual or group to be positive (or enthusiastic, as Britt and Dickinson suggest) about assigned objectives. Combatants can be highly motivated to carry out tasks that they don't necessarily want to engage with, that they are not hopeful, optimistic, or confident will succeed, because they are disciplined or even coerced into action. As John Keegan has pointed out, '"Kill or be killed" is the logic of battle – to which military law adds the rider, "risk being killed by the enemy or else risk being killed by your own provost-marshal".'[26] This coercive dynamic (or 'negative' motivation), Keegan claims, has been present in warfare throughout the ages.[27] Soldiers have to accept 'the basic philosophy governing human relationships within an army',[28] said S.L.A. Marshall, or take what Samuel A. Stouffer et al. referred to as 'the institutionally sanctioned consequences'.[29] Brigadier A.B. McPherson, who, in 1950, compiled the British War Office monograph on discipline, wrote that 'in the inculcation of "morale" discipline is an indispensable factor. Self-respect, self-control and obedience to authority, which go hand in hand in training in discipline, are sturdy elements also in the foundation of morale.'[30]

There is evidence, then, to suggest that military professionals closely link morale with motivation but explicitly distance it from associations with positive affective states or the group. For example, General Sir Bernard Law Montgomery, in a paper he wrote on the subject in April 1946, defined morale as 'endurance and courage in supporting fatigue and danger... the quality which makes men go forward in an attack and hold their ground in defence'.[31] He stated categorically that high morale 'is not contentment or satisfaction' or 'happiness'. Happiness, according to Montgomery, 'may be a contributory factor in the maintenance of morale over a long period, but it is no more than that. A man can be unhappy but can still, regularly and without complaining, advance and defend.'[32] He saw group dynamics in very much the same light.[33] Morale, according to Montgomery, is clearly like an overall causative influence on a soldier's conduct; indeed some psychologists use the term 'motivation' in similar contexts.[34]

Following the broad thrust of Montgomery's approach, morale is here defined as the willingness of an individual or group to engage in an action required by an authority or institution; this willingness may be engendered by a positive desire for action and/or by the discipline to accept orders to take such action. The degree of morale of an individual or group relates to the extent of their willingness or discipline to act, or their determination to see an action through. This is the broad approach to the conceptualization of morale that is taken in this chapter. This definition does not conflate morale with emotion or interpersonal loyalty, although it excludes neither as potential sources of motivation. Instead it recognizes that military institutions require their personnel, first and foremost, to be willing to carry out orders. If troops are willing to carry out orders, any military organization will have a chance of success irrespective of the mood of their men and women or the strength of primary group bonds.[35] Indeed, as King discusses in Chapter 5 of this volume on discipline, it seems almost certain that the severe, even fanatical, discipline of the Wehrmacht positively encouraged German soldiers to fight, even though they can scarcely have been happy about the sanctions with which they were constantly threatened.

Assessment of Morale

There are relatively few reliable primary or secondary sources that assess levels of morale in armies. Those that do, such as the works of Morris Janowitz and Edward Shils, S.L.A. Marshall and Samuel A. Stouffer et al. in the Second World War, and Leonard Wong et al. in the 2003 Iraq War, make extensive use of contemporaneously recorded attitudinal surveys. These studies typically link morale with motivation; for example, S.L.A. Marshall contended that morale had 'no meaning' unless it was understood as 'the body of thought' that 'disposes the thinker to high endeavor or toward failure',[36] while Wong et al. explicity engaged with the problem of 'combat motivation'.[37] Nevertheless, these studies are sufficiently wide-ranging that they can also be used to assess morale as a positive affective state or aspect of group dynamics. They provide scholars with valuable attitudinal information for military organizations at particular places and periods of time. However, their findings are often highly contextualized; their relevance, therefore, to other combatant nations, wars, and environments must be understood in this light.

In the majority of cases where attitudinal surveys have not taken place, scholars attempting to assess morale are typically dependent on sources such as personal recollections and memoirs. These sources use varying conceptualizations of morale and also suffer from serious methodological shortcomings, not least the fallibility of individuals' memories, especially where interviews take place decades after the event, and the impact of prevailing cultural and social interpretations of the meaning of events on the recollections of historical actors. Contemporaneously recorded diaries or letters are more reliable as historical sources, but, again, they lack consistent definitional clarity and it is often difficult to amass a representative sample of such sources for an army. Recent studies, such as Alexander Watson's *Enduring the Great War*, this author's *Combat and Morale in the North African Campaign*, and Jonathan Boff's *Winning and Losing on the Western Front* have tried to circumvent this problem by making use of newly discovered or underused official sources, such as censorship summaries of soldiers' mail and morale reports.[38] Such sources cover morale widely and deeply and tend to express views that represent a considerable body of opinion among troops. They are so extensive and detailed that the appropriate definitional focus can relatively easily be mined within them. However, again, these sources are available in only a minority of cases, and, thus, may be limited as a tool to assess morale in armies across the twentieth century.

In addition to the use of attitudinal surveys, combatant letters and diaries, censorship summaries, and morale reports, military professionals (particularly commanders and medical officers), theorists, and scholars have also used rates of sickness, battle exhaustion, desertion, absence without leave (AWOL), and self-inflicted wounds (SIW) in units as a useful proxy for morale. These metrics do not assess morale as an affective state or as an aspect of group dynamics but accurately, albeit indirectly, assess the willingness of troops to engage in actions required by the authority or institution. In this regard, the functional definition used here not only clarifies the meaning of morale but also facilitates its measurement. Militaries tend to record and keep this kind of data, and, therefore, this method may prove generally useful to assess morale across armies and across time periods. Historians of the British and Commonwealth Armies in the Second World War have been particularly proactive in utilizing these sources as a tool to assess morale. Indeed, both David French and John Buckley have used battle exhaustion, desertion, AWOL, and SIW statistics to argue that perceptions of weakness in British morale in the Normandy campaign have been misplaced and misguided.[39] This author has also used sickness, battle exhaustion, desertion, and AWOL and surrender statistics to plot a morale crisis and recovery in the Eighth Army in the North African campaign in 1942.[40]

There is a significant body of contemporaneous evidence that confirms the use of this approach in the assessment of morale. F.A.E. Crew, the official historian of the British Army Medical Services in the Northwest Europe campaign of the Second World War, wrote that sickness in the campaign had been kept low in no small measure due to morale. The fact that the campaign had lasted less than a year, been 'filled with victory', and that 'no deepening gloom born of disaster' had set in played a role.[41] Montgomery, the Commander of 21st Army Group, of which Second Army was a part, was emphatic that the 'truly remarkable success of the medical organisation' in Northwest Europe had been built upon two factors: the quality of the manpower available and the morale of the Army. Montgomery noted the 'exhilarating effect of success' on 'reducing the rates of sickness' in Second Army. It was quite clear to him that 'the medical state' of the army had not been 'dependent on the doctors alone'. Their efforts had been 'immeasurably facilitated when morale' had been 'at its highest'.[42]

The number of battle exhaustion cases in a unit was also seen as an accurate indicator of morale. Major D.J. Watterson, the psychiatrist attached to Second Army in Normandy, wrote in his report for June 1944 that the number of exhaustion cases in a unit was an excellent 'index to that unit's quality of men and of its wellbeing and morale'. In fact, he said, 'it is as good a guide to the unit's state of mental health as is the temperature chart in a case of fever'.[43] A similar view was present in the Canadian Army. Major R.A. Gregory, Neuropsychiatrist to 3 Canadian Infantry Division, wrote during the Battle of the Scheldt in October 1944 that 'there was one thing of note among all troops admitted for exhaustion, [that is] lack of morale or lack of volition to carry on'. The psychiatrist's work, he wrote, 'is first to advise on things which tend to lower morale and increase the incidence [of] exhaustion'.[44]

According to F.A.E. Crew, there was a close association between battle exhaustion and certain military crimes, such as desertion, AWOL, and SIW.[45] In fact, he argued that these issues could be viewed as two sides of the same coin. 'Whereas some men went sick and were evacuated, others suffering from much the same condition ran away, were charged and awarded penal servitude. The psychological escape of the former and the physical escape of the latter were expressions of the same mechanism.'[46] A report produced in June 1945 on '"Soldiers Under Sentence" for Such Offences as Desertion, Cowardice, Mutiny etc., Whose Cases Have Been Reviewed in British Second Army' lends support to this contention.[47] During the period 17 November 1944 to 22 January 1945, 596 men who had committed offences in June, July, and August 1944 and served three months of their sentences were interviewed with the intention of ascertaining their suitability for further service in the Army. The 'great majority of the prisoners' were found to be 'good personality types, only too anxious to be given the opportunity to redeem their characters'. Their crimes were, according to the report, 'in a large majority of cases not premeditated, but occurred on the "spur of the moment" when under great stress'. The men gave a number of reasons for their behaviour, most of which are known influences on morale in battle:

> "I was browned off at the time", "I was very tired, we got no sleep", "I was worried about bad news from home"... Resentment at being transferred to infantry after long service in other arms of the service was often evident. Here... the plea put forward as an excuse for behaviour was "lack of training" and in many cases... recent reinforcements... had obviously not been long enough with their units to develop that group spirit and friendship which is the best bond to security.[48]

It is conceivable, of course, that the inmates were telling the military authorities what they wanted to hear. Nevertheless, the conclusions of the report are supported by evidence from other theatres during the war and the behaviour of the men themselves once they returned to combat.[49] Of the 596 men who had their cases reviewed, 435, or 73 per cent, were returned to full duties in the line.[50] By the end of the war, these men had seen a number of months of active duty, the majority taking part in the Reichswald Forest battle, which, according to the report, was 'more grim' than the period in which the men had committed their offences. Overall, almost three-quarters of the men returned to full duties in the line gave satisfactory and meaningful military service. The report concluded that 'the percentage of "real bad eggs" has been small' and that 'the outstanding impression gained' was 'the great similarity there exists between many of the cases reviewed and those that are referred through medical channels for psychiatric opinion'. The 'natural conclusion' to be made from this study was, therefore, 'that the majority of deserters' were not 'true cowards', and, in fact, that issues relating to morale, just as in cases of battle exhaustion, had been the key driver in the soldiers' behaviour.[51]

Morale, according to Q.V.B. Wallace, the Deputy Director of Medical Services 1 Corps, Second Army, was also central to keeping instances of SIW in an army at a low level. He wrote that the welfare of the soldier, leave, the provision of newspapers, the speed of delivery of letters, and so on had all 'been much more thoroughly planned for' in Northwest Europe and this 'had a very obvious effect on morale', which, in turn, affected SIW.[52] Lieutenant-Colonel A.T. A. Browne, who was given the task of investigating the causes of SIW during the Normandy campaign, painted a very similar picture. He argued that SIW were the product of the interplay of five factors: (1) the cultural and social conditions prevalent at the time; (2) the appalling physical conditions experienced on the battlefields of Normandy; (3) the heavy casualties; (4) poor quality training and manpower selection; (5) the need for rest and recuperation.[53] All of these factors were key influences on morale.[54]

Thus, an examination of historical sources suggests that morale affects and is affected by rates of sickness, battle exhaustion, desertion, AWOL, and SIW. It may additionally be suggested that the relationship between these factors is so close that a knowledge of rates of sickness, battle exhaustion, desertion, AWOL, and SIW can provide a useful indicator of the levels of morale at different stages of a campaign. The extent and strength of these associations may be explored by use of quantitative approaches. This requires a systematic assessment of available primary sources, such as censorship summaries of soldiers' mail, to describe and 'quantify' levels of morale (through the use of a numerical morale scale) as well as the tabulation of rates of sickness, battle exhaustion, desertion, AWOL, and SIW. If the tabulated rates are indeed indicators of morale, they should, broadly speaking, fluctuate along the lines of the shifts in morale levels as encapsulated in the morale scale.

Quantitative Assessment of Morale

In order to carry out this quantitative assessment, the case of British Second Army in the Northwest Europe campaign of the Second World War was explored. Thirty-six bi-weekly censorship summaries, covering the period between 4 April 1944 and 15 October 1945, were found at Library and Archives Canada and the Directorate of History and Heritage, Ottawa.[55] These summaries were compiled from 1,494,479 letters sent by officers, noncommissioned officers, and other ranks in the British Army and have never before, as far as this author is aware, been used in an historical study. These summaries proved a most useful source for the building of the morale scale. Each summary was based on the censorship of an average of over 40,000 letters, and, thus, gave a reliable appraisal of changes in morale in Second Army. A typical modern national opinion survey uses a sample of 1100 to give an error margin of ± 3 per cent; if a similar methodology were used on a sample of 40,000, the error would be tiny (± 0.5 per cent). The censorship reports used a fairly consistent language and usually described the level of morale in a summary at the start. In order to turn a qualitative assessment of morale into a scale, all typical descriptions of morale were first graded from the 'best' to the 'worst' morale, taking into account the definitional approach being used here. The different descriptions were then put in categories of meaning that were as far as possible of equal intervals apart. Then each description of morale was given a score. Where morale was described as 'excellent', it was awarded a score of 3. 'High' morale was given a score of 2 and 'good' morale was scored 1. 'Satisfactory' morale was given a score of 0 (neither positive or negative). Morale described as 'severely tried' was scored –1, while 'low' and 'very low' morale were scored –2 and –3 respectively.[56]

Statistics relating to weekly levels of sickness, battle exhaustion, and SIW were found in the War Diaries of the Deputy Director of Medical Services, Second Army.[57] The War Diaries reported sick admissions per 1000 per week (which also included battle exhaustion cases) in the 'Weekly Hygiene Reports' for Second Army and the incidence of battle exhaustion per 1000 for each weekly period in the 'Monthly Hygiene Reports' for Second Army.[58] The number of battle exhaustion cases was calculated by applying the battle exhaustion rate per 1000 to the strength of Second Army (which was provided in each 'Weekly Hygiene Report'). The overall weekly sick number was also computed in this manner. By then subtracting the weekly battle exhaustion number from the overall weekly sick number, thus obtained, it was possible to calculate the true sick number (i.e. without battle exhaustion). The weekly figures for SIW were outlined in a 'Report on Surgery in Second Army in the Northwest Europe Campaign, 1944–45'.[59] Monthly convictions by Courts Martial for desertion and AWOL were recorded in the 'Administrative History of 21st Army Group' and in the War Diaries of the Deputy Judge Advocate General, 21st Army Group.[60] While some of these figures have been presented before,[61] this chapter offers, for the first time, as far as this author is aware, a comprehensive picture of the number of sick, battle exhaustion, desertion and AWOL, and SIW in Second Army during the Northwest Europe campaign of the Second World War (see Tables 7.1 and 7.2).

This detailed presentation of the facts should allow, through the use of a correlation analysis, a clearer picture to be formed of the relationship between morale and those factors historically associated with it.[62] The available statistics for desertion and AWOL, which were presented in a monthly fashion, could not be meaningfully broken down into bi-weekly figures (to be correlated with bi-weekly reported morale levels contained in the censorship summaries). Thus, it was decided to carry out the analysis on the eleven monthly periods of the Northwest Europe campaign, June 1944 to April 1945.[63] The two bi-weekly censorship summaries in every month had, therefore, to be combined to create a morale score for each month as a whole (see Table 7.3). In cases where the censorship summaries described different levels of morale in the first and second half of a month, a best effort was made to find a balanced score taking fully into account the complexity of the issues addressed in both summaries. In many cases, this decision was not overly problematic. For example, it was sensible in instances where morale was described as 'good' (scored 1) in one half of the month and 'excellent' (scored 3) in the other half of the month to record an overall morale score of 2 (or

Table 7.1. Strength of the fighting portion of Second Army plus weekly admissions for casualties (cas)*, sick, battle exhaustion (BE), and self-inflicted wounds (SIW), 6 June 1944 to 26 May 1945.[64]

Week Ending	Strength	Cas	Sick	BE	SIW	Week Ending	Strength	Cas	Sick	BE	SIW
6–10 Jun-44		5259^				09-Dec-44	268,000	590	2007	54	3
17-Jun-44	217,437	7194	1155	563	16	16-Dec-44	261,000	287	1976	21	6
24-Jun-44	294,393	2554	1027	365	18	23-Dec-44	286,000	474	1599	34	2
01-Jul-44	355,000	6020	2329	925	18	30-Dec-44	297,000	471	1719	39	6
08-Jul-44	424,922	3739	2855	999	30	06-Jan-45	304,000	907	2290	73	8
15-Jul-44	399,003	4225	2131	1365	31	13-Jan-45	299,000	993	2805	72	6
22-Jul-44	421,576	6615	6281	2373	28	20-Jan-45	299,000	742	2177	54	5
29-Jul-44	343,839	2670	2225	622	20	27-Jan-45	210,000	945	1889	92	8
05-Aug-44	296,000	3968	2717	681	19	03-Feb-45	175,000	269	1562	9	4
12-Aug-44	274,000	5173	1984	923	21	10-Feb-45	166,000	234	1264	27	1
19-Aug-44	238,000	1730	2468	231	8	17-Feb-45	130,000	194	1267	29	1
26-Aug-44	221,190	492	1803	53	5	24-Feb-45	102,000	86	883	26	0
02-Sep-44	234,000	893	1114	87	3	03-Mar-45	101,000	110	816	20	3
09-Sep-44	244,000	1419	686	39	2	10-Mar-45	228,000	883	1913	82	1
16-Sep-44	245,000	1220	1186	186	5	17-Mar-45	300,000	469	1900	15	0
23-Sep-44	250,000	2005	933	190	3	24-Mar-45	346,000	1314	1597	69	0
30-Sep-44	255,000	1554	1760	260	4	31-Mar-45	316,000	5124	2473	411	13
07-Oct-44	290,000	1599	1717	165	3	07-Apr-45	325,000	2657	1577	156	11
14-Oct-44	274,000	876	1540	219	3	14-Apr-45	356,000	2214	1751	125	9
21-Oct-44	265,000	1496	1582	334	5	21-Apr-45	357,000	2360	2289	136	6
28-Oct-44	303,000	1616	2018	182	1	28-Apr-45	357,000	1469	2366	143	9
04-Nov-44	268,000	1956	1729	295	8	05-May-45	373,000	1522	2355	112	18
11-Nov-44	269,000	1092	1552	108	16	12-May-45	401,000	714	2327	20	6
18-Nov-44	275,000	1232	1601	96	2	19-May-45	347,000	479	2461	7	1
25-Nov-44	262,000	1534	2261	128	2	26-May-45	311,000	303	1473	0	0
02-Dec-44	266,000	931	1819	82	0						

* This includes all casualties admitted for reasons of battle wounds, battle accidents and burns (but does not include killed and captured).
^ Admissions to medical units were not classified during this period. Thus, it is probable that this figure includes sick and battle exhaustion.

Table 7.2. Number of convictions by courts martial for desertion and AWOL, Second Army, July 1944 to April 1945.[65]

Month	Desertion	Absence	Total
July 1944	141*	96*	237
August 1944	393*	82*	475
September 1944	303	169	472
October 1944	343	143	486
November 1944	353	304	657
December 1944	380	261	641
January 1945	321	353	674
February 1945	310	327	637
March 1945	451	333	784
April 1945	410	264	674

* Figures for 103 and 105 Reinforcement Groups have been excluded as they were situated in the UK at this time.[66]

Table 7.3. Morale descriptions and scores based on the censorship summaries, Second Army, June 1944 to April 1945.

Month	Morale description	Morale score
June 1944	'Excellent'	3
July 1944	'Good'	1
August 1944	'Good'	1
September 1944	'Excellent'	3
October 1944	'High'	2
November 1944	'High'	2
December 1944	'High'	2
January 1945	'High'	2
February 1945	'High'	2
March 1945	'High'	2
April 1945	'Excellent'	3

LAC T 17924 21st Army Group Censorship Reports, April to August 1944; LAC RG24 Volume 10706 File No. 46-3-6/Int Vol.1 21st Army Group Censorship Reports, August to November 1944; LAC RG24 Volume 10706 File No. 46-3-6/Int Vol.2 21st Army Group Censorship Reports, December 1944 to February 1945; LAC RG24 Volume 10706 File No. 46-3-6/Int Vol.3 21st Army Group Censorship Reports, February to April 1945; LAC RG24 Volume 10547 File No. 153/Censor Reps/1/3 21 Army Group Censorship Reports, 1944 to 1945.

'high'). The sick, battle exhaustion, and SIW figures, recorded on a weekly basis, had also to be added together to produce monthly statistics.[67] The absolute figures for each month were reconverted into rates per 1000 by taking an average from the size of army statistics contained in each 'Weekly Hygiene Report' (see Table 7.4); this ensured that the varying size of the army at different times did not distort the relationships. The rates for each category were then combined to produce a composite measure that could be correlated with the morale scale (see total/1000 in Table 7.4).

The results of the correlation analysis showed that the composite measure (the combined rate of sickness, battle exhaustion, desertion, AWOL, and SIW)

Table 7.4. Monthly sick, battle exhaustion (BE), desertion/AWOL, and SIW rates per 1000 in Second Army, June 1944 to April 1945.

Month	Sick/1000	BE/1000	Desertion*/1000	SIW/1000	Total/1000
June 1944	14.42	5.96	0.82	0.18	21.39
July 1944	36.99	14.22	1.20	0.30	52.72
August 1944	35.72	6.49	1.87	0.19	44.26
September 1944	19.64	2.81	1.96	0.06	24.46
October 1944	26.93	3.70	2.32	0.06	33.01
November 1944	28.79	2.09	2.39	0.09	33.36
December 1944	29.31	0.65	2.42	0.07	32.45
January 1945	36.12	1.09	2.29	0.11	39.62
February 1945	33.70	0.77	2.82	0.04	37.32
March 1945	30.75	2.00	2.61	0.06	35.41
April 1945	22.80	1.70	Not available	0.11	24.61**

* Includes AWOL. As Courts Martial usually tried cases in arrears, the figures for each month are based on the Courts Martial from the following month, i.e. the figure for July 1944 is derived from the number of Courts Martial carried out in August 1944.
** Does not include desertion/AWOL.

had an extremely strong negative correlation with morale (−0.949, $P<0.001$). This is a remarkably strong relationship and shows that these factors when taken together can be used as a quantitative method to assess levels of morale, at the very least for the Army and campaign under discussion. Coincidentally, it offers a quantitative justification of the approach used by this author in *Combat and Morale in the North African Campaign*. Of some interest also is the finding that of the factors taken singly, sickness had by far the strongest negative correlation with morale (−0.875, $P<0.001$).[68]

In Search of a Theory to Explain Combat Cohesion: The Case of British Second Army in Northwest Europe

Thus, the evidence suggests, that for Second Army, at this time, in the conditions prevalent in Northwest Europe, morale can be assessed quantitatively. This methodological innovation makes it possible, using figures for sickness, battle exhaustion, desertion, AWOL, and SIW in Second Army to assess and graph levels of morale during the campaign on a monthly basis (see Figure 7.1).

As one would expect from the high level of correlation, Figure 7.1 shows that the tabulated figures for sickness, battle exhaustion, desertion, AWOL, and SIW coincide with the morale scale presented in Table 7.3, with the exception of July 1944, and, to a lesser extent, January and February 1945. In fact, it is suggested that where there is a discrepancy between the tabulated figures and the morale scale, the tabulated figures should be used. The bi-weekly censorship summaries, although an excellent source, are sometimes limited in their utility as a measure of morale due to the fact that they occasionally omit mail from key units during

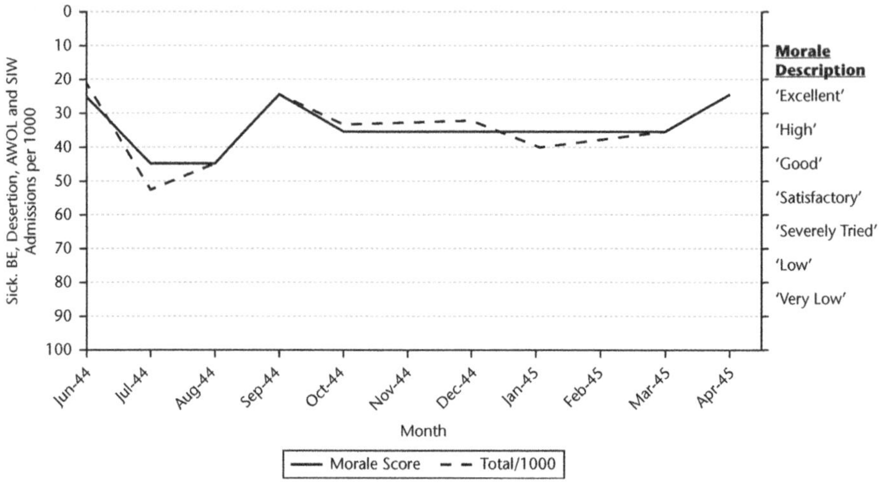

Figure 7.1. Second Army, monthly admissions per 1000 to general hospitals and casualty clearing stations for sickness, battle exhaustion (BE), desertion, AWOL, and SIW as compared with the morale score derived from the censorship summaries, June 1944 to April 1945.[69]

critical phases of the campaign. For example, the summary for 16 to 31 July 1944 did not assess mail from the main assault divisions (Guards Armoured Division, 7th Armoured Division, and 11th Armoured Division) during Operation Goodwood, the key battle during the report period.[70] In fact, only 326 of the 27,366 letters assessed in this bi-weekly summary came from these units. The tabulated figures include sick, battle exhaustion, desertion, AWOL, and SIW for all units during all phases of the campaign. Furthermore, by assessing these factors together, they render irrelevant reporting irregularities such as the fact that units often classified similar battle behaviours in very different manners (for example, some units classified breakdowns in battle as exhaustion while others classified them as military crimes, such as desertion). Thus, it can be argued that the tabulated figures capture fluctuations in morale even more accurately than the morale scale derived from the censorship summaries.

The picture of morale that emerges from the tabulated figures shows clearly that, for most of the Northwest Europe campaign, morale in Second Army was either high or excellent, as indeed one would expect during a successful period of fighting. In fact, morale appears to have been lower only during the static and arguably less successful fighting in Normandy, in July and August 1944. This can add considerably to our understanding of the Normandy campaign and the impact of morale on combat cohesion more generally. Much has been written on the state of morale in Second Army during this period, and, while an army with satisfactory morale can hardly be described as in crisis, this assessment of morale may go some way to explain Second Army's less than spectacular performance in the critical Battles around Caen and Falaise in the summer of 1944.

To understand the role of morale in combat cohesion in Second Army in Northwest Europe better, however, an even more granular picture can be compiled (see Figure 7.2) by presenting weekly fluctuations in the rates of sickness, battle exhaustion, and SIW (weekly figures for desertion and AWOL are not available). These factors taken together have almost as strong a negative correlation with morale (–0.936, $P < 0.001$) as the total tabulated figures presented in Table 7.4.

Figure 7.2 shows that while morale was, on the whole, high to excellent for much of the campaign, it did dip at some critical moments in July and November 1944 and January and February 1945. It could barely be described as 'good' during operations in the first half of August 1944, notably during Operation Bluecoat and the attempt to close the Falaise Pocket, and was seriously problematic on one occasion—during Operation Goodwood in July 1944.

Indeed, the graphical analysis of morale presented in Figures 7.1 and 7.2 supports the conclusions of much of the recent historiography on the British Army in Northwest Europe: morale was a necessary component of combat cohesion (morale in Second Army was broadly speaking high throughout the victorious campaign); however, morale was not a sufficient explanation for Second Army's successes and failures on the battlefield.[71] For example, morale would appear to have been at its highest before and during Operation Market Garden. But Market Garden was a failure. It is likely, as John Buckley has argued, that Market Garden was a conceptual failure rather than a morale one.[72] Morale would also appear to have been mostly high during operations in the Low Countries and Germany, but these operations were beset with setbacks and delays.

In fact, one of the key advantages of the quantitative assessment of morale presented in this chapter may be that it allows the relationship between levels of morale and combat cohesion to be quantitatively evaluated. Because the 'Colossal Cracks' operational approach employed by Second Army was essentially attrition-based,[73] an assessment of German casualties would be an excellent proxy for combat effectiveness. However, while such figures are available in ten-day periods for the German Army as a whole, they do not differentiate between casualties inflicted by Allied nationality, for example British, Canadian, or American, making an accurate assessment of Second Army's effectiveness on this basis problematic. As the number of enemy prisoners captured by Second Army per week is available, it may be used as a crude but useful alternative proxy for combat success or cohesion. As can be seen in Figure 7.3, there is a weak negative correlation (–0.236, $P = 0.179$) between the level of morale in each week of the campaign (assessed using figures for sickness, battle exhaustion, and SIW) and the number of enemy prisoners captured per week. Insofar, therefore, as the number of prisoners captured might indicate combat

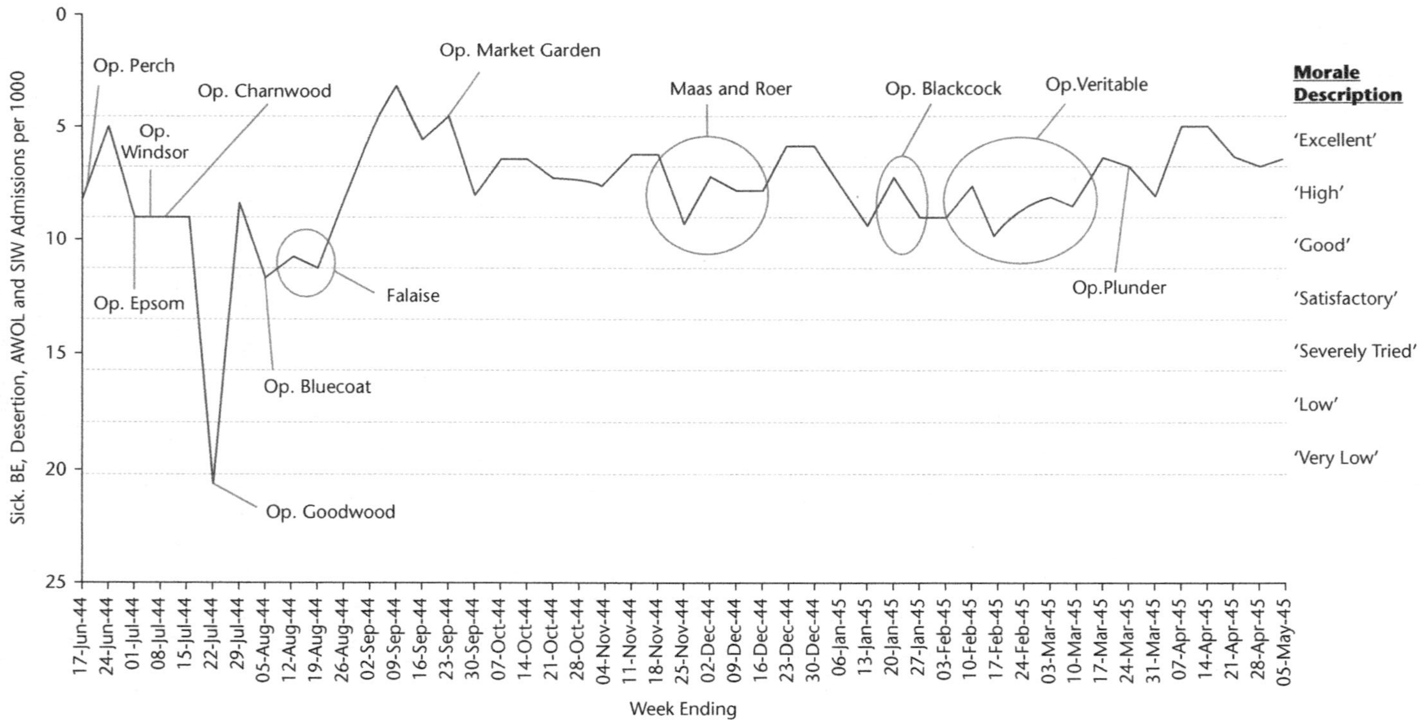

Figure 7.2. Second Army, weekly admissions per 1000 to general hospitals and casualty clearing stations for sickness, battle exhaustion, and SIW, 11 June 1944 to 5 May 1945. Morale scale equivalents are presented on the right-hand Y-axis.[74]

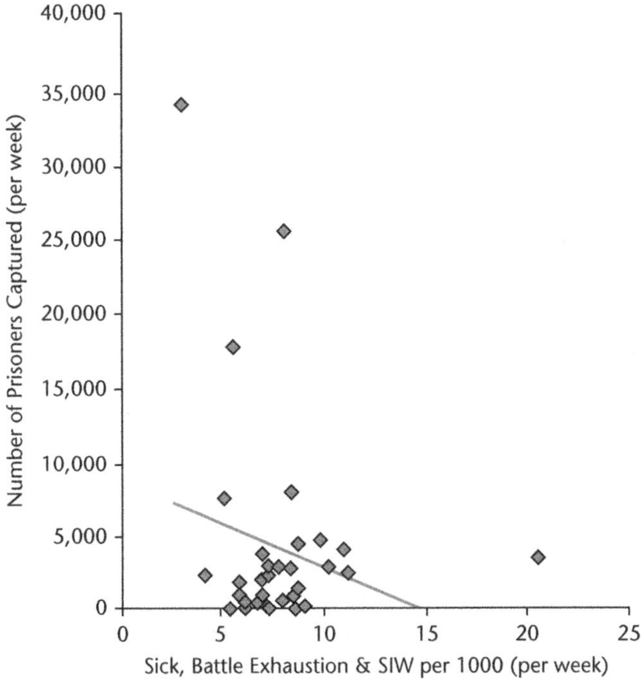

Figure 7.3. Second Army, rate of sick, battle exhaustion, and SIW per 1000 and number of prisoners captured per week.[75]

success, this success is only weakly associated with the strength of morale in Second Army.

The empirical evidence presented in Figures 7.2 and 7.3 may have implications for our understanding of combat cohesion in the twentieth century. It clearly indicates that the 'morale narrative', as outlined by King, may, while correctly emphasizing the important role of morale in combat cohesion, tend to overestimate that importance to the exclusion of other key factors.

Additionally, any suggestion that Second Army's combat performance was the product primarily of the application of overwhelming firepower (the 'brute force narrative') appears to be undermined also by the available evidence. Army Operational Research Group Memorandum No. E20, 'Some Statistics on the North West Europe Campaign', clearly states that there 'was no significant correlation between German casualties and either Allied [Anglo-Canadian 21st Army Group and US 12th Army Group] casualties or Allied ammunition expenditure'.[76] Furthermore, Figure 7.4 demonstrates that there was not a strong relationship (–0.234, P=0.179) between the firepower employed by 21 Army Group, of which Second Army was a key component, and the area of ground captured from the enemy (again, a crude but useful

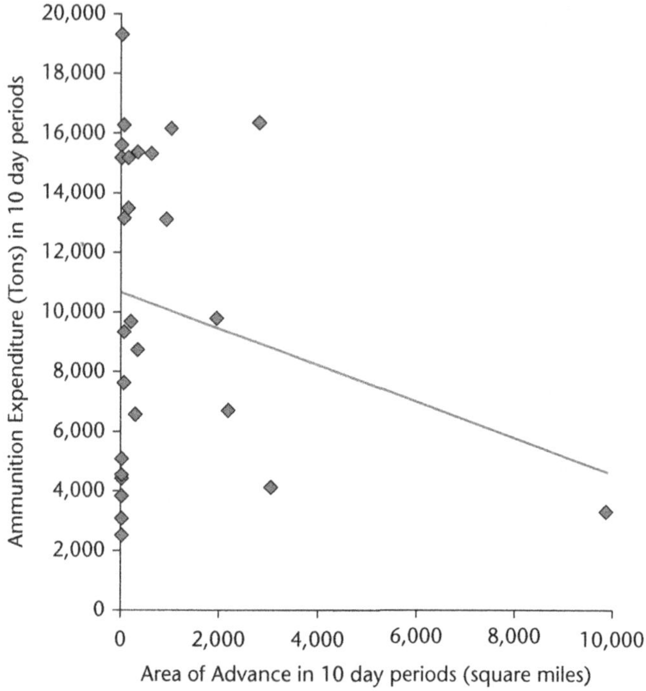

Figure 7.4. 21st Army Group, area of advance in ten-day periods to ammunition expended (1 July 1944 to 28 March 1945).[77]

proxy for combat success). In fact, there was a similarly weak correlation (–0.160, P = 0.445) between the firepower employed by 21 Army Group and the number of enemy prisoners captured.[78]

Furthermore, any suggestion that the performance of Second Army was driven primarily by conceptual considerations (the 'tactics narrative') appears to be ill-founded. David French, Stephen Hart, and Stephen Biddle have all argued that Montgomery employed a systematic, or even formulaic, approach to battle in the Northwest Europe campaign.[79] Indeed, by 1944, Montgomery had devised an operational approach that gave the British Army the best chance possible to tackle the Wehrmacht in battle. As John Buckley has argued, Montgomery's 'greatest achievement' was 'correctly adopting an operational approach that emphasized British strengths (firepower, planning, logistics and intelligence) and avoided weaknesses (close-combat tactics and manoeuvre)'.[80] However, both theoretically and practically, there are limits to the effectiveness of systemic approaches to battle. Both Buckley and Copp have argued that by 1944, German tactics had become so predictable that the Allies were able to devise their own strategies around them. The same can be said for the British approach. By 1944, the Germans were fully aware, for example, that the direction of creeping barrages indicated the thrust lines of impending advances. This made it easy for the Wehrmacht to place reserves at the decisive point at the key moment. Normandy did not go exactly according to plan; no military plan ever survives first contact with the enemy unchanged.[81] In fact, as Buckley and Copp have intimated, the Northwest Europe campaign was characterized far more by adaptation and innovation than heretofore recognized.[82] It can be argued, therefore, that British success was not built primarily on an operational, or conceptual, approach to the problem (however much this may have helped), but rather on the ability of individuals (including Montgomery) and units at every level of the military organization to balance the available means (be they material, tactical, or morale) with the objectives and ends at hand, as indeed they did over the course of the successful campaign. This interactive, dynamic process required a nuanced, flexible, and strategic approach to combat cohesion.

Combat Cohesion and Strategy

Further consideration, therefore, of the relationship between cohesion and strategy may shed light on the issue. According to Clausewitz, 'war is an act of force to compel our enemy to do our will'.[83] This statement suggests that, more often than not, belligerents fight for a goal, or an end, that they try to achieve or impose upon an enemy. That goal does not have to be rational or particularly well thought out; it can be driven by reason, passion, hatred, and pure chance. Nevertheless, the challenge of devising a successful strategy to achieve a goal dominates military theory and practice. Indeed strategy was the central and unifying theme of Clausewitz's *On War*.[84]

Clausewitz defined strategy as the 'use of the engagement for the purpose of the war'.[85] More recently, Colin Gray defined strategy as 'the use that is made of force and the threat of force for the ends of policy',[86] while Liddell Hart described strategy as 'the art of distributing and applying military means to fulfill the ends of policy'.[87] Thus, it appears clear that the relationship between military means and policy objectives (or ends) is at the heart of the strategic process.[88] Military decision makers have constantly to align their means to coincide with policy; vice versa, policymakers have to create policy in line with the available means. As Strachan puts it

> In the ideal model of civil–military relations, the democratic head of state sets out his or her policy, and armed forces coordinate the means to enable its achievement. The reality is that this process – a process called strategy – is iterative, a dialogue where ends also reflect means, and where the result – also called strategy – is a compromise between the ends of policy and the military means available to implement it.[89]

To achieve policy by use of violent means, a belligerent has typically to match his effort against what Clausewitz referred to as an enemy's 'power of resistance'. This he expressed as 'the product of two inseparable factors, viz. the total means at his disposal and the strength of his will'.[90] To put it another way, all belligerents in a conflict engage with strategy and attempt to balance ends with means. Military means are a product of the interplay between the material capability to fight and the will to fight (morale, as defined above).[91] When a belligerent can no longer continue to fight, because their material strength has been whittled away through attrition, or they are no longer willing to fight, and they desert or surrender en masse, that belligerent must eventually, by engaging in the strategic process, also alter policy (ends must reflect means). Victory ensues when a belligerent comes to the conclusion that they no longer have the means, either physical or psychological, or both, to resist the will of the enemy and they alter policy (to, for example, surrender or enter negotiations for a ceasefire).[92]

As Michael Howard has put it

> The military means used to obtain the purposes of the war were divided by Clausewitz and his successors into two: *Vernichtungsstrategie*, 'strategy of annihilation' – the destruction of the enemy capacity to defend himself by destroying his armed forces on the battlefield; and *Ermattungsstrategie*, the use of attrition to wear down his will to resist. The first disarms the adversary, leaving him literally at the mercy of the victor. The second persuades him that victory is, if not impossible, only obtainable at an unacceptable price.[93]

While strategy, in the current dominant usage, is 'designed to make war useable by the state, so that it can, if need be, use force to fulfill its political objectives',[94] it can also be used in the sense that one might have a 'strategy' to cross the fire-swept zone and capture, for example, a pill box on the other side of a field. Militaries tend to refer to this type of activity as 'tactics' or 'drill'. However, 'tactics' or 'drill', it can be argued, represent nothing more than formalized pre-packaged strategy. They provide junior leaders with ready-made solutions to balance ends and means in oft-repeated military scenarios. Strategy, understood as a process, appears, therefore, to play a key role, in different guises, at different levels of military activity and may be understood to increase in complexity as one climbs the levels of war. Strategy at the tactical level can almost always be formalized in drill. Strategy at the operational level can be formalized in operational doctrine. However, at the military strategic, grand strategic, or political levels, strategy becomes so complex and contingent that it is beyond formalization or doctrine and becomes truly the art of the military or political genius.

Thus, it is fair to argue that decision makers at all levels in war have to act strategically (balance ends with means). Indeed, S.L.A. Marshall insisted that they do.[95] As the strategies devised by political leaders and senior officers influence the goals and objectives (policies) and means available to those below them, strategy cascades in an interactive fashion from the top of the state apparatus to the activities of junior officers on the frontline. To take the Second World War (and particularly the climactic 'end-game' in Northwest Europe) as an example, those in charge of the state (the politicians) decided that the only strategy for realizing the end of protecting British long-term interests was to wage war with Germany. As a consequence, decision makers (at the grand strategic level) were tasked with devising strategies to balance the ends of winning the war with the military, economic, diplomatic, and intelligence means available. These strategies influenced the policy, or the goal, and means available at the military strategic level. Those tasked with strategy at the military strategic level then had to balance these ends and means, or, if this proved impossible, ask their superiors to rein in ends or ramp up the means. In the case of Northwest Europe, commanders decided that the best way to bring about the defeat of the Wehrmacht on the Continent was to launch an invasion across the Channel to the shores of Normandy. This strategy outlined the ends and means for operational level commanders in the British Army who had to develop their own 'strategies' to get ashore at Normandy. One element of the operational challenge was the requirement for 3 British Infantry Division to storm Sword Beach. Thus, 'strategy' at the operational level influenced policy (or the ends) and the means available at the tactical level, and junior officers had to develop their own 'strategies' to succeed with the specific assignments tasked to them on 6 June 1944.

If we understand strategy in this manner, as an interactive process that both cascades down and feeds back up the levels of war, rather than a level of war or activity situated only as a bridge between national policy and tactics,[96] we may begin to see inadequacies in the three broad meta-narratives of combat cohesion outlined in this chapter. Each 'narrative' explains combat cohesion as the product mainly of one element of strategy, rather than the emergent outcome of a complex multidimensional process. The 'brute force narrative' focuses on the 'material capability' element of military means. The 'morale narrative' focuses on the 'will' element of military means. The 'tactics narrative' tends to interpret strategy as a fixed doctrine (for example Biddle's 'modern system of force employment') rather than a dynamic interactive process where means constantly affect policy as much as policy affects means. Thus, the 'tactics narrative' assumes that formalized responses to military problems are adequate when, in fact, war is infinitely complex and decisions and activities have to be made in light of constantly shifting and evolving events (as S.L.A. Marshall put it, in war 'the unusual is met usually and the abnormal becomes the normal').[97] Furthermore, the 'tactics narrative', as Strachan has pointed out, can obscure the dynamic interlinkages between the different levels of war. For example, it can treat tactics and operations as a 'politics free zone' and can, therefore, encourage military practitioners to 'diverge dangerously' from the policies devised at the political, grand strategic, and military strategic levels, as indeed it did with the German Army in the Second World War, and, arguably, in the counter-insurgency campaigns undertaken by the British and American Armies in more recent conflicts in Iraq and Afghanistan.[98] Strategy can, and perhaps should, be understood as an iterative multi-level decision-making continuum where decisions on means and ends at each level can affect decisions on means and ends at all other levels.

The three broad meta-narratives of combat cohesion unquestionably add to our understanding of the role of firepower, tactics, and morale in battle. However, while it is certainly possible that firepower, tactics, or morale might have played a dominant role in specific Battles in the Northwest Europe campaign, the evidence derived from this study suggests that patterns of cohesion in British Second Army cannot be explained by reference to one 'narrative' on its own. In the case of Second Army, it appears that there was no ready-made solution to the problems of cohesion on the battlefield (see Figures 7.2, 7.3, and 7.4). The best that commanders could do, as Clausewitz argued in *On War*, was to recognize the components of the problem and manage and balance them appropriately in each battle scenario, that is, behave strategically.[99] The individuals in Second Army, including army commanders and junior officers, made decisions and behaved in ways that were highly contextualized, contingent, and interlinked. While all-encompassing theories may be attractive, the challenge for scholars trying to search for patterns is to recognize complexity and embrace its methodological challenges rather than shy away from it in the search for simplicity. Indeed empiricism demands such a scholarly approach.

Conclusion

This chapter has re-evaluated the role of morale in the military strategic process. It has proposed a functional conceptualization of morale, which focuses its meaning and relevance on motivation and the willingness to act in a manner required by an authority or institution. This conceptualization of morale has facilitated a first step towards developing a methodologically sound quantitative approach to assessing morale in an historical battle context. Additionally, the conceptual and methodological innovations explored here have facilitated a more thorough interrogation of the complex web of interrelationships that shape military performance. The chapter offers evidence to suggest that the dominant narratives on combat cohesion have either overestimated or underestimated the importance of morale, not least, perhaps, owing to the fact that it is typically conceptually misunderstood and inaccurately assessed. Thus, the findings point to the need to re-evaluate the dominant theories on combat cohesion and put strategy, understood as an interactive process that both cascades down and feeds back up the levels of war, at the heart of any understanding of combat cohesion.

These conclusions resonate with the broader themes of this volume. Indeed, the irregular wars of the twenty-first century are more similar in nature to the conventional ones of the twentieth than is generally perceived.[100] The use of 'brute force' by the British Army in Helmand in Afghanistan from 2006 did not produce the outcomes desired, nor, in recent years, has a tactical and operational approach which focused on 'hearts and minds', or the willingness (morale) of combatants and citizens to fight, led to a peaceful and prosperous Afghanistan. As Hew Strachan has argued in relation to the German Army in 1940, 'tactical and operational successes' have not been 'given the shape which strategy could have bestowed'.[101] In Iraq and Afghanistan, operational doctrine, the 'tactics' narrative, has filled the void left by strategy, but, in spite of its sophistication, has failed to do duty for effective strategy.[102] Thus, cohesion at the tactical and operational levels in Iraq and Afghanistan cannot be understood outside the context of strategy. Indeed, it would appear that combat cohesion is contingent, contextual, multidimensional, and inherently complex, and any attempt to understand it must acknowledge that reality. The challenge for military practitioners in the twenty-first century, as surely it was in the twentieth, is to balance and coordinate ends and means at all levels of war and in all military scenarios. This (strategic) process requires intelligent, well-trained, highly motivated, and well-provisioned personnel to identify the character of the war they are fighting and act appropriately. As Clausewitz wrote:

> Theory cannot equip the mind with formulas for solving problems, nor can it mark the narrow path on which the sole solution is supposed to lie by planting a hedge of principles on either side. But it can give the mind insight into the great mass of phenomena and of their relationships, then leave it free to rise into the higher realms of action.[103]

Notes

1. Stephen Biddle, *Military Power: Explaining Victory and Defeat in Modern Battle* (Princeton, NJ: Princeton University Press, 2004), 207. See also Jonathan Boff, 'The Morale Maze: the German Army in late 1918'; Jonathan Fennell, 'In Search of the "X" Factor: Morale and the Study of Strategy'; and Vanda Wilcox, 'Morale and Battlefield Performance at Caporetto, 1917', in a special issue of the *Journal of Strategic Studies* 37(6–7) (2014): 799–878.

2. For just a few examples see: Paul Addison and Angus Calder (eds.), *Time to Kill: The Soldier's Experience of War in the West 1939–1945* (London: Pimlico, 1997); John Baynes, *Morale: A Study of Men and Courage* (London: Cassell, 1967); Omer Bartov, *The Eastern Front, 1941–45: German Troops and the Barbarisation of Warfare* (Oxford: Macmillan, 1985); Stephen Biddle, *Military Power: Explaining Victory and Defeat in Modern Battle* (Princeton, NJ: Princeton University Press, 2004); Thomas W. Britt, Carl Andrew Castro, and Amy B. Adler (eds.), *Military Life: The Psychology of Serving in Peace and Combat*, vol. 1: *Military Performance* (Westport, CT: Greenwood, 2006); Trevor N. Dupuy, 'Theory of Combat', in Franklin D. Margiotta (ed.), *Brassey's Encyclopaedia of Military History and Biography* (Washington, DC: Brassey's 1994); John Ellis, *The Sharp End: The Fighting Man in World War II* (London: Pimlico 1993); Stephen G. Fritz, *Frontsoldaten: The German Soldier in World War II* (Lexington: University Press of Kentucky, 1995); Reuven Gal and A. David Mangelsdorff (eds.), *Handbook of Military Psychology* (Chichester: John Wiley, 1991); Richard Holmes, *Acts of War: The Behaviour of Men in Battle* (London: Cassell, 2004); S.L.A. Marshall, *Men Against Fire: The Problem of Battle Command in Future War* (Alexandria, VA: Byrdd Enterprises, Inc., 1947); Allan R. Millet and Williamson Murray (eds.), *Military Effectiveness*, vol. 1: *The First World War* (London: Allen and Unwin, 1988); Charles Moskos, *The American Enlisted Man: The Rank and File in Today's Military* (New York: Russell Sage Foundation, 1970); Sam C. Sarkesian (ed.), *Combat Effectiveness: Cohesion, Stress, and the Volunteer Military* (London: Sage Publications, 1980); Edward A. Shils and Morris Janowitz, 'Cohesion and Disintegration in the Wehrmacht in World War II', *Public Opinion Quarterly* 12(2) (1948): 280–315; Guy L. Siebold, 'The Essence of Military Group Cohesion', *Armed Forces and Society* 33(2) (2007): 286–295; Nora Kinzer Stewart, *Mates and Muchachos, Unit Cohesion in the Falklands/Malvinas War* (New York: Brassey's, 1991); Samuel A. Stouffer et al., *The American Soldier: Combat and Its Aftermath* (New York: John Wiley, 1965); Leonard Wong, Thomas A. Kolditz, Raymond A. Millen, and Terrence A. Potter, 'Why They Fight: Combat Motivation in the Iraq War', *Report for the Strategic Studies Institute* (July 2003).
3. Anthony King, *The Combat Soldier: Infantry Tactics and Cohesion in the Twentieth and Twenty-First Centuries* (Oxford: Oxford University Press, 2013), 36.
4. Sam C. Sarkesian, 'Combat Effectiveness', in Sarkesian (ed.), *Combat Effectiveness: Cohesion, Stress, and the Volunteer Military*.
5. Allan R. Millet, Williamson Murray, and Kenneth H. Watman, 'The Effectiveness of Military Organizations', in Millet and Murray (eds.), *Military Effectiveness Vol. I: The First World War*, 2.
6. Millet, Murray, and Watman, 'The Effectiveness of Military Organizations', 3.
7. Millet, Murray, and Watman identified two approaches, a sociological approach and an operational approach.
8. Gerd Hardach, *The First World War, 1914–1918* (Berkeley: University of California Press, 1977), 80.
9. Marshall, *Men Against Fire*, 23.
10. Marshall, *Men Against Fire*, 66–7.

11. Hew Strachan, 'Training, Morale and Modern War', *The Journal of Contemporary History*, 41(2) (2006): 225.
12. John Buckley, *Monty's Men: The British Army and the Liberation of Europe* (London: Yale University Press, 2013), 9; Basil Liddell Hart, *The Other Side of the Hill* (London: Cassells, 1951); Chester Wilmot, *The Struggle for Europe* (London: The Reprint Society, 1954, first published 1952).
13. John Ellis, *Brute Force: Allied Strategy and Tactics in the Second World War* (London: André Deutsch, 1990), 538.
14. See, for example, Biddle, *Military Power*; Buckley, *Monty's Men*; Terry Copp, *Fields of Fire: The Canadians in Normandy* (London: University of Toronto Press, 2003); David French, *Raising Churchill's Army: The British Army and the War against Germany 1919–1945* (Oxford: Oxford University Press, 2000); Stephen Ashley Hart, *Colossal Cracks: Montgomery's 21st Army Group in Northwest Europe, 1944–45* (Mechanicsburg, PA: Stackpole, 2007); Daniel P. Marston, *Phoenix from the Ashes: The Indian Army in the Burma Campaign* (London: Praeger, 2003); T.R. Moreman, *The Jungle, The Japanese and the British Commonwealth Armies at War, 1941–45* (Abingdon: Frank Cass, 2005).
15. Biddle, *Military Power*, 190.
16. Biddle, *Military Power*, 3.
17. Biddle, *Military Power*, 190.
18. King, *The Combat Soldier*, 204. See also, John Baynes, *Morale*; John Keegan, *The Face of Battle* (London: Pimlico, 1991); Ardant du Picq, *Battle Studies: Ancient and Modern Battle* (Charleston, SC: BiblioBazaar, 2006).
19. King, *The Combat Soldier*, 41. See also Baynes, *Morale* and Leonard V. Smith, *Between Mutiny and Obedience*, (Princeton, NJ: Princeton University Press, 1994) on this matter.
20. Biddle, *Military Power*, 207. For some exceptions see: Boff, 'The Morale Maze'; Fennell, 'In Search of the "X" Factor'; J-F. Fils, 'The Measurement of Morale among Belgian Military Personnel Deployed in Crisis Response Operations: A Longitudinal Survey Design', in *Human Dimensions in Military Operations—Military Leaders' Strategies for Addressing Stress and Psychological Support.* Meeting Proceedings RTO-MP-HFM-134, Paper 28. Neuilly-sur-Seine, France, April 2006: 1-8; J.M. Foret, 'Measuring Morale within the French Army', in *Human Dimensions in Military Operations—Military Leaders' Strategies for Addressing Stress and Psychological Support.* Meeting Proceedings RTO-MP-HFM-134, Paper 29. Neuilly-sur-Seine, France, April 2006: 1-10; Melvin J. Kimmel and Francis E. O'Mara, 'The Measurement of Morale', US Army Research Institute for the Behavioural and Social Sciences, 1981; Melvin J. Kimmel, Francis E. O'Mara, and Nehama Babin, 'The Development of a Unit Morale Measure for Army Battalions', Technical Report 617, US Army Research Institute for the Behavioural and Social Sciences, March 1984; S.J. Motowidlo, B. E. Dowell, W.C. Borman, P.D. Johnson, and M.D. Dunnette, *Motivation, Satisfaction, and Morale in Army Careers: A Review of Theory and Measurement* (Arlington, VA: US Army Research Institute for Behavioral and Social Sciences, 1976); Wilcox, 'Morale and Battlefield Performance at Caporetto, 1917'.
21. Michael I. Handel, *Masters of War: Classical Strategic Thought* (London: Frank Cass, 2001), 15.

22. Thomas W. Britt and James M. Dickinson, 'Morale during Military Operations: A Positive Psychology Approach', in Britt, Castro, and Adler (eds.), *Military Life*, 162; Reuven Gal and Frederick J. Manning, 'Morale and its Components: A Cross National Comparison', *Journal of Applied Social Psychology* 17(4) (1987): 369–391; Robert M. Guion, 'Industrial Morale (A Symposium): 1. The Problem of Terminology', *Personnel Psychology* 11(1) (1958): 60–2; Motowidlo, Dowell, Borman, Johnson, and Dunnette, *Motivation, Satisfaction, and Morale in Army Careers*, 49–52.
23. Elisabeth Kier, 'Homosexuality in the US Military: Open Integration and Combat Effectiveness', *International Security* 23(2) (1998); Robert MacCoun, 'What is Known about Unit Cohesion and Military Performance', in National Defence Research Institute *Sexual Orientation and US Military Personnel Policy: Options and Assessment* (Washington, DC: RAND, 1993).
24. John A. Lynn, *The Bayonets of the Republic: Motivation and Tactics in the Army of Revolutionary France* (Oxford: Westview Press 1996), 34–5.
25. Stephen D. Wesbrook, 'The Potential for Military Disintegration', in Sarkesian (ed.), *Combat Effectiveness*, 257.
26. John Keegan, 'Towards a Theory of Combat Motivation', in Addison and Calder (eds.), *Time to Kill*, 6.
27. Keegan, 'Towards a Theory of Combat Motivation'. Hew Strachan has also argued that coercion is not always given enough recognition as a motivational tool. Hew Strachan, 'The Soldier's Experience in Two World Wars: Some Historiographical Comparisons', in Addison and Calder (eds.), *Time to Kill*, 374–5.
28. Marshall, *Men Against Fire*, 165.
29. Stouffer et al., *The American Soldier*, 101.
30. National Archives (NA) WO 277/7 A.B. McPherson, 'The Second World War 1939–1945, Army Discipline', 2.
31. Imperial War Museum (IWM) 99/1/2 Major General Raymond Briggs Papers, Paper by Field-Marshal Montgomery, 'Morale in Battle: Analysis', 30 April 1946, 43.
32. Montgomery, 'Morale in Battle: Analysis', 47–8.
33. Montgomery, 'Morale in Battle: Analysis', 51–3.
34. Douglas A. Bernstein, Alison Clarke-Stewart, Edward J. Roy, and Christopher D. Wickens, *Psychology* (Boston, MA: Houghton Mifflin,1997), 337.
35. In fact, one of Britt's other studies supports this view. He asked a small sample of soldiers to indicate what they thought were the key characteristics of morale. He found that soldiers were most likely to indicate the attributes of motivation and drive in their views of morale (Britt and Dickinson, 'Morale During Military Operations', in Britt, Castro, and Adler (eds.), *Military Life*, 163).
36. Marshall, *Men Against Fire*, 157–8.
37. Leonard Wong, Thomas A. Kolditz, Raymond A. Millen and Terrence A. Potter, 'Why They Fight: Combat Motivation in the Iraq War', Report for the Strategic Studies Institute, July 2003.
38. Jonathan Boff, *Winning and Losing on the Western Front: The British Third Army and the Defeat of Germany in 1918* (Cambridge: Cambridge University Press, 2012); Jonathan Fennell, *Combat and Morale in the North African Campaign: The Eighth Army and the Path to El Alamein* (Cambridge: Cambridge University Press, 2011);

Alexander Watson, *Enduring the Great War: Combat, Morale and Collapse in the German and British Armies, 1914–1918* (Cambridge: Cambridge University Press, 2008).

39. John Buckley, *British Armour in the Normandy Campaign 1944* (London: Frank Cass, 2004); Buckley, *Monty's Men*; David French, '"Tommy is no soldier": The Morale of the Second British Army in Normandy, June–August 1944', *Journal of Strategic Studies* 19(4) (1994): 154–78. See also, Terry Copp and Bill McAndrew, *Battle Exhaustion: Soldiers and Psychiatrists in the Canadian Army, 1939–1945* (London: McGill-Queen's University Press, 1990).
40. Fennell, *Combat and Morale*, Chapter 1.
41. Francis Albert Eley Crew, *The Army Medical Services, Campaigns*, vol. 4, *North-West Europe* (London: Her Majesty's Stationery Office, 1962), 563.
42. Field Marshal Montgomery, 'Despatch', *The London Gazette*, 3 September 1946, in Crew, *The Army Medical Services, North-West Europe*, 560.
43. NA WO 177/321 Monthly Report for June 1944 by Psychiatrist Attached 2nd Army, 7 July 1944.
44. Library and Archives Canada (LAC) Volume 12631 File No. 11/Psychiatry/5 Reports, Major R.A. Gregory, Neuro-psychiatrist 3 Canadian Infantry Division, 'Psychiatric Report', October 1944.
45. SIW were punished with two years' imprisonment with hard labour. Desertion, by comparison, was punished with three years' penal servitude—Australian War Memorial (AWM) 54 624/2/6 Report on Functions of 'AG's' Branch During Invasion of Western Europe, 1944, 237–8.
46. Crew, *The Army Medical Services, North-West Europe*, 326.
47. NA WO 285/25 Report on '"Soldiers Under Sentence" for Such Offences as Desertion, Cowardice, Mutiny etc., Whose Cases Have Been Reviewed in British Second Army', June 1945.
48. NA WO 285/25 Report, '"Soldiers Under Sentence"'.
49. See, for example, NA WO 204/6701 Morale British Troops in Italy, 1943 to 1945; AWM 54 423/11/98 Middle East Military Weekly Censorship Summaries, Covering Dominion, Colonial and Allied Forces, Allied Air Forces, Women's Forces and British Prisoners of War in Enemy hands, 1942.
50. Of the rest, 135 were transferred for service in other parts of the army, thirteen were identified as 'irreconcilable types' and kept in prison, twelve were admitted to a psychiatric hospital, and one was discharged from the army altogether.
51. NA WO 285/25 Report, '"Soldiers Under Sentence"'.
52. WO 177/336 War Diary H.Q. (Medical) 1 British Corps, January 1945, Appendix IV: Quarterly Report by Brigadier Q.V.B. Wallace, D.D.M.S. 1 Corps, Quarter Ending 31 December 1944, Appendix B: Some Notes on Medical Ops.
53. Lieutenant-Colonel A.T.A. Browne, 'A Study of the Anatomy of Fear and Courage in War', *The Army Quarterly and Defence Journal* 106(3) (1976): 3–4.
54. See Fennell, *Combat and Morale* and 'In Search of the "X" Factor'.
55. LAC T 17924 21st Army Group Censorship Reports, April to August 1944; LAC RG24 Volume 10706 File No. 46-3-6/Int Vol.1 21st Army Group Censorship Reports, August to November 1944; LAC RG24 Volume 10706 File No. 46-3-6/Int Vol.2 21st Army Group Censorship Reports, December 1944 to February 1945; LAC

RG24 Volume 10706 File No. 46-3-6/Int Vol.3 21st Army Group Censorship Reports, February to April 1945; LAC RG24 Volume 10547 File No. 153/Censor Reps/1/3 21 Army Group Censorship Reports, 1944 to 1945; Directorate of History and Heritage File No. 581.023 (D1) Censorship Reports, Canadian Forces in Netherlands and Germany, August to October 1945.

56. The scale, therefore, approximates to what is known as an 'equal appearing interval scale'.
57. NA WO 177/321 D.D.M.S. 2nd Army, War Diaries, 1943–44; NA WO 177/322 D.D. M.S. 2nd Army, War Diaries, 1945. Precise weekly figures for Battle casualties (broken down into Battle wounds, Battle accidents, SIW, and burns) are contained in WO 177/322 in an appendix to a 'Report on Surgery in Second Army in the Northwest Europe Campaign, 1944–45'.
58. Due attention was given to the difference between 'admissions' and 'incidence' in order to avoid double counting. Admissions refer to the total number of new cases in each period; 'incidence' refers to the total number of cases being treated during each period (thus, it could include admissions from a previous period). However, in the case of Battle exhaustion, the absolute figures for 'admissions' and 'incidence' were extremely similar. For example, the Battle exhaustion figures for the period of the Normandy campaign calculated here tally almost exactly with the figures presented as 'exhaustion cases' in French, '"Tommy is no soldier"', 173. They also tally almost exactly with absolute figures presented elsewhere in the primary sources. Accordingly, there appears to be little danger of double counting. The similarity in the 'admissions' and 'incidence' figures may be explained by the quick turnaround experienced by those who suffered from Battle exhaustion (five to seven days in July 1944). See, for example, NA WO 177/321 Report by Psychiatrist Attached to 2nd Army for Month of July 1944. In some cases where exact numbers for Battle exhaustion were found in the sources, they were used in preference to the computed figure.
59. WO 177/322 Report on Surgery in Second Army in Northwest Europe, 1944–1945.
60. WO 171/182 War Diaries, Deputy Judge Advocate General, 21st Army Group, April to December 1944; Liddell Hart Centre for Military Archives (LHCMA) De Guingand Papers, Administrative History of 21 Army Group, 6 June 1944 to 8 May 1945.
61. French, "'Tommy is no Soldier"'; Buckley, *British Armour in the Normandy Campaign*, 200–2.
62. The findings from this research are explored further in the author's upcoming book, *The People's Armies: The British and Commonwealth Armies and the Second World War* (Cambridge: Cambridge University Press, forthcoming 2015).
63. The figures for May were not included, as the German Army surrendered to 21st Army Group on 6 May 1945.
64. WO 177/321 D.D.M.S. 2nd Army, War Diaries, 1943–44; WO 177/322 D.D.M. S. 2nd Army, War Diaries, 1945; LAC Volume 12630 File No. 11/Psychiatry/1/2 Lt-Col. J.C. Richardson, 'Memorandum about Current Neuropsychiatric Problems in the Canadian Army in the European Theatre of Operations', 27 October 1944; WO 285/18 2nd Army, Health Reports, December 1944 to June 1945. Figures for casualties (which include Battle wounds, Battle accidents, and burns, but not killed

in action or captured) have been included in case a reader wishes to calculate the percentage of, for example, battle exhaustion to wounded.

65. LHCMA De Guingand Papers, Administrative History of 21 Army Group, 6 June 1944 to 8 May 1945; NA WO 171/182 War Diaries, Deputy Judge Advocate General, 21st Army Group, April to December 1944. The figures presented here are higher than those in French, '"Tommy is no Soldier"', 172. This may be because French does not include the independent armoured brigades and line of command troops in his figures.
66. 103 and 105 Reinforcement Groups accounted for 76 courts martial for desertion and 244 for absence in July 1944 and 112 for desertion and 417 for absence in August 1944. 103 and 105 Reinforcement Groups moved to the continent in September 1944 and, therefore, figures from these units have been included from this period onwards.
67. Where a week straddled two months, sick, Battle exhaustion, and SIW figures were assigned to each month in proportion to the number of days of the week that fell in each month.
68. A strong correlation between two variables would produce an r-value in excess of + 0.9 or – 0.9. Terry Lucey, *Quantitative Techniques* (Continuum: London, 1996), 113.
69. Incidences of excellent morale coincided with figures of between twenty and thirty sick, Battle exhaustion, desertion, AWOL, and SIW per 1000. Incidences of high morale coincided with figures between thirty and forty per thousand. These findings were verified statistically by use of principal coordinate analysis and a T-test.
70. The summary did, however, include a very small number of letters from Guards Armoured Division on 18 July 1944.
71. See, for example, Buckley, *Monty's Men*; Hart, *Colossal Cracks*.
72. Buckley, *Monty's Men*, 208–31.
73. Hart, *Colossal Cracks*.
74. These weekly morale scale equivalents were computed by dividing the monthly levels from Table 8.1 by 31 and multiplying by 7.
75. WO 219/1531 Casualty Reports, British, Canadian and US Forces Daily Summaries, including enemy prisoners taken, June to October 1944; WO 219/1532 Casualty Reports, British, Canadian and US Forces Daily Summaries, including enemy prisoners taken, Oct 1944 to Mar 1945; WO 177/316 Medical Diaries, D.D.M.S. 21st Army Group, 1944; LHCMA Papers of Gen Sir Harold English Pyman 5/21, 'A Short Account of the Operations of the Second British Army'.
76. NA CAB 106/1084 Army Operational Research Group Memorandum No. E20, 'Some Statistics of the North West European Campaign, June 1944 to May 1945', 27.
77. Derived from statistics presented in NA CAB 106/1084, Some Statistics of the North West European Campaign, June 1944 to May 1945. The extent to which the ten-day periods coincide with bombardments and advances could possibly distort these relationships, but probably not sufficiently to take from the overall thrust of the argument outlined here.
78. WO 219/1531 Casualty Reports, British, Canadian and US Forces Daily Summaries, including enemy prisoners taken, June to October 1944; WO 219/1532 Casualty

Reports, British, Canadian and US Forces Daily Summaries, including enemy prisoners taken, Oct 1944 to Mar 1945; WO 177/316 Medical Diaries, D.D.M.S. 21st Army Group, 1944; LHCMA Papers of Gen Sir Harold English Pyman 5/21, 'A Short Account of the Operations of the Second British Army'; NA CAB 106/1084 Some Statistics of the North West European Campaign, June 1944 to May 1945.

79. Biddle, *Military Power*; French, *Raising Churchill's Army*; Hart, *Colossal Cracks*.
80. Buckley, *Monty's Men*, 29.
81. Daniel J. Hughes, *Moltke on the Art of War: Selected Writings* (Novato, CA: Presidio, 1993).
82. Copp, *Fields of Fire*; Buckley, *Monty's Men*.
83. Clausewitz, *On War*, 83.
84. Hew Strachan, 'A Clausewitz for Every Season', *American Interest* (July/August 2007), 33.
85. Carl Von Clausewitz, *On War* (London: Everyman Library, 1993, first published in 1832), 207.
86. Colin S. Gray, *Modern Strategy* (Oxford: Oxford University Press, 1999), 17.
87. Basil H. Liddell Hart, *Strategy: The Indirect Approach* (London: Faber and Faber, 1967), 335.
88. Hew Strachan, 'Strategy or Alibi? Obama, McChrystal and the Operational Level of War', *Survival* 52(5) (2010): 158.
89. Hew Strachan, 'The Lost Meaning of Strategy', *Survival* 47(3) (Autumn 2005): 52.
90. Clausewitz, *On War*, 86.
91. For other studies that have made use of Clausewitz's perspective of the relationship between 'means' and 'will', see Mark Clodfelter, 'Aiming to Break Will: America's World War II Bombing of German Morale and its Ramifications', *Journal of Strategic Studies* 33(3) (2010): 401–35; Niall Ferguson, 'Prisoner Taking and Prisoner Killing in the Age of Total War: Towards a Political Economy of Military Defeat', *War in History* 11(2) (2004): 148–92; Michael Howard, 'When are Wars Decisive?', *Survival* 41(1) (1999): 126–35.
92. This understanding of strategy appears to apply also to nuclear wars and limited 'small' wars. See Fennell, 'In Search of the "X" Factor'.
93. Michael Howard, 'When are Wars Decisive?', *Survival* 41(1) (1999): 128.
94. Strachan, 'The Lost Meaning of Strategy', 49.
95. Marshall, *Men Against Fire*, 190.
96. Strachan, 'The Lost Meaning of Strategy', 36–40.
97. Marshall, *Men Against Fire*, 184.
98. Strachan, 'The Lost Meaning of Strategy', 47.
99. Clausewitz, *On War*.
100. See for example, Colin M. Fleming, 'New or Old Wars? Debating a Clausewitzian Future', *Journal of Strategic Studies* 32(2) (2009): 213–41; Bart Shuurman, 'Clausewitz and the "New Wars" Scholars', *Parameters* (Spring 2010): 89–100.
101. Strachan, 'The Lost Meaning of Strategy', 46.
102. Strachan, 'Strategy or Alibi?', 157–82.
103. Clausewitz, *On War*, 698.

Part IV
Case Studies

8

New Model Diggers

Australian Identity, Motivation, and Cohesion in Afghanistan

Garth Pratten

Introducing the post-deployment book for Australia's Second Reconstruction Task Force in Afghanistan, Lieutenant Colonel Harry Jarvie wrote:

> War and military service has always been central to the Australian character, and the Aussie Digger has become a byword for mateship, courage and sacrifice. The ANZAC tradition is the core of our military heritage, and sets the standards by which all Australian forces will be judged. Having had the rare privilege to command these extraordinary young diggers on this most difficult and dangerous operation, I am confident that the ANZAC tradition is in safe hands.[1]

By evoking Anzac and the digger—the archetypal Australian soldier—Jarvie explicitly linked his troops and their endeavours in Afghanistan in 2006 to a set of values, standards, and expectations founded in a war almost one hundred years distant. In keeping with the central theme of *Frontline*—why do soldiers fight together—this chapter focuses on Australian combat soldiers—infantry, armour, engineers, and special forces—in Afghanistan. It will explore their views of themselves and their operations, and their relationship with society, in order to offer some observations about who are the modern diggers, what motivates and binds them together, and, indeed, whether the very idea of the digger is still relevant to them.

Although the digger stereotype is still important to the identity of contemporary Australian soldiers, its role as a source of motivation and cohesion should not be overstated. It is neither a universal, nor a unitary, concept. Soldiers' behaviour has been subject to a variety of influences including the Army's own subcultures, broader Australian cultural trends and expectations, and the character of the war itself. While some commentators have emphasized traditional digger attributes as a source of Australian military prowess in Afghanistan, others in the Army hierarchy regard the digger as a crude and anachronistic caricature, incompatible with the modern profession of arms, which has undermined discipline and standards. Indeed, some Australians in Afghanistan have asserted a more generic combat ethos, founded on the idea of a warrior code, rather than a distinctly Australian ideal. Ultimately, in the Australian Army's reflections on Afghanistan we see a modern professional force grappling with its citizen soldier heritage.

The Changing Face of the Digger

There is not the scope in this chapter to chart, in detail, the evolution of the idea of the digger, nor examine its historical veracity.[2] It shall suffice to say that a distinct stereotype of the typical Australian citizen soldier emerged from the First World War, and it remains a potent symbol of Australian identity. The original digger stereotype emphasized an aversion to formal discipline and a disdain for the hierarchy of the army; a live-hard, play-hard attitude in which raucous and often riotous behaviour in rear areas was justified by effective performance at the front; an ability to overcome the worst conditions with a mixture of initiative, resolve, and humour; and above all a commitment to standing beside one's mates until the very last. The digger was distinctively Australian, robustly masculine, and confident in his superiority over allies and enemies alike.

In his study of Australian First World War soldiers, historian Bill Gammage provided a snapshot of the attitudes that spawned the idea of the digger: 'they considered themselves unalterably civilian, fighters, not soldiers, volunteers for a job, not subjects to a medley of archaic impositions'.[3] It is pertinent to note the use of the term 'fighters' because the original digger stereotype emphasized brutally effective martial prowess. Thus, one of the great Australian heroes of the First World War—hailed as Australia's 'greatest fighting soldier' and thus the archetypal digger—was Albert Jacka.[4] He was awarded the Victoria Cross (VC) for leading a counter-attack at Gallipoli in May 1915 during which he shot and bayoneted seven Turks. At Pozières, in France, the following year Jacka's trench was overrun. Surviving a grenade attack on his dugout, Jacka led a counter-attack that plunged headlong into a party of Germans rounding up prisoners. Jacka was shot seven times but still managed to kill a dozen Germans with his revolver and a bayonet and thwart the German attack. Jacka's image adorned recruiting posters during the war and he was mobbed on his return to Australia in 1919. His story, repeated in numerous formats, thrilled many throughout the interwar years, and 30,000 people lined the streets of Melbourne for his funeral procession in 1932.[5]

In more recent times, the archetypal Australian military hero was transformed. Jacka's was no longer a household name; men who killed with bayonet and revolver were not in keeping with the sensibilities of modern Australia. At the Australian War Memorial only two individuals are commemorated with individual statues, both of which were erected in the 1990s. One is Private John Simpson Kirkpatrick, an Anzac icon famed for rescuing wounded soldiers at Gallipoli with donkeys, the other Lieutenant Colonel Edward 'Weary' Dunlop, a prisoner-of-war surgeon remembered for his efforts to keep Australians alive on the Burma–Thailand railway. Both are notable for saving lives rather than taking them.[6] As if sensing the broader mood of society, but also reflecting its main activities at the time, Australian Defence Force (ADF) recruiting in the 1990s emphasized its humanitarian roles.[7]

Tales of past glories had long been used to inculcate morale and cohesion in the Australian Army, but an initiative was launched in 2006 to establish an official ideal of the Australian soldier, which was, ironically, derivative of a similar campaign by the United States Army. The 'I'm an Australian Soldier' initiative set forth nine core behaviours to serve as goals for individual achievement and as guiding principles for the development of Army education, training, and employment. One core behaviour—every soldier demonstrates compassion—seems to have been developed to satisfy public expectations and reflect the Army's increasing role responding to political instability and humanitarian emergencies. However, the bulk of the core behaviours could be considered traditional digger virtues: every soldier an expert in close combat; every soldier a leader; every soldier physically tough; every soldier courageous; every soldier takes the initiative; every soldier works for the team.[8] Like the digger stereotype, the initiative asserted the value of human characteristics and a distinctive, superior, Australian approach on battlefields dominated by technology. Announcing the initiative, the Chief of Army, Lieutenant General Peter Leahy wrote:

> If you look at what other armies are doing to equip their soldiers for future operations you will notice how similar the technology is: ballistic helmets, body armour, better weapon sights, improved personal radios... to use a motor racing analogy, it's as though all soldiers are going to be driving the same cars. The best way to keep our advantage over other soldiers it to become 'better drivers'.[9]

Two of the core behaviours—every soldier mentally prepared, and every soldier committed to continuous learning and self-development—recognized that military prowess is as much a product of institutional learning and adaptation as of individual characteristics. They thus reflected a major historical factor in the success of Australian units and formations,[10] as well as contemporaneous discussions about the responsiveness of Western armies to the changing character of war.

The initiative overtly sought to link a new generation of Australian soldiers to their forebears, but it also asserted a professional ethos. References to Australian military professionalism can be found as early as the 1930s, but it became an institutionalized virtue with the creation of the Australian Regular Army after the Second World War. Professionalism subsequently served as a unifying ideal for the regulars and conscripts with whom Australia fought Confrontation and the Vietnam War. In *The Soldier and the State*, Samuel Huntington defined the distinguishing characteristics of a profession as expertise, responsibility, and corporateness,[11] all of which are readily identifiable in the inscription on the medallion issued to promote the 'I'm an Australian Soldier' initiative:

> I'm proud to share traditions built on the foundation of the Anzac legend.
>
> I'm proud to build on the heritage of service, mateship and sacrifice of previous generations.
>
> I'm proud to serve the Australian people by securing their future.
>
> I'm proud to secure peace and safety for the people of other countries when they need help.[12]

Thus, the Army articulated an official standard for its soldiers. The modern digger was a product of folklore, history, aspiration, and public expectation, and these factors have all shaped how Australian soldiers have experienced and understood their service in Afghanistan.

Australians in Afghanistan

Australia's contribution to the coalition effort in Afghanistan was relatively small. It never exceeded 1600 personnel in-country, which at the peak of the International Security Assistance Force's (ISAF) expansion in 2011 represented a little over 1 per cent of the force.[13] Australia's initial deployment to Afghanistan was a special forces task group committed to Operation Enduring Freedom from November 2001 to December 2002. United States' pressure brought the ADF back to Afghanistan in strength in September 2005. Special forces deployed into the southern province of Uruzgan to set the conditions for the subsequent deployment of Australian conventional troops in a reconstruction role. The first Reconstruction Task Force (RTF) commenced operations in August 2006 as part of the Dutch-led Task Force Uruzgan (TFU) and included both engineering and security elements. Uruzgan would remain the focus of Australian operations in Afghanistan until the end of 2013. Australia made other contributions to ISAF,[14] but its main effort was its two task forces—one conventional and one special forces—in Uruzgan.

The evolution of the composition and roles of the conventional task forces reflected the changing ISAF concept of operations. In October 2008 the RTF took on a mentoring role and became the Mentoring and Reconstruction Task Force (MRTF), and then in February 2010 the reconstruction component was wound up leaving a Mentoring Task Force (MTF) to concentrate solely on preparing the 4th Brigade of the Afghan National Army for independent operations. Australian forces switched to an advisory role at the end of 2012, before withdrawing from Uruzgan in December 2013.

The experience of Australian combat soldiers in Afghanistan was defined by their roles, tasks, and the limitations imposed on their activities. Concerns about casualties and a desire to limit the scope of the deployment meant that until 2010 any deployment of Australian troops outside Uruzgan required permission from Canberra, causing much frustration among alliance partners and embarrassment to the officers having to deal with them.[15] Even when the restriction was modified, only the special forces operated routinely outside of the province. Australia demurred from taking the lead in Uruzgan following the Dutch withdrawal in 2010, but in 2012, with the countdown to withdrawal already underway, reversed this decision. In general, Australian strategic direction regarding Afghanistan was confused, and often contradictory, leaving tactical commanders to negotiate a web of national caveats and risk management procedures while seeking to provide clear direction to their troops.[16]

The infantry attached to the RTFs were essentially restricted to providing security for the engineers' reconstruction projects, while the infantry deployed with the task forces with a mentoring role experienced more active operations as the Afghan National Security Force presence was pushed out into Uruzgan's Valleys. Major engagements with insurgent forces, however, remained the exception rather than the rule. Throughout the Australian commitment to Uruzgan the special forces played the most active role targeting insurgent networks, which routinely resulted in close-quarter engagements. Although they operated largely in Uruzgan, they were not under the direct command of TFU or its successor Combined Force Uruzgan and conducted operations largely independent of the conventional forces.

In January 2014 Australian fatalities in Afghanistan numbered forty, evenly split between the conventional and the special forces; non-fatal casualties amounted to 261.[17] Although relatively low compared with the casualties of the major troop contributors, Australian casualties were in proportion to the relative size of the Australian contribution to ISAF. While operations in Uruzgan were less intensive than elsewhere in southern Afghanistan—Uruzgan was always a supporting effort behind the main effort in Kandahar and Helmand—it was still a dangerous place.

Upholding the Traditions of the Anzacs

Correspondent Chris Masters observed in *Uncommon Soldier* that 'in the broad Army there is a general suspicion of the Anzac industry's amplification of the digger'.[18] Nevertheless, many Australian soldiers have seen their service in Afghanistan as part of a broader historical tradition and make reference to the attributes of the digger stereotype in writings, interviews, and eulogies for their fallen comrades. Ideas of mateship, irreverence, endurance in the face of adversity, and a distinct Australian military identity predominate. A eulogy for Sapper Jamie Larcombe, killed during an engagement with insurgents in February 2011,[19] highlights the manner in which the digger stereotype was embraced in the ranks:

> You were a truly outstanding bloke... You never shied away from a difficult situation and pushed through the hard times with that renowned Aussie attitude ... You are an immortal Anzac now mate, up there with Akka [Cpl Richard Atkinson] looking over us and especially our mates overseas...[20]

One soldier reflected 'from what I saw in Afghanistan... the Anzac spirit is still alive and important'. He continued that being called an Anzac is a 'privilege' that has to be earned and that the defining characteristics of the modern Australian soldier and the Anzac are synonymous: 'Willing[ness] to take the enemy on. Battle discipline. Initiative. Work hard, play hard'.[21] Private Damien Tomlinson, who lost both of his legs to an improvised explosive device (IED) strike in April 2009, similarly commented that 'Aussie mateship and the spirit of Anzac are indeed very special and is something that not too many people can fully appreciate until they experience what I have.'[22]

It is not surprising that Australians serving in Afghanistan have discussed their service in these terms. As Major Jim Hammett has observed, 'the past achievements of Australian Infantrymen are acclaimed across the military community: barracks are named after famous battles, bases are named after famous commanders, and battle honours are commemorated annually'.[23] Indeed, both the perceived characteristics and achievements of prior generations of Australian soldiers served as a standard to emulate, and be measured against, in Afghanistan. Writing of the achievements of his RTF 2 platoon, Lieutenant Glenn Neilson observed that they could 'proudly display their Infantry and Army Combat Badges to the hardest veterans on ANZAC Day'.[24] In his post-operations report, Lieutenant Colonel Chris Smith, the Commanding Officer (CO) of MTF 3, noted a different standard that some of his soldiers sought to emulate. They 'believed quite passionately that an Australian soldier is expected to "muck up" on operations' and that they were 'obliged to live up to a rogue, irreverent and scruffy stereotype (a distorted view of the larrikin)'.[25] 'Larrikin' is a word that routinely appears in homages to Australian soldiers in Afghanistan and, a long way from its origins as a term of abuse for the members of Sydney street gangs in the 1870s,[26] is generally used as a term of endearment for a mischievous, but good-natured, scallywag.[27]

'Don't Knock Off until the Job is Done': Motivation and Cohesion

Although ideas derived from the Anzac legend and the digger stereotype seem to have played an important role in the way Australian soldiers in Afghanistan constructed their identity and described their service, they appear to have been much less significant as factors contributing to motivation and cohesion in-country. Motivation here is defined as the influences that cause individuals to serve, fight, and risk wounding or death, while cohesion is the uniting bonds that enable combat troops to act collectively in pursuit of their assigned mission and not fragment under pressure.[28] The two concepts are intertwined and cannot be easily separated; shared sources of personal motivation facilitate group cohesion. The sources of motivation and cohesion that Australian soldiers describe in their interviews and writings—comradeship, altruism, adventure, professional commitment, and organizational standards—are not peculiarly Australian in origin, although the first two are often couched in the Australian vernacular as 'mateship' and providing a 'fair go'. They point to a more universal theory of combat motivation and cohesion. This universal combat experience also raises the question of to what degree the public rhetoric of Anzac has actually been internalized by modern Australian soldiers.

Concluding his recent memoir, Corporal Mark Donaldson, Australia's first VC recipient in Afghanistan, equivocated in his assessment of whether the nation's involvement had been 'worth it'. Regarding his personal contribution he was more definite: 'as a soldier, you can look at your own role. I wanted to go because that is what I trained for. If your country wants you to go somewhere, 99 per cent of soldiers will say yes.'[29] In an earlier interview, Donaldson referred to his actions during the engagement in the Ana Kalay Valley as just doing his job.[30] Some were surprised that the question of motivation is even an issue. Corporal Nicole Spohn responded to media questioning about her deployment with MRTF 1 in 2009 thus: 'It's why we join the forces . . . There's no use in pretending otherwise.'[31] One officer similarly stated that he was 'happy to fight because I was told to'. He continued: 'I kept on going through culturally not knowing anything different. I've been well indoctrinated that you don't knock off until [the] job is done, hence I've never had to call upon a higher power or sense of national pride.'[32] 'Doing my job' was a common sentiment used by Australian soldiers to explain their motivation for service in Afghanistan.[33] This sentiment emphasizes not only the incorporation of professionalism into the Australian military ideal, but also the role of what has been referred to as 'functional interdependence' or 'task cohesion'—the phenomena of various individuals and elements in a military force being united by their commitment to a shared mission and the different, interdependent capabilities they bring to it. Artillery joint fires observer Bombardier David Robertson was awarded the Medal of Gallantry for his efforts to deliver indirect fire support to an MTF 2 infantry patrol. He later recalled: 'When the time came, everyone else was relying on me to do my job and I really didn't want to let anyone down.'[34]

Inherent in the concept of doing a job is being paid for it—an age-old motivation that held true for some Australians in Afghanistan[35]—and also having the opportunity to employ skills developed during training 'for real'. Deployment on operations is seen as the practical test of a soldier's skills and character and as the most significant achievement of the military profession. One infantryman, still awaiting his opportunity to deploy, remarked that troops 'go stir crazy' after being stuck in training scenarios in Australia: 'You also get shown more respect if you have done a few tours.'[36] His sentiments are echoed by Private Thomlinson, who did deploy in 2009: 'I was excited to be finally going. After all, this is what you join for. And it is also about earning some respect amongst the unit. At a normal unit, it's about getting some time up, whereas at my unit it's like, "Well, what have you done?"'[37] For reservists the desire to be involved in active operations was even more pronounced. They had put their civilian careers on hold and volunteered to serve in Afghanistan for what may have been their only chance to 'play in the big game' and have the 'honour and the pleasure of... mixing it up in the dasht and villages of AFG'.[38]

For fighting soldiers, combat was the defining experience, both for the application of practical skills in extremis and the accompanying adrenalin rush. Corporal Leon Gray of MRTF 1 described his first contact thus: 'I had been waiting a long time to fire my weapon. When it happens it seems the most natural thing in the world. It is surreal to be actually doing it, with the bullets coming past. Part of the brain is terrified, but as an infantryman it is the pinnacle of what we do.'[39] Lieutenant Colonel Smith observed that some soldiers chased the adrenalin rush—and also the opportunity for the YouTube video filmed with the ubiquitous helmet cams—often with a degree of naivety as to the potential consequences.[40] The exhilaration of engaging in, and surviving, the first contact is palpable in Trooper Aaron Healy's description of contact near Langar: 'We ceased firing[,] checked everyone was ok and we had no wounded, resupplied with ammo and called in the contact over the radio, everyone was cheering at the experience, for most it was their first contact and there was a lot of celebrations and talk of how "sick that was".'[41]

As Masters has noted, the ardour for combat was most pronounced among the special forces, some of whom completed up to nine operational rotations, and even engineered transfers between subunits to ensure an earlier return to Afghanistan.[42] Discussing the early stages of the Battle of Tizak, which he spent orbiting in a helicopter providing airborne fire support, Corporal Ben Roberts-Smith commented: 'It sounds funny but, honestly, we were depressed at the prospect of missing out on the action.'[43]

The motivation for action, however, does not seem to have been solely self-actualization but rather one linked to a collective experience. Here we again glimpse the nexus between motivation and cohesion. Corporal Ryan Avery from MTF 1 has reflected that the engagement at Derapet on 4 December 2010 was a real 'eye opener': 'I realised the insurgents weren't playing around and didn't care if they went home.' He concluded that the 'contact was exhilarating, *fighting with my mates* [my emphasis]'.[44] Comradeship, and the code of behaviour attached to it, is the most frequently evoked source of motivation and cohesion, both explicitly and implicitly, for Australian troops in Afghanistan. The Officer Commanding Combat Team Spear, the security component of RTF 3, invoked mateship as the foundation of his troops' superiority over their enemy:

> above all else we look after our mates for they are the ones who will stand beside us when all hell breaks loose; they are the ones who will watch our backs when the enemy hits us from all sides. They are the one who will pick us up when we are down for we leave no one behind[45]

Individual soldiers and officers have variously identified mateship as the principal source of their team's cohesion and effectiveness: 'Within the Platoon we had time to bond as strong teams which helped enormously'; their individual motivation: 'It got to the point where it was like "f*** this, I'm not going to sit here while one of the boys gets hit" ... I thought I'd just have a crack'; and even their personal survival: 'I wouldn't be here today if it wasn't for the boys with me'.[46] The mateship ideal was summed up by Lieutenant Ashley Judd, an MRTF 1 platoon commander, thus: 'Sounds like big talk but most would find the thought of letting down a team member worse than themselves getting harmed.'[47]

The commitment to mateship also resulted in a more visceral motivation for combat—revenge. For some it was a clearly expressed desire,[48] for others it manifested a change in general attitude and behaviour in response to casualties. Reflecting on his experiences with MRTF 2 in 2009–10, Captain Matt Proud noted the appearance of a 'thirst for revenge' when casualties were sustained and the more aggressive behaviour that resulted.[49] The motivation for revenge was bound up with notions of honour and a perception that Australian troops needed to strike back to maintain both personal morale and the initiative in the campaign. Briefing the media following the death of

Signaller Sean McCarthy in an IED explosion in July 2008, the Chief of the ADF, Air Chief Marshall Angus Houston, stated that the incident would 'harden our resolve to keep these Taliban leaders, these Taliban bomb makers under pressure.'[50] In a similar vein, he congratulated the special forces troops who killed the Taliban leader who orchestrated the rocket attack that killed Private Gregory Sher in January 2009: 'It is quite clear you made them pay for the death of a comrade.'[51] These sentiments were echoed by those who took the fight to the Taliban. Mark Donaldson has recalled that he and his Special Air Service Regiment (SASR) comrades 'were wild with frustration' when they were denied the opportunity to track down the Afghan soldier responsible for the death of Lance Corporal Andrew Jones in May 2011: 'It was a matter of honour and urgency. An Australian soldier had been killed; we thought there should be no debate... showing people what would happen if they treated us like that should override whatever managerial and political issues were going on.'[52]

Sustaining casualties also had the effect of hardening the professional commitment to seeing the mission through and thus ensuring that casualties had not been suffered in vain. Again, official pronouncements, largely made for the consumption of the media and the general public, rhymed with the attitudes of deployed soldiers. In his July 2008 press conference Houston also expressed his confidence that Australians in Afghanistan were 'supporting each other' and were 'more determined than every [*sic*] to continue their important mission in honour of their fallen comrades'.[53] Major General John Cantwell, Australian national commander in the Middle East Area of Operations 2009–10, has recalled a conversation with a soldier after the deaths of Sappers Darren Smith and Jacob Moreland in an IED explosion in June 2010: 'One bloke says to me, "We gotta keep goin', hey sir, otherwise Snowy and Smithy died for nothing".' Seemingly conscious of accusations that the soldier was simply parroting official platitudes, Cantwell concluded the account thus: 'He's not trying to impress me: this is what he believes.'[54] Masters noted similar sentiments during his attachment to MTF 1, exemplified by the reflections of Private James Smith: 'They need help here... If we leave, all the work done, for all those years, will go to nothing.'[55]

Smith highlighted the predominant higher order motivation for Australian troops in Afghanistan expressed in available sources. In general, they have not linked their own efforts to national strategic goals. References to fighting terrorism or making the world more secure are relatively few, and these have generally been made by special forces, which is unsurprising given their combat mission.[56] It has been more common for Australians expressing a higher motivation for their service in Afghanistan to invoke a spirit of altruism—helping an oppressed and impoverished population towards a better life—described by Captain Liam Hansen, an RTF 1 engineer, as 'doing the righty'.[57]

Many Australian soldiers apparently internalized the expectations of their community, the Army's recent history of stability and humanitarian operations, and the altruism inherent in the 'I'm an Australian soldier' mantra. Sapper James Martin, killed by a rogue Afghan soldier in August 2012, explained his motivations for army service in a letter to his parents: 'I want to wake up every morning and be proud of what I do and who I am. I want to... fight for civil liberties that most Australians take for granted. I want to be an Australian soldier.'[58] Others have similarly made reference to fighting 'bullies' in Afghanistan and to helping 'others who can't help themselves', ideas linked to the Australian cultural ideal of the fair go: 'If there's someone being oppressed by someone else, you want to fight the oppressor. It's an Aussie thing. You want to back the underdog.'[59] Reporting from Afghanistan in 2012, ABC journalist Kathy McLeish observed that the soldiers she interviewed 'believed passionately' in what they were doing: 'They have a strong belief they are supporting their country and they hope to make a difference for the people of Afghanistan, where conditions in health, education and employment are very poor.'[60] McCleish's ABC colleague Masters has echoed her observations. His interviews, conducted across multiple visits to Afghanistan, include Australian soldiers with an urge to help others, soldiers motivated by the change they saw resulting from their operations, and soldiers with a desire to return because of these achievements.[61]

For some soldiers individual motivations were mutually reinforcing. They desired service in Afghanistan, indeed revelled in it, because of the experience as a whole. Private Nathan Mullins from 1 Commando Regiment described his experiences during a 2010 patrol in Kandahar province as 'awesome!':

> This was exactly why I joined the Army. I was in a strange situation in a far-off country, fighting an enemy with incredible resolve, pushing boulders out of the way to help the progress of a column of armoured vehicles... it was an incredible experience, thrilling and filled with an exciting buzz of danger, physical effort, guns, a mate beside me, an exotic locale.[62]

A member of RTF 3's Light Armoured Vehicle (LAV) troop similarly observed that he and his comrades had 'experienced what most people only ever witness on television and movies'; it was this experience that brought the 'troop together when fatigue, cold and boredom' set in.[63] Trooper Healy's emails home speak of excitement, boredom, fear, and fatigue. He writes about the 'insane' mountains he drives his LAV through, how he loves learning about Islam and Afghan culture, of 'peaking' while driving on steep vertical paths to avoid IEDs, and being 'wrecked' but still 'enjoying the adventure'.[64] The overwhelming impression is of a young man embracing a life experience for better or worse—'what we're doing and what I'm learning really is a once in a lifetime adventure and a chance to be part of something bigger than life'—and perhaps in this regard the motivations of 'Az' and his comrades are not far removed from those of their historical forebears.

Team Australia?

The combined military and civilian effort in Afghanistan was often referred to as Team Australia.[65] It was a title that evoked the ideal of mateship and national solidarity as well as the whole-of-government effort that was deemed essential to achieving stabilization objectives in Afghanistan. 'Team Australia', however, belied the differences in organizational culture, outlook, capabilities, and resourcing that posed a challenge to the implementation of a whole-of-government approach. Similarly, in exploring the identity and motivations of the Australian Army in Afghanistan we have to be careful not to ascribe to it a homogeneous culture and ethos and a universal set of motivations. Australian soldiers in Afghanistan belonged to a collection of subcultures based around their roles and experience with attendant suspicions, rivalries, and frustrations. As Masters suggests, although mateship may be a strong factor in motivation and cohesion within small groups, its role as a unifying influence across the contingent should not be overestimated: 'there are many non-mates. I could not help but notice how many disaffected soldiers there were, how much contempt they held for some, and how much rivalry existed between others.'[66] The observation of these internal fissures and fractures, in a force that was generally regarded to have performed effectively in the field, adds weight to the notion that functional interdependence and commitment to a shared mission functioned as sources of cohesion for a number of small groups with their own internal codes of behaviour.

With combat widely seen as the pinnacle of military endeavour, the basic distinction maintained by many was between those who routinely ventured beyond the wire and those who worked inside fortified bases. This is an historical distinction, based on perceptions of unequal sacrifice and privileges, which has persisted despite the critical role of command, intelligence, maintenance, and support functions in maintaining troops in the field.[67] In Afghanistan, the 'base wallah' of the Second World War, or the REMF or POGO of the Vietnam era,[68] became a 'fobbit'—a conjunction of the abbreviation for the ubiquitous forward operating base (FOB) and the hobbits of J.R.R. Tolkien's *Lord of the Rings*, who rarely ventured beyond their bunker-like homes in the Shire.[69] The role of the fobbits was readily dismissed by combat troops—they are waved away in one of Trooper Healy's emails with 'enough about fobbits'[70]—and the number of days spent beyond the wire was worn as a badge of honour.[71]

Further distinctions were maintained by those who did regularly venture outside the FOB. The special forces, particularly the SASR, form a small, insular community within the Army, and did so in Afghanistan. The insularity is emphasized in Major General Cantwell's reflection on his first briefing on Special Operations Task Group (SOTG) operations in Afghanistan in January 2010: 'It is a rare opportunity for someone like me, with a conventional forces background, to be admitted to the shadowy and lethal world of Special Forces.'[72] As mentioned, the SOTG's more aggressive combat role largely separated it from the conventional infantry in the field, and operational security concerns did the same in the Australian base at Tarin Kowt where it occupied an 'island of secrecy behind locked gates'.[73] Ironically, for troops for whom secrecy is a key consideration, they were often relatively easy to identify in the FOB environment. Chris Masters observed: 'They overstate the understatement. With their advanced weaponry, looser clothing and longer hair, they stand out like Jedi Knights on a lunar landscape.'[74]

The special forces seem to have eschewed the digger identity and embraced what is referred to as a warrior ethos, or code, in its place. There was little of the boastfulness of the traditional digger in this code, and it was inextricably linked to the concept of a military profession with an upright sense of self-discipline at its core. It emphasized physical and mental robustness, controlled aggression, tight-knit comradeship, humility, and discretion. An indication of the code's pervasiveness is the frequent references to it in the eulogies of special forces soldiers killed in action.[75] The code was epitomized by the words of Warrant Officer 'H' at the funeral of Sergeant Brett Wood from 2 Commando Regiment, killed by an IED in May 2011:

> For a warrior it is a disgrace for the men around him to show more valour than he. He will strive to surpass his fellow operators time and again. Is it his lust for glory? No! It is for the silent respect of the man next to him – the nod we give as we pass each other.[76]

Standards were enforced from within, and the only accolade that mattered was that bestowed by another warrior. As VC recipient Corporal Ben Roberts-Smith pointed out, 'the most meaningful recognition for us in the SAS comes from within, not from the public. When your fellow operators tell you that you've done something really well, it's the greatest feeling in the world.'[77] The warrior ethos celebrated lethality as a hallmark of military aptitude. Special forces elements adopted unofficial insignia featuring Spartan and Viking helmets, wolves, death's heads, or daggers, and in their downtime sported T-shirts carrying slogans such as '72 virgins dating service—we'll hook you up', 'I kill terrorists', 'Trample the weak, hurdle the dead', and 'There are problems only Heckler and Koch can solve'.[78]

The SOTG camp was not without its own internal divisions—the commando and SASR force elements have been described as 'head-butting bikie gangs'.[79] Some in the SASR regarded the commando company as a rather blunt instrument full of ostentatious cowboys. This perhaps reflected the commandos' role of conducting shock action at troop or company scale as opposed to the more surgical patrol level activity that is the SASR's specialty. There is evidence of tensions between the two in 2009, and Masters records that in 2010 some in the SASR expressed a reluctance to work with the commandos.[80]

The SOTG was the only component of the Australian force apart from the medical elements to routinely deploy significant numbers of reservists. This introduced another fault line, with regular troops expressing scepticism regarding the reservists' competence, particularly after several incidents in 2009.[81] Although reservists exhibited similar base motivations and sources of cohesion to their regular counterparts, some seem to have clung to a citizen soldier identity, maintaining a unique subculture based around non-conformism—which could be seen as diggeresque irreverence—and on asserting a perception of intellectual superiority stemming from their lives and careers outside of the Army. Private Mullins, an NGO programme manager in civilian life, served with the SOTG in 2010, and has recounted how his company used the Elton John song 'Rocketman' as the title for an operation targeting insurgent rocketeers:

> There is a definite difference in how we approach these things to how a unit of regular soldiers would approach them. I think it would be inconceivable for a regular army unit to name an operation after a homosexual pop artist's song. It just wouldn't happen. They would probably call the operation 'Kill Rocket' or 'Protect Base'... Boring, predictable and macho as can be.[82]

The relationship between the Australian conventional and special forces elements in Afghanistan was marked at times by suspicion, frustration, and professional jealousy. Although instances of direct cooperation were relatively infrequent, some in the SOTG considered combined operations a waste of their capability and the conventional forces a millstone that deprived them of speed and mobility, along with the surprise that resulted from it.[83] At least one special forces soldier, however, has attributed these frustrations in part to a sense of superiority, and perhaps arrogance, maintained by the special forces in regard to their conventional comrades.[84] Conversely, some in the conventional force elements were often sceptical of the efficacy of the special forces' counter-leadership operations. Killed or detained leaders could be replaced, and the raids often angered local populations, with whom the conventional forces were trying to build relationships.[85] Elements of the infantry, in particular, seem to have been frustrated by their tasks in Afghanistan and the restrictions imposed upon their operations, and felt that their stated role of closing with, and then killing or capturing the enemy, had been usurped by the special forces.[86] Perceptions of better food in the special forces' enclave for a time caused dissatisfaction, as did their access to alcohol, which was on occasion flaunted, in an otherwise dry camp.[87]

Golfing in the Green Zone

Asked what the defining characteristics of the Australian soldier were, one officer responded 'I love the Larrikin attribute. In conflict it translates to "how can I best screw the other guy over?", which consistently results in innovative solutions that give us a comparative advantage.'[88] The view that a slightly unorthodox approach to soldiering surprises the enemy and thus brings success on the battlefield is commonly held. The television mini-series *Anzacs* portrayed the genesis of the highly successful 'peaceful penetration' operations of April 1918 as a single-man raid on a German-held farmhouse by the archetypal larrikin Pat Cleary (Paul Hogan) in search of booze and souvenirs.[89] In the 1998 edition of *The Fundamentals of Land Warfare*, the Army sought to institutionalize this idea with the inclusion of the concept of 'battle cunning', which was said to result from the Australian cultural traits of pragmatism and anti-authoritarianism. The latter was said to lead to a 'disinclination to blindly accept orders. It does not imply a lack of discipline or professionalism, but rather a desire to positively contribute to operational success.'[90] 'Battle cunning' as a concept was rejected by many within the Army, and did not appear in the next edition of *The Fundamentals of Land Warfare*.[91] Experience in Afghanistan, however, has demonstrated that the contribution of the digger virtues to Australian military effectiveness is still a debated issue.

Most assessments of the effectiveness of Australian troops in Afghanistan have been positive. They have been lauded for their professionalism, courage, and endurance by their own military and political leadership, as well as that of their coalition partners. The assessment of Lieutenant Colonel Stuart Yeaman, the Commanding Officer of RTF 4, is typical of the general perception of what Afghanistan has meant for the Army: 'The Army . . . has become much harder, more resilient and more capable after the years of commitment to Afghanistan . . . There is a new era of soldiers and young officers who have experienced first hand the harsh realities of the battle field.'[92] Some accounts have, predictably, emphasized Australian superiority over their coalition partners, particularly the Americans and the Dutch. Others have underlined what are considered traditional Australian strengths, such as dispersed small unit operations and the maintenance of a strong security posture, and good situational awareness through active patrolling.[93]

Australia's involvement in Afghanistan, however, has been long, complex, and involved 26,000 troops. We would be at risk of perpetuating another stereotype if we were to assume that all Australian actions in Afghanistan were characterized by the highest standards of professionalism, courage, and endurance. Detailed analysis of Australian operations in Afghanistan awaits, but a survey of press reporting, correspondence, memoirs, and official documentation indicates that the Australian Army in Afghanistan was as prone to failings, oversights, and the frictions of war as any other. Several investigations into civilian casualties and other allegations of misconduct in the field were ongoing in January 2014;[94] Australian troops were returned to Australia for disciplinary reasons;[95] deployed troops made allegations that they had been inadequately supported or equipped;[96] and inquiry officers' reports highlighted poor personal discipline; inappropriate or ignored tactics, techniques, and procedures; and inadequate preparation as contributing factors to casualties.[97]

Concerns about the potential negative influence of the digger stereotype on the Army's standards have been expressed at the highest level; in March 2013 the Chief of Army, Lieutenant General David Morrison, addressing the United Nations International Women's Day Conference, stated that it was a 'pantomime caricature' that 'breeds a dangerous complacency about how professional and sophisticated soldiering really is'.[98] The tension between the digger stereotype and the standards of professional behaviour expected by many of the Army's senior leadership is highlighted by two accounts of Australians in Afghanistan. One, from Masters' *Uncommon Soldier* celebrates an Army cook who, dressed in footy shorts, a T-shirt, and thongs,[99] intervened in a tense stand-off between coalition troops and Afghan police over the fate of an insurgent prisoner.[100] The other, from Lieutenant-Colonel Smith's MTF 3 post-operations report despairs over examples 'throughout Australia's commitment in Afghanistan' of soldiers sunbathing in tactical positions, hitting golf balls from over watch positions into the green zone, kicking footballs in tactical positions, and meeting helicopters at landing zones clad in thongs and T-shirts.[101]

Such behaviour, however, was born of not just a larrikin spirit; the digger stereotype was one of several discernible influences on the behaviour of Australian troops in Afghanistan. Clausewitz wrote that 'war and its forms result from ideas, emotions and conditions prevailing at the time' and by extension so do the motivations and actions of its participants.[102] The behaviour of Australian troops was not only shaped by interpretations of the digger stereotype, but also by broader cultural trends and expectations, and the character of the operations in which they were involved. Reflections on the performance of the Army in Afghanistan reveal an active discussion about the relative significance of these factors and suggest an ongoing reshaping, or at least re-examination, of Army culture. The impact of the soldiers of the 'iPod generation' on the Army and their effectiveness on operations is central to such discussions.

The relative affluence and narcissistic tendencies of contemporary Australian youth culture clearly shaped the expectations of many Australian soldiers, and these ideas sat awkwardly beside traditional military ideals of conformity and subordination of the self. Modern soldiers have been praised for their technological aptitude[103]—vital on an increasingly digital battlefield—but uniforms and equipment have also become as much a personal statement as a means of doing the job. An observation by Masters is revealing: 'they seem to love their *toys* [my emphasis] and all the accoutrements'.[104] Particular looks were considered 'cool', sought after, and remarked upon.[105] Highlighting both the pervasiveness of popular culture and the inherent challenges presented by multiple subcultures and standards within one army, some conventional troops sought to imitate the behaviour and appearance of the special forces—the epitome of the modern warrior as celebrated in the computer games they played in their downtime.[106] As mentioned previously, helmet cams provided the opportunity for them to star in their own first-person shooter.

Discussions about appearance and conformity among Australian troops in Afghanistan followed traditional paths. Some soldiers argued that short hair and uniformity did not make a soldier fight better and that adapting to their operational environment was most important, particularly when issued clothing and equipment were deemed inadequate.[107] The attitude of the modern soldier to their kit, however, has made it 'difficult to distinguish a good pragmatic idea from a fashionable desire'.[108] Soldiers throughout history have modified their uniforms and equipment to cope with the situation at hand, and often armies, as institutions, have been slow to catch up. In Afghanistan, early concerns about issued uniforms and equipment by the conventional force elements were progressively addressed with the Army eventually adopting the approach 'if the equipment was good enough for... SOTG... then it was good enough... for our Tier Two combatants'.[109] The historical irony is that an Army traditionally known for scrounging to make up its deficiencies ended up with kit that was the envy of its allies in Afghanistan.[110] A distinction should be drawn, however, between the type of clothing and equipment used and the manner in which they were carried and worn. Lieutenant-Colonel Smith noted that there was usually a 'correlation between soldiers' attitudes to matters of appearance and general behaviour and their attitude to important things such as protective clothing, field craft and equipment checks'.[111] Thus, while one Australian officer wrote of being 'consistently disappointed' at the organisational attempts to 'snuff out such [individual] behaviour through a "one size fits all approach"',[112] Smith maintained that uniformity has a more fundamental role of reinforcing the subservience of the individual to the army, the team, and the mission.

Contrary to the digger virtue of living hard and rough, Australian troops in Afghanistan were provided with an historically unprecedented level of amenities in many bases, including air conditioning, gym equipment, and uninterrupted internet connectivity.[113] Although Australian soldiers in Afghanistan frequently demonstrated the ability to go without,[114] there still seems to have been the expectation that personal comfort was important and should be provided for, a desire that was often facilitated by the prevalence of vehicle-mounted operations. This inventory of comforts from Private Mullins' Bushmaster[115] is indicative: 'I had a swag, cooked meals, soft drink, camp chairs, energy bars, a book to read, movies on a laptop, a short-wave radio.'[116] In keeping with previously discussed ideas about soldiering being a 'job' and deployment to Afghanistan being an 'experience', it has also been observed that soldiers had further expectations of periods of downtime and enjoyable recreation even when outside the wire.[117] This partly explains the phenomena of soldiers hunched over laptops under tarpaulins stretched between their vehicles, bonfires with an unrestricted line of sight (and fire) to uncontrolled territory, the abandonment of uniforms and body armour for shorts and T-shirts, and the now infamous golfers of the Green Zone.

Although some may have thought such sporting shenanigans carried on a digger tradition, they could also be seen as contradicting it, for they represented 'an erosion of standards of field discipline and fieldcraft' upon which the Australian Army has traditionally prided itself.[118] One can only guess why soldiers felt safe enough to disregard their training. Stories of a one-man piquet in the night vision–equipped remote weapon station of a Bushmaster suggest overconfidence in Australian technological dominance as one reason.[119] Smith has surmised that complacency born of the relatively low level of insurgent activity in Uruzgan, combined with poor insurgent fieldcraft and marksmanship, fostered such practices. The fact that they were allowed to persist, and have been commented upon by two Australian contingent COs and at least one board of inquiry, also suggests problems in junior command culture that lie beyond the scope of this chapter.

The Digger as Victim

A cornerstone of the digger identity that emerged from the First and Second World Wars was that the digger was an everyman. As volunteer citizens' forces, the Australian armies of these two great conflicts were said to be representative of the culture and outlook of the society from which they were drawn and that this was their great strength.[120] Despite the modern Australian Army being only 10 per cent of the maximum size of its Second World War predecessor, and consisting of two-thirds regular soldiers,[121] links with society and the presence of the everyman in its ranks are still considered to be a factor in its strength; there is no distinct 'warrior caste'.[122] Lieutenant Judd observed: 'The thing of soldiers doing amazing things is not that they are super heroic, but they are normal blokes. They come from normal society.'[123]

The Australian soldier in Afghanistan, however, was not an everyman. During the war an Australian warrior caste of sorts emerged, and a large gulf of understanding and outlook opened up between Australian soldiers and the community. Despite the diversity of RTF 4 documented by Lieutenant Colonel Yeaman,[124] the Army is not representative of contemporary Australian society. Australians born overseas, or from non-English speaking backgrounds are considerably underrepresented, as are women, particularly in the combat elements where gender-based restrictions have yet to be fully removed.[125] Moreover, as discussed previously, the bulk of the combat operations in Afghanistan were conducted by the special forces, which constitute roughly 5 per cent of the Army.[126] Within that cohort, some operators completed multiple deployments to Afghanistan. Given the closed world of the special forces and the intensity of their experience, it is not surprising that they viewed themselves as part of a warrior brotherhood, or caste, separate from both the rest of the Army and the wider community.[127]

The phenomenon of special forces soldiers volunteering for back-to-back deployments to Afghanistan raises a fundamental contradiction in popular views of Australian soldiers. The idealized digger is a man of action, and Australian soldiers in Afghanistan, both conventional and special forces, conceived of themselves as such. Yet popular views of the digger—from Gallipoli to Vietnam—have also portrayed him as a victim whose fighting prowess has been squandered in pursuit of unworthy objectives by uncaring and incompetent politicians and alliance partners. The digger as victim is a powerful and recurring Australian cultural symbol.[128]

Although opinion polls suggested that two-thirds of Australians were in favour of the initial deployment to Afghanistan, from 2007 they consistently revealed a majority of the population to be opposed to the continuing mission.[129] In the relatively limited media coverage accorded to the deployment the greatest attention was devoted to casualties, PTSD, fatalities, ramp ceremonies, and funeral services; what one journalist called a 'narrative of loss'.[130] This imagery of victims, however, did not generally accord with the motivations or identities of Australian soldiers discussed above. In 2010, Masters asked a group of Afghanistan veterans if they would be willing to return: 'It should not be thought they don't grasp the evil of war; they know it better than most. But ask them if they would go again, and they say: "Tomorrow", "In a heartbeat", "Can't fucking wait".'[131] The widening gulf between soldiers and their society is epitomized by Private Mullins: 'I told people I was going to Afghanistan and they acted as if I had cancer! – I wanted to go!' Mullins wrote his memoir in part to counteract the victim narrative:

> The guys I went with and I enjoyed our time in AFG... that's the largest misconception people have about going to AFG. Everyone I speak to seems to believe that I would've hated my time in AFG and somehow got forced into going... I wasn't trying to paint war as a rosy and persistently enjoyable pastime... but I wanted people to see that it can be fulfilling, challenging and affirming, and therefore enjoyable on another level.[132]

Soldiers have lamented that the story of Australia in Afghanistan has not reflected the sense of achievement discussed earlier in the chapter, whether that was providing the infrastructure and security to foster economic development and a better standard of life, or aggressively targeting insurgent networks and their leadership to the point that they were incapable of mounting coherent operations.[133]

The use of lethality as a measure of military prowess may provide another reason as to why Australian society has a predilection to portray its soldiers as victims. As was discussed earlier, contemporary Australian society seems ill at ease with the violence committed in its name. Two of Australia's four VC recipients earned their award serving with the SASR; both have made it clear that their main job in Afghanistan was to kill the enemy and both have become public celebrities. In the process of creating celebrities, however, the warrior has been dusted down, airbrushed, and sanitized. Accounts of Corporal Mark Donaldson's action focused on his recovery of a wounded interpreter under fire rather than the vicious ambush and firefight that surrounded it. Given the close-range fighting during the assault at Tizak, in which seventy insurgents were killed, stories of Corporal Ben Roberts-Smith's award could not hide the grim reality of his actions. There has been a trend in subsequent reporting, however, to focus on his impressive physical stature and his role as a husband, father, and, most recently, charity patron and veterans' advocate. Donaldson was invited to appear on the home renovation show *Backyard Blitz*; both he and Roberts-Smith have appeared as cover models for *Men's Fitness* magazine.[134]

A mixed reaction in the blogosphere to an account of a commando who, on a raid, strangled an insurgent during a building entry provides further evidence that part of Australian society is no longer comfortable with heroes from the Jacka mould.[135] As several overseas scholars have observed, it is difficult for contemporary society to accept combatant soldiers as the heroes of the battlefield—they are too violent and lack moral purity; sacrifice and victimhood are 'safer heroic terrain' than killing the enemy.[136] Reflecting on his relationship with the society he serves, Donaldson has pondered that perhaps it is better off not knowing all the details: 'What we do can be so brutal, it will shock people, and maybe turn them against us, unless they understand the full context. And it takes so long to explain it, you begin to wonder whether only those who were there and trained for it can understand.'[137]

It is telling that one of the most celebrated heroes of Australia's commitment to Afghanistan was not a soldier but rather an explosive detection dog named Sarbi. Lost during a contact and found by an American patrol fourteen months later, Sarbi was the subject of multiple media stories and a best-selling 366-page book, had her own website, and was decorated in an awards ceremony at the Australian War Memorial.[138] Major General Cantwell has expressed his frustration at the traction gained by the Sarbi story, which was more 'than all of... [his]... other media engagements combined'. He attributes this situation partly to the media but 'largely due to the draconian control of information by the Department of Defence Public Affairs Office and the Defence Minister's Office'.[139] The Department of Defence's media management in Afghanistan has been roundly criticized, with the general assessment being that media access to the deployed forces and the release of information was tightly controlled in order to limit potential damage to the ADF's reputation, support for the mission, and the strategic communications objectives of the broader campaign.[140]

It is almost impossible to disentangle the relationship between public expectations, media reporting, and community attitudes towards the Australian operations in Afghanistan. Have public attitudes towards Australian operations in Afghanistan been shaped by media reporting, or the lack of it? Or have preconceived views, or the perception of them, informed what has been reported? Conceiving of the digger as a victim may explain why the community could still voice support for the troops while disapproving of their war. Alternatively, the lack of a concerted campaign for Australia to end the commitment to Afghanistan, despite its unpopularity, is perhaps the greatest evidence of the emergence of a professional warrior caste separated from the rest of society—with the cost of the war limited to a small section of the community, the rest of Australia seems to have been not that bothered by it all.[141]

Conclusion

Welcoming home Australia's last rotation of conventional troops from Uruzgan, Lieutenant General Morrison remarked that they had added their 'own distinctive chapter to the legend of the ANZACs': 'You have joined that long, loping column stretching back in the memory's eye through the mist of time, of those who have worn the slouch had and Rising Sun badge abroad in the service of their country.'[142] Morrison's words, just like the reflections of lower-ranking soldiers discussed earlier, are recognition that the legend of Anzac remains Australia's dominant military narrative into which the experience of Afghanistan will be subsumed. The Anzacs of Morrison's speech, however, are not diggers. There is no talk of larrikins, or egalitarianism, or adventure but rather of 'the profession of arms' and 'service to the nation', along with a warning against a 'false sense of self-importance'. Pride is derived from 'duty faithfully executed with integrity'.[143] We have seen how soldiers have discussed their experience in similar terms: of being motivated by an unwillingness to let the team down, of their commitment to the arbitrary concept of the mission, and of their adherence to internal standards of tactical and technical competence. In an historical juxtaposition, Anzac packaging has been used to articulate, and legitimate, a clearly professional ethos.

In emphasizing professional achievement over strategic effect, and rejecting the victim tag, both individual soldiers and the larger Army are perhaps also steeling themselves for the inevitable post-mortems of the ISAF campaign. Legacies can be questioned; the earnestness with which they were pursued cannot. Morrison enjoined his audience to 'never let anyone tell you that you did not make a difference in Afghanistan . . . to defend the vulnerable is always noble'. The question of emphasis in the national narrative of Afghanistan has implications beyond personal worth, or institutional reputation. As Christopher Coker has pointed out, 'whether a soldier is traumatised or not will depend in part on how he interprets that experience or how he is allowed to'.[144]

The Australian experience of Afghanistan, however, must not be massaged into a tidy narrative in which the irreverent digger is replaced by the noble professional. As has been illustrated in this chapter, Australian troops in Afghanistan were motivated, united, and divided by a myriad of factors stemming from their own expectations of military service, organizational culture, societal influences, and the operational environment. Mateship is mentioned too frequently to be discounted as a source of both motivation and cohesion, but as the discussion of the rivalries and jealousies inside the Australian task forces demonstrated, its influence should not be overstated. The commitment to a professional ideal was clearly another unifying factor, but one that some chose to ignore in pursuit of personal expression and combat fashion. Although this chapter has made reference to the emergence of an Australian warrior caste, and Lieutenant General Morrison has marked out Australian soldiers as being set apart from society for having eschewed 'the path of ease, comfort, affluence, prestige, and fame', they were still a product of that society and expected an unprecedented level of comfort and amenity, which stood in stark contrast to that enjoyed by the people they were operating among.

We must also be wary of the exceptionalism that bedevils a great deal of Australian military historiography. By 2013 it was difficult for the untrained eye to differentiate between the troops of the United States, the United Kingdom, Australia, and many other national contingents in Afghanistan. Derivations of Crye Precision Multicam had become the camouflage pattern of choice, and soldiers were similarly bundled up in ballistic eyewear, enhanced combat helmets, and lightened body armour festooned with pouches. This observation points to the universality of the combat experience in Afghanistan, particularly in the southern provinces. Although the level of activity varied, combat troops essentially faced the same enemy and the same challenges. Thirteen years of coalition operations influenced national military cultures, doctrine, and operational practices, and fused ideas and values.[145] Even a cursory survey reveals a common experience of the war in Afghanistan, and perhaps the emergence of an international military culture. 'Fobbits' was used by British, Canadian, and American soldiers as well as Australians; in 2006–8 the British military establishment struggled to enforce standards of dress and bearing amidst a desire for a fashionably unkempt frontline appearance known as 'looking ally'; Australian special forces share their warrior ethos with United States operators; and United States veterans have been left wondering why society has a predilection for portraying them as victims. In Afghanistan, Australians formed part of a coalition that at its zenith was over 100,000 strong. To fully exploit what that coalition has to tell us about the experience of war it needs to be examined not just in national chunks, but as a whole.[146]

Notes

1. Harry Jarvie, in *Rebuilding Afghanistan: The Story of Australia's 2nd Reconstruction Task Force* (Salisbury: Australian Defence Force, 2008), vii.
2. Those interested in scholarly examinations of the idea of the digger should consult L.L. Robson, 'The Australian Soldier: Formation of a Stereotype', in Michael McKernan and Margaret Browne (eds.), *Australia: Two Centuries of War and Peace* (Canberra: Australian War Memorial, 1988), 313–37; Jane Ross, *The Myth of the Digger: The Australian Soldier in Two World Wars* (Sydney: Hale & Iremonger, 1985); Joan Beaumont, 'Remembering the Heroes of Australia's Wars: From Heroic to Post-Heroic Memory', in Sibylle Scheipers (ed.), *Heroism and the Changing Character of War* (Basingstoke: Palgrave Macmillan, 2014), 334–48; or Graham Seal, *Inventing Anzac: The Digger and National Mythology* (St Lucia: University of Queensland Press, 2004).
3. Bill Gammage, *The Broken Years: Australian Soldiers in the Great War* (Richmond: Penguin, 1985), 231.
4. Michael Evans, 'Close Combat: Lessons from the Cases of Albert Jacka and Audie Murphy' in Michael Evans and Alan Ryan (ed.), *The Human Face of Warfare: Killing, Fear and Chaos in Battle* (St Leonards: Allen and Unwin, 2000), 42; Kevin Fewster,

'Jacka, Albert (1893–1932)', *Australian Dictionary of Biography*, National Centre of Biography, Australian National University, <http://adb.anu.edu.au/biography/jacka-albert-6808/text11779>. Accessed 5 December 2013.

5. Robert Macklin, *Jacka VC* (Crows Nest: Allen & Unwin, 2006), 355; Evans, 'Close Combat', 42.
6. Both statues were sculpted by Peter Corlett. His theme of soldiers who save was continued by his statue 'Cobbers' erected at Fromelles in northern France in 1998 and depicts a soldier carrying a wounded comrade to safety.
7. The iconic defence image of this period was Trooper John Church from the Special Air Service Regiment carrying a Rwandan child into a first-aid post following the massacre at Kibheo in April 1995; AWM Image P04111.030 available at <http://www.awm.gov.au/collection/P04111.030>. Accessed 29 January 2015.
8. 'I'm an Australian Soldier', *Army*, 3 May 2007, 16, 21.
9. Peter Leahy, 'I'm an Australian Soldier', *Army*, 3 May 2007, 16.
10. See, for example, Peter Pedersen, 'The AIF on the Western Front: The Role of Training and Command', in McKernan and M. Browne (eds.), *Australia Two Centuries of War and Peace*, 167–93; Robert Stevenson, *To Win the Battle: The 1st Australian Division in the Great War* (Melbourne: Cambridge University Press, 2013); Garth Pratten, *Australian Battalion Commanders in the Second World War* (Melbourne: Cambridge University Press, 2009); Richard Bushby, *'Educating an Army': Australian Army Doctrinal Development and the Operational Experience in South Vietnam*, 1965–72, Canberra Papers on Strategy and Defence No. 126 (Canberra: Strategic and Defence Studies Centre, 1998); and John Blaxland, *The Australian Army from Whitlam to Howard* (Melbourne: Cambridge University Press, 2013).
11. Samuel Huntington, *The Soldier and the State: The Theory and Politics of Civil-Military Relations* (Cambridge, MA: Harvard University Press, 1985), 8–10.
12. Inscription on 'I'm an Australian Medallion', issued to Garth Pratten, November 2007.
13. It is often claimed that the Australian contribution to Afghanistan was the largest by a non-NATO nation, but this is incorrect. French troop strength peaked at 2890 in January 2009; France did not rejoin NATO until April 2009.
14. Other contributions made by Australia included an aviation detachment of two CH-47 Chinooks each summer from 2006 to 2013, AP-3C Orion surveillance aircraft operating out of the United Arab Emirates from 2003 to 2012, air traffic controllers, medical personnel, and staff officers in various ISAF headquarters.
15. David Hook, interview, 15 July 2009; author's observations as War Diarist/Senior War Diarist RC-S, January–July 2010; John Cantwell, 'A Price Too High', *The Monthly*, October 2012, <http://www.themonthly.com.au/issue/2012/october/1349327194/john-cantwell/price-too-high>. Accessed 22 December 2013.
16. James Brown, 'What did we learn from the war in Afghanistan?', *The Age*, 30 October 2013, <http://www.theage.com.au/comment/what-did-we-learn-from-the-war-in-afghanistan-20131029-2we2w.html>. Accessed 20 December 2012; Stuart Yeaman, *Afghan Sun: Defence, Diplomacy, Development and the Taliban* (Brisbane: Boolarong Press, 2013), 56–7; Chris Smith, 'Commanding Officer's Observations Mentoring Task Force Three'.

17. 'Battle Casualties in Afghanistan', <http://www.defence.gov.au/operations/afghanistan/personnel.asp>. Accessed 18 November 2013. The breakdown of Australian fatalities in Afghanistan in January 2014 was as follows: mines and IEDs, 15; gunshot wounds, 10; 'green on blue', 7; helicopter accidents, 6; rocket-propelled grenades, 1; large-calibre rockets, 1.
18. Chris Masters, *Uncommon Soldier: Brave, Compassionate and Tough, the Making of Australia's Modern Diggers* (Sydney: Allen & Unwin, 2012), 39.
19. The exact cause of death of Sapper Larcombe has not been made public; the Inquiry Officer Report was not released to the public at the request of Sapper Larcombe's family. 'Inquiry into Sapper Jamie Larcombe's death finalised', Department of Defence Media Release, 13 April 2012, <http://news.defence.gov.au/2012/04/13/inquiry-into-sapper-jamie-larcombes-death-finalised/>. Accessed 8 December 2013.
20. Robert Armand in Andrew Hetherington, 'A Faithful Mate, a Valued Soldier', *Army*, 3 March 2011, 2.
21. Soldier 1, questionnaire, 19 March 2013. The RSM of the Australian Army has also stated that 'There is a reasonable view that the Army works hard and plays hard': David Ashley, 'Army's Spirit', *Australian Army Journal*, 10(3) (2013): 209.
22. Michael Brooke, 'Mates Saved my Life: Cdo', *Army*, 10 December 2009, 2.
23. Jim Hammett, 'We Were Soldiers Once...The Decline of the Royal Australian Infantry Corps?', *Australian Army Journal*, 5(1) (2008): 44–5.
24. Glenn Neilson, '12 Platoon, 1 RAR – Call Sign 43', in *Rebuilding Afghanistan: The Story of Australia's 2nd Reconstruction Task Force*, 87. See also Yeaman, *Afghan Sun*, 140.
25. Smith, 'Commanding Officer's Observations Mentoring Task Force Three', 7.
26. The history of the concept of the larrikin is discussed in Melissa Bellanta, *Larrikins: A History* (St Lucia: University of Queensland Press, 2012).
27. Greg Stolz, 'Town Mourns Larrikin and Warrior', *The Courier-Mail*, 18 June 2011, <http://www.couriermail.com.au/ipad/town-mourns-larrikin-and-warrior/story-fn6ck51p-1226077350801>. Accessed 11 December 2013; Neilson, '12 Platoon, 1 RAR', 81; Charlie Lynn, contribution to 'Death of Australian Servicemen in Afghanistan' Debate, NSW Parliamentary Debates, Legislative Council, 13 September 2012, 15099; Tyron Butson, 'Farewell for a "soldier, a friend, a larrikin"', *The Age*, 28 March 2009, <http://www.theage.com.au/national/farewell-for-a-soldier-a-friend-a-larrikin-20090327-9eb7.html>. Accessed 11 December 2013; Melanie Schinkel, 'Loyal Larrikin Farewelled', *Army*, 2 September 2010, 2; 'Statement by Sergeant Michael Lyddiard', Department of Defence Media Release MECC 6/08, 16 January 2008.
28. These definitions are in keeping with the way military historians have traditionally understood these terms but run contrary to the work of many sociologists who have written of cohesion solely as the emotional bonds between soldiers and thus as the primary source of individual combat motivation. A good summary of the sociological literature on military cohesion can be found in Anthony King, *The Combat Soldier: Infantry Tactics and Cohesion in the Twentieth and Twenty-First Centuries* (Oxford: Oxford University Press, 2013), 24–39.

29. Mark Donaldson, *The Crossroad* (Sydney: Pan Macmillan, 2013), 414.
30. Peter Pedersen, '"Recock and Refigure": Trooper Mark Donaldson VC Speaks', *Wartime* 46, (2009): 15.
31. Nicole Spohn in Patrick Walters, 'Shrinking Support for Afghan War at Odds with Troops: Newspoll', *The Australian*, 24 March 2009, <http://www.theaustralian.com.au/news/shrinking-support-for-afghan-war-poll/story-e6frg6n6-1225690653333>. Accessed 13 December 2013.
32. Soldier 2, questionnaire, 2. Similar sentiments are expressed by Private Tom Williams from MTF 1: 'I went from gung-ho to scared shitless to back to normal in three days . . . I don't know whether it is an army thing or a bloke thing, but you just want to finish the job': Tom Williams in Konrad Marshall, 'Gone to War', *The Age*, 24 June 2013, <http://www.theage.com.au/national/gone-to-war-20130623-2oq8d.html >. Accessed 25 June 2013.
33. See, for example, Ben Roberts-Smith in Andrew Hetherington, 'Recognised for Heroism', *Army*, 3 February 2011, 3.
34. David Robertson in Andrew Hetherington, 'I Really Didn't Want to Let Anyone Down', *Army*, 27 September 2012, 20.
35. Soldier 1, questionnaire; 'The Jolly Satan', blog post, 6 June 2012, <http://www.reddit.com/r/australia/comments/unb9w/abc_reporter_kathy_mcleish_is_heading_to/>. Accessed 13 December 2013.
36. 'The Jolly Satan', blog post, 6 June 2012.
37. Damien Thomlinson, *Without Warning: A Soldier's Extraordinary Journey* (Sydney: Harper Collins Publishers, 2013), 75. Similar sentiments were expressed by Private Andrew Duff, 3 RAR, in Andrew Hetherington, 'Carrying on the Family Tradition', *Army*, 7 June 2012, 18.
38. Nathan Mullins, *Keep Your Head Down: One Commando's Brutally Honest Account of Fighting in Afghanistan* (Crows Nest: Allen & Unwin, 2011), 149.
39. Leon Gray in Masters, *Uncommon Soldier*, 183.
40. Chris Smith, telephone interview, 17 March 2013.
41. Aaron Healy, email, undated, AWM PR, item 05190, 6. Ashley Judd, a platoon commander from 7 RAR, has similarly remarked on the excitement of combat: 'I had been on combat operations overseas which, while dangerous, was quite a rush': Ashley Judd in Nick Wiseman, 'It wasn't Until Later that I Realised there was a Problem', *Army*, 13 September 2012, 6.
42. Masters, *Uncommon Soldier*, 275. See also Keith Fennell, *Warrior Brothers* (Sydney: Bantam, 2008), xii, 125, 167; Donaldson, *The Crossroad*, 322–3, 361.
43. Ben Roberts-Smith, in Peter Pedersen, 'The Falling Leaves of Tizak', *Wartime* 57 (2012): 11.
44. Ryan Avery in Andrew Hetherington, 'Fearless Actions', *Army*, 21 June 2012, 3.
45. Michael Bassingthwaite, 'Combat Team Spear – Security Task Group – Officer Commanding's Comments', *Reconstruction Task Force 3*, 93.
46. Soldier 1, questionnaire, 2; Ben Roberts-Smith in Andrew Hetherington, 'Extreme Peril', *Army*, 3 February 2011, 19; Daniel Keighran in Dave Morley, 'Facing Down the Enemy', *Army*, 8 November 2012, 2. See also *Reconstruction Task Force 3*, 49.
47. Ashley Judd in Masters, *Uncommon Soldier*, 184.

48. One of reserve commando Nathan Mullin's stated motivations for volunteering for service in Afghanistan was to avenge the death of his friend Private Greg Sher, who was killed in a rocket attack in January 2009. Mullins, *Keep Your Head Down*, 23–4.
49. Matt Proud, 'The COIN Environment: The Integration and Execution of Lethal and Non-Lethal Effects at the Tactical Level – Platoon Team Lessons from MRTF-2 in Afghanistan', *Australian Army Journal*, 8(1) (2011): 37. See also the reflections of Tom Williams on the cumulative effect of attacks on a patrol base by IED, rockets, and small arms—Tom Williams in Marshall, 'Gone to War'.
50. 'Media conference by the Chief of Defence Force regarding the death of an Australian soldier in Afghanistan', Defence Media, MECC 80709/08, 9 July 2008, 3.
51. Angus Houston in Stephen Lunn and Nicola Berkovic, 'Troops avenge comrade', *The Australian*, 12 January 2009, <http://www.theaustralian.com.au/national-affairs/policy/troops-avenge-comrade/story-e6frg8yo-1111118536349>. Accessed 13 December 2013.
52. Donaldson, *The Crossroad*, 368–9.
53. 'Media conference by the Chief of Defence Force regarding the death of an Australian soldier in Afghanistan', Defence Media, MECC 80709/08, 9 July 2008, 1.
54. John Cantwell, *Exit Wounds: One Australian's War on Terror* (Melbourne: Melbourne University Press, 2012) 301.
55. James Smith in Masters, *Uncommon Soldier*, 251. See also 15.
56. See, for example, Donaldson, *The Crossroad*, 414; Thomlinson, *Without Warning*, 86; and the attitudes of Private Scott Palmer in Peter Munro and Tom Hyland, 'Families speak out for the men they loved and lost', *Sydney Morning Herald*, 17 October 2010, <http://www.smh.com.au/national/families-speak-out-for-the-men-they-loved-and-lost-20101016-16ohi.html>. Accessed 17 December 2013.
57. Liam Hansen in Masters, *Uncommon Soldier*, 118.
58. James Martin in, 'James was Well Suited to be a Sapper from Day One', *Army*, 27 September 2012 [pages incorrectly dated 13 September 2012], 3.
59. Mullins, *Keep Your Head Down*, 28; Masters, *Uncommon Soldier*, 173; Douglas McGuire, 'Archaelogist [*sic*] Digging for Answers', *Army*, 24 July 2008, 16; Craig Reedman in, *Reconstruction Task Force 3*, 47; Thomlinson, *Without Warning*, 22.
60. Kathy McLeish, 'Your questions answered', 15 June 2012, <http://www.abc.net.au/news/2012-06-11/afghan-assignment-embedded-with-the-troops/4063890#qjun15>. Accessed 14 December 2013.
61. Masters, *Uncommon Soldier*, 15, 81, 118, 173; Jamie Tanner, interview with Chris Masters, <http://www.abc.net.au/news/2011-06-05/beyond-the-wire/2746760>. Accessed 14 December 2013.
62. Mullins, *Keep Your Head Down*, 226.
63. *Reconstruction Task Force 3*, 98.
64. Healy, 'Afghanistan', 'Barracks Life', 'ANA Contact', 'Recon Patrol', 'First Hit', 'Nearlly there' [*sic*], undated emails, AWM, item PR05190.
65. Along with ADF members, Australia also deployed personnel from its intelligence agencies, Australian Federal Police officers, and public servants from the Australian Agency for International Development, the Department of Foreign Affairs and Trade, and the Department of Defence to Afghanistan.

66. Masters, *Uncommon Soldier*, 342.
67. See, for example, Mark Johnston, *At the Frontline: Experiences of Australian Soldiers in World War II* (Melbourne: Cambridge University Press, 1996), 130–8.
68. REMF: rear echelon mother fucker. The origins of Pogo are obscure. Bruce Picken in *Viet Nam Diggers' Language* (Loftus: Australian Military History Publications, 2006) identifies it as an initialism derived from 'personnel on garrison operations'; Bruce Moore in *A Lexicon of Cadet Language* (Canberra: Australian National Dictionary Centre, 1993) states it is a shortened form of pogo stick, which is rhyming slang for prick (penis)—a generic Australian insult.
69. Donaldson, *The Crossroad*, 381; Masters, *Uncommon Soldier*, 118.
70. Healy, 'Afghanistan', undated email, AWM, item PR05190.
71. Masters, *Uncommon Soldier*, 118, 173; J. Sergeant and [no initial] Fitzgerald, '10 Section', *Rebuilding Afghanistan*, 94; Yeaman, *Afghan Sun*, 312–14.
72. Cantwell, *Exit Wounds*, 254.
73. Cantwell, *Exit Wounds*, 254.
74. Masters, *Uncommon Soldier*, 122.
75. See Andrew Hetherington, 'Salute to Ultimate Warrior', *Army*, 15 November 2007, 3; Andrew Hetherington and Dave Morley, 'Warriors Mourned', *Army*, 8 July 2010, 2; Peter Martinelli, 'Farewell to a Warrior', *Army*, 22 July 2010, 3; Schinkel, 'Loyal larrikin farewelled'; Melanie Schinkel, 'Thousands Gather to Farewell Warrior', *Army*, 9 June 2011, 2; Melanie Schinkel, 'A Tenacious Leader', *Army*, 21 July 2011, 2; Nick Wiseman, 'Saying Goodbye to Fallen Warrior', *Army*, 18 July 2013, 2.
76. 'WO2 H' in Masters, *Uncommon Soldier*, 280.
77. Roberts-Smith in Pedersen, 'The Falling Leaves of Tizak', 15. See also Thomlinson, *Without Warning*, 59–60.
78. Masters, *Uncommon Soldier*, 308, 319. See also: <http://images.defence.gov.au/20100419_909_0264.jpg>; <http://images.defence.gov.au/20100419_909_0301.jpg>; <http://images.defence.gov.au/20131005adf8094997_036-31.jpg>; <http://images.defence.gov.au/20131003adf8094997_214_2.jpg>; <http://i1091.photobucket.com/albums/i394/MyZone2/018.jpg>.
79. Masters, *Uncommon Soldier*, 101, 295.
80. Donaldson, *The Crossroad*, 335–6; Masters, *Uncommon Soldier*, 231–3.
81. Masters, *Uncommon Soldier*, 172, 180–3; Tom Hyland, 'Death and doubt in Afghanistan', *The Sydney Morning Herald*, 21 March 2010, <http://www.smh.com.au/world/death-and-doubt-in-afghanistan-20100320-qneg.html>. Accessed 28 December 2013.
82. Mullins, *Keep Your Head Down*, 127–8.
83. Donaldson, *The Crossroad*, 249, 324–7.
84. Mullins, *Keep Your Head Down*, 290.
85. Masters, *Uncommon Soldier*, 169, 194.
86. Greg Cotton, 'Enhancing Operational Capability', *Australian Army Journal*, 5(1): 52–3; J. Hammett, 'We were Soldiers Once', 5(1): 41–4; Yeaman, *Afghan Sun*, 56–5, 313; Will Langdon in *Rebuilding Afghanistan*, 72; *Reconstruction Task Force 3*, 62; Rajeshni Naidu, 'Australian troops labelled "lighthouses"', 29 May 2008, ninemsn.com.au,

<http://milnewstbay.pbworks.com/f/CAN-AUS-AFG-ninemsn-290427May08.pdf>. Accessed 12 October 2013.

87. Brendan Nicholson, 'Special forces on top tucker while the others go Dutch', *The Age*, 4 June 2009, <http://www.theage.com.au/national/special-forces-on-top-tucker-while-the-others-go-dutch-20090603-bvp8.html>. Accessed 17 December 2013; 'Aussie tucker for diggers in Afghanistan', <http://www.abc.net.au/news/2009-06-03/aussie-tucker-for-diggers-in-afghanistan/1702540>. Accessed 3 June 2009; Masters, *Uncommon Soldier*, 232–3.
88. Soldier 2, questionnaire, 2.
89. *Anzacs*, Burrowes-Dixon Production Company, 1985.
90. *Land Warfare Doctrine 1: The Fundamentals of Land Warfare* (Australian Army, Puckapunyal, 1998), 6–7.
91. Richard Hughes, 'On the Culture of the Australian Army', *Australian Army Journal*, 10(3): 228.
92. Yeaman, *Afghan Sun*, 334.
93. Masters, *Uncommon Soldier*, 117, 119, 177; Michael Hindmarsh, foreword to Yeaman, *Afghan Sun*, iv, 161; I. Langford, 'Australian Special Forces in Afghanistan: Supporting Australia in the 'Long War'', *Australian Army Journal*, 7:1, 22, 25, 29–30.
94. 'Operations 06: ADF Combat Incidents Resulting in Allegations of Death or Injury to Afghan Civilians and Subject to Administrative Inquiry as at 17 May 2013', Defence Senate Estimates Brief SB13-000511, 24 May 2013; Michael Brissenden, 'Australian special forces troops under investigation for cutting off hands of dead Afghan insurgent', <http://www.abc.net.au/news/2013-08-30/claims-sas-troops-cut-hands-off-afghan-insurgent/4924694>. Accessed 30 August 2013.
95. 'Soldiers return to Australia following disciplinary action', Department of Defence media release, 16 January 2012; Alex White, 'Thirty-four Diggers sent home from Middle East in disgrace', *Herald-Sun*, 5 June 2013, <http://www.heraldsun.com.au/news/thirty-four-diggers-sent-home-from-middle-east-in-disgrace/story-fni0fiyv-1226658035702>. Accessed 21 December 2013; Sally Sara, 'Australian soldiers facing disciplinary action', ABC Radio Australia, 25 March 2011, <http://www.radioaustralia.net.au/international/2011-03-25/australian-soldiers-facing-disciplinary-action/224080>. Accessed 21 December 2013; Masters, *Uncommon Soldier*, 233.
96. Ian McPhedran, 'Our Diggers betrayed in the hell of Afghanistan', *The Telegraph*, 21 September 2010, <http://www.dailytelegraph.com.au/our-diggers-betrayed-in-hell-of-afghanistan/story-e6frewp9-1225926985259>. Accessed 12 October 2013; Bryan Littlely, 'Soldiers forced to pay for equipment', *The Advertiser*, 25 September 2010, <http://www.adelaidenow.com.au/news/national/soldiers-forced-to-pay-for-equipmentl/story-e6frea8c-1225929106776>. Accessed 12 October 2013; Linton Besser and Dan Oakes, 'Letting the troops down', *The Sydney Morning Herald*, 8 May 2010, <http://www.smh.com.au/action/printArticle?id=1425187>. Accessed 8 May 2010.
97. 'Inquiry Officer's Report into the facts and circumstances surrounding a shooting incident in Afghanistan that resulted in the deaths of LCPL. S. Milosevic, SPR J. T. Martin and PTE R. H. F Poate and injuries to other Australian soldiers on or about 29 AUG 12', 5; 'Inquiry Officer's Report into matters relating to the force

preparation, training, certification and leadership associated with the force element identified in the Inquiry Officer's report into the death of Lieutenant M. K. H. Fusell', 30.

98. David Morrison, speech to United Nations International Women's Day Conference, 8 March 2013, <http://www.army.gov.au/Our-work/Speeches-and-transcripts/~/media/Files/Speeches/CA_UNIWD_8-9MAR2013.pdf>. Accessed 15 October 2013, 5.
99. 'Flip flops' for non-Australian readers. Masters, *Uncommon Soldier*, 118.
100. Masters, *Uncommon Soldier*, 176–7.
101. Smith, 'Commanding Officer's Observations Mentoring Task Force Three', 10.
102. Carl von Clausewitz, *On War*, trans. M. Howard and P. Paret (Princeton, NJ: Princeton University Press, 1989), 580.
103. Yeaman, *Afghan Sun*, 175–6.
104. Masters, *Uncommon Soldier*, 207.
105. Yeaman, *Afghan Sun*, 169; Smith, 'Commanding Officer's Observations Mentoring Task Force Three', 7.
106. Smith, 'Commanding Officer's Observations Mentoring Task Force Three', 7.
107. Mullins, *Keep Your Head Down*, 176–7; Smith, 'Commanding Officer's Observations Mentoring Task Force Three', 7.
108. Smith, 'Commanding Officer's Observations Mentoring Task Force Three', 8.
109. Mark Jennings in Andrew Hetherington, 'Life-Saving Upgrades', *Army* Uruzgan Special, 5 December 2013, 8.
110. Blaxland, *The Australian Army*, 274.
111. Smith, 'Commanding Officer's Observations Mentoring Task Force Three', 7.
112. Soldier 2, questionnaire, 2.
113. 'Life at a patrol base', JTF 633 Facebook page, 24 June 2012, <https://www.facebook.com/media/set/?set=a.436160176418755.104363.115155888519187&type=1>. Accessed 23 December 2013; Kathy McLeish, 'Keeping in touch with home', ABC News, 27 June 2012, <http://www.abc.net.au/news/2012-06-11/afghan-assignment-embedded-with-the-troops/4063890>. Accessed 23 December 2013; Masters, *Uncommon Soldier*, 226.
114. See, for example, the account of the construction of Patrol Base Qudus in Yeaman, *Afghan Sun*, 153–80.
115. Australian designed and manufactured mine-resistant vehicle.
116. Mullins, *Keep Your Head Down*, 79. See also Tom Hyland, 'In the shadow of the valley', *The Age*, 8 March 2009, <http://www.theage.com.au/world/in-the-shadow-of-the-valley-20090307-8rwo.html>. Accessed 23 December 2013; 'Inquiry Officer's Report into the facts and circumstances surrounding a shooting incident in Afghanistan that resulted in the deaths of LCPL. S. Milosevic, SPR J. T. Martin and PTE R. H. F Poate...', 32.
117. Smith, 'Commanding Officer's Observations Mentoring Task Force Three', 7, 10.
118. Smith, 'Commanding Officer's Observations Mentoring Task Force Three', 7.
119. Mullins, *Keep Your Head Down*, 82.
120. See, for example, John Laffin, *Digger* (London: Cassell, 1959), 233–4.

121. At the height of its expansion in the Second World War the Australian Army comprised 466,593 soldiers. The 2011 ADF census lists the strength of the Army as 46,450, composed of 30,020 regulars and 16,430 active reservists. Paul Hasluck, *The Government and the People 1942–1945* (Canberra: Australian War Memorial, 1970), 333; *Department of Defence Census 2011—Public Report*, (Canberra: Department of Defence, 2012), 2.
122. Blaxland, *The Australian Army*, 13–14; Yeaman, *Afghan Sun*, 139–40.
123. Judd in Masters, *Uncommon Soldier*, 183–4.
124. Yeaman, *Afghan Sun*, 64–5.
125. *Review into the Treatment of Women in the Australian Defence Force—Phase 2 Report* (Canberra: Australian Human Rights Commission, 2012), 46–7; *Department of Defence Census 2011—Public Report* (Canberra: Department of Defence, 2012), 10.
126. Donaldson, *The Crossroad*, 321. Donaldson estimates that the Special Forces suffered 90 per cent of the Australian casualties in Afghanistan. This estimate cannot be verified due to ADF policy of not releasing the details of non-fatal casualties. Given the more aggressive role of the Special Forces it is possible, but the even split of fatalities between the conventional and Special Forces would suggest it is unlikely.
127. Kira Harris et al., 'Military Retirement: Reflections from Former Members of Special Operations Forces', *Australian Army Journal* 10(3): 102.
128. The genesis of the victimized digger lies in the Boer War with lieutenants Peter Hancock and Harry 'The Breaker' Morant, who were Australian enlistees (Morant was British) executed by British authorities for shooting Boer prisoners. They were subsequently dubbed as 'scapegoats of the Empire' and Morant in particular has become somewhat of a folk hero. John Laffin epitomized this sentiment when he railed against the 'British military tradition and inept leadership' that in the First World War 'sentenced thousands of Diggers to death' (Laffin, *Digger*, 131), and one of the most poignant scenes of modern Australian cinema is the climax of the film *Gallipoli* where the lives of Australian soldiers at the Nek are blasted away by Turkish machine guns while British officers at Suvla Bay are 'sitting on the beach drinking cups of tea' (*Gallipoli*, R & R Films, 1981). During the Second World War, the Australian soldiers taken prisoner at Singapore were said to have been betrayed by the British, leading to three years of Japanese captivity; books on the prisoner-of-war experience are legion; accounts of the fighting that preceded it are not. The traumatized, forgotten, and alienated Vietnam veteran—always a conscript never a volunteer—is a more recent example of the victimized digger.
129. The Lowy Institute provided the most consistent polling on the Australian commitment to Afghanistan. Asked 'Should Australia continue to be involved militarily in Afghanistan?', participants responded: 2007 46 per cent no/46 per cent yes; 2008 56/42; 2009 51/46; 2010 54/43; 2011 59/40. Lowy polling in 2013 asked if, considering the costs and benefits to Australia, the war in Afghanistan had been worth fighting; 61 per cent of participants responded 'no' and 35 per cent 'yes'. Fergus Hanson, *Australia and New Zealand in the World: Public Opinion and Foreign Policy* (Sydney: Lowy Institute, 2012), 7; Alex Oliver, *Australia in the World: Public*

Opinion and Foreign Policy (Sydney: Lowy Institute, 2013), 8. See also: 'Should Australia Pull its Troops Out of Afghanistan?', *The Sydney Morning Herald*, 26 November 2011, <http://www.smh.com.au/federal-politics/the-question/should–australia-pull-its-troops-out-of-afghanistan-20111125-1nyy6.html>. Accessed 13 December 2013; Patrick Walters, 'Shrinking Support for Afghan War at Odds with Troops: Newspoll', *The Australian*, 24 March 2009, <http://www.theaustralian.com.au/news/shrinking-support-for-afghan-war-poll/story-e6frg6n6-1225690653333>. Accessed 13 December 2013.

130. Chris Masters, 'Australia in Afghanistan: Longest war draws to a "bittersweet" end after 12 years', 11 November 2013, ABC News, <http://www.abc.net.au/news/2013-11-11/australias-war-in-afghanistan-draws-to-an-end/5082976>. Accessed 28 December 2013.
131. Masters, *Uncommon Soldier*, 259.
132. Mullins, *Keep Your Head Down*, 334–5.
133. Yeaman, *Afghan Sun*, 334–5; Masters, 308–9.
134. *Men's Fitness*, April 2012 and December 2013.
135. Kristin Shorten, 'Killer Commando: America poaches Australia's toughest man', news.com.au, 7 March 2013, <http://www.news.com.au/national/killer-commando-america-poaches-australias-toughest-man/story-fncynjr2-1226592161560>. Accessed 13 March 2013; Kristin Shorten, 'Diggers defend killer Commando Sergeant Paul Cale', news.com.au, 8 March 2013, <http://www.news.com.au/national/diggers-defend-killer-commando-who-strangled-taliban-commander/story-fncynjr2-1226593075852>. Accessed 13 March 2013.
136. Christopher Coker, *The Warrior Ethos: Military culture and the War on Terror* (Abingdon: Routledge, 2007), 9, 98–102; Michael Gisick, 'Citizen-Soldier, Warrior Caste: American soldiers in the "War on Terror"', PhD proposal, Australian National University, 2013, 10. I must thank Mike for drawing my attention to this literature.
137. Donaldson, *The Crossroad*, 291.
138. See, for example, Sandra Lee, *Saving Private Sarbi: The True Story of Australia's Canine War Hero* (Crows Nest: Allen & Unwin, 2011); 'RSPCA awards Sarbi the Purple Cross', Department of Defence media release, 5 April 2011. The Sarbi webpage was hosted on the Australian Department of Defence website at http://www.defence.gov.au/sarbi/index.htm 2011–2014.
139. Cantwell, *Exit Wounds*, 326.
140. See, for example, Michael Brissenden, 'Afghanistan: The War We Hardly Knew', *The Drum*, 14 November 2013, <http://www.abc.net.au/news/2013-11-14/brissenden-afghanistan-the-war-we-hardly-knew/5090826>. Accessed 28 December 2013 and Kevin Foster, 'Don't Mention the War: Lessons for the Media from Afghanistan', *The Conversation*, 4 December 2013, <https://theconversation.com/dont-mention-the-war-lessons-for-the-media-from-afghanistan-20310>. Accessed 28 December 2013. The most detailed explorations of Australian media management in Afghanistan are provided by Jason Logue, *Herding Cats: The Evolution of the ADF's Media Embedding Program in Operational Areas* (Canberra: Land Warfare Studies Centre, 2013) and

Kevin Foster, *Don't Mention the War: The Australian Defence Force, the Media and the Afghan Conflict* (Clayton: Monash University Publishing, 2013).

141. The genesis of this idea needs to be attributed to a discussion with UNSW Canberra (ADFA) PhD scholar Tom Richardson.
142. David Morrison, speech to welcome home parade for 2nd Cavalry Regiment Task Group and Combined Team Uruzgan Four and Five, 1 March 2014, <http://www.army.gov.au/Our-work/Speeches-and-transcripts/~/media/Files/Speeches/CA_2CAVREGTWelcomeHome Parade_1MAR2014.pdf>. Accessed 3 March 2014.
143. Morrison, speech, 1 March 2014.
144. Coker, *The Warrior Ethos*, 102.
145. James Brown, 'Fifty Shades of Grey: Officer Culture in the Australian Army', *Australian Army Journal*, 10(3) (2013): 247.
146. I would like to thank my fiancé Carol Holmes for her forbearance during the writing of this chapter. Carol, along with Joan Beaumont, Mark Johnston, and Anthony King, also provided helpful comments on my first draft. Thanks also go to Miesje de Vogel for her assistance with some of the research.

9

A Profession of Love

Cohesion in a British Platoon in Afghanistan

Patrick Bury and Anthony King

Introduction

British infantry platoons have been engaged in high-intensity combat and counter-insurgency operations in Afghanistan since 2006. During this time, the British infantry in particular have taken some of the heaviest casualties per number of serving soldiers in NATO.[1] Yet despite these relatively high numbers of casualties, a complex and arguably vague counter-insurgency mission, and a constantly shifting UK strategic narrative for involvement in Afghanistan, the combat performance of the British infantry has remained remarkably effective. Along with the US infantry and Marines, the British infantry has been deployed to the most violent regions of the country and consistently involved in some of the heaviest fighting of the war. Moreover, some engagements have been so intense and protracted that they have been described by commanders as the heaviest fighting undertaken by the British army since the Korean War.[2] Indeed, despite being initially deployed to isolated bases in dangerously small numbers whilst being chronically under-resourced,[3] a British infantry platoon has never been overrun, unlike their predecessors in Korea, or some of their NATO allies' infantry platoons in Afghanistan.[4] Of course, while the Korean War had its own specific conditions in terms of enemy force size and tactics, superior air power and advances in weapons technology have acted as force multipliers for infantry platoons in a way not possible in Korea. Yet, as we shall see, even in intense close-quarters engagements in Afghanistan, where the benefits of these assets have been negated and the insurgents have often held the initiative, British infantry platoons have remained highly effective. These engagements have often occurred in difficult terrain and in extreme climatic conditions, where platoons have remained resolute when faced with an equal or even superior enemy in conventional and non-conventional combat situations. Although the availability of airpower, especially the Apache helicopter—which the Taliban greatly feared—in many firefights has to be acknowledged, evidence suggests that even the Taliban insurgency in Helmand province learnt to respect certain British infantry units.[5] Indeed, the insurgency's switch to asymmetric tactics, as evidenced by the massive increase in the improvised explosive device (IED) threat from 2008 onwards, represented a tacit acknowledgement of the combat superiority of the professional British infantry platoon at the tactical level. While it is true that—like its retreat from Basra in 2007—the British Army's withdrawal from Sangin in 2010 is seen by many as a stain on its reputation, it is important to note that this was the result of a lack of political appetite for casualties rather than imminent tactical defeat. Indeed, despite coming close on numerous occasions, to date the British infantry platoon remains tactically undefeated in Afghanistan. Sociologically, in terms of the academic literature on military unit cohesion, it may be interesting to examine some of the factors that explain why this is so.

While the strong combat performance of the professional British infantry platoon in Afghanistan is by no means unique when compared with the performance of other nations' infantry, case studies solely focusing on the micro-interactions of the British infantry platoon remain rare. While the British example has been used to address the question of how professional Western infantry platoons have remained so effective in recent conflicts, and specifically in Afghanistan,[6] this chapter seeks to add to the recent literature on military cohesion by exclusively focusing on the micro-interactions of one British platoon in combat in Afghanistan. The platoon that is examined was an air-assault infantry platoon in Ranger Company of 1st Battalion, The Royal Irish Regiment, attached to 2nd Battalion, The Parachute Regiment (2 PARA). It served in Sangin, Helmand province over the summer of 2008, during Operation Herrick 8, where it conducted counter-insurgent patrols and strike operations. In studying the combat interactions of this platoon, the chapter uses the case-study approach to analyse two combat situations that one of the authors (Patrick Bury) and his platoon experienced in Helmand province within six days of each other, to address some of the fundamental arguments about cohesion its components in modern professional forces.

Drawing on personal experience of combat in Afghanistan in 2008,[7] this chapter seeks to build upon and extend this 'practical' or professional paradigm. It utilizes two firefights involving 7 Platoon of the Royal Irish Regiment in Sangin in June 2008 as an evidential base from which to contribute to contemporary debates about cohesion and to the advance of the professional paradigm described in Chapter 1. In *The Combat Soldier*, King singles out the use of mnemonics in infantry battle drills that remind commanders of the procedures that they should go through when faced with different tactical problems.[8] As we shall see, these are precisely the procedures tactical commanders and those under their command use when reacting to contact with the enemy; not only do they engender effective individual and collective performance, they also distract from fear by occupying the minds of those executing the drills and stop them focusing on the chaos and danger of combat. However, while substantially supporting the perspective developed by King, Strachan, and Ben-Ari that training constitutes the central explanation of cohesion in combat in a professional force, the chapter also explores the emotional bonds between the soldiers to explain their actions in combat. In this way, the chapter seeks to address a potential oversight in the emerging paradigm and, perhaps, to reconcile it partially with the classical literature where interpersonal bonds in the primary group were prioritized. Nevertheless, as the evidence clearly demonstrates, even these intense emotional and apparently purely personal bonds are constituted through, and indivisible from, shared professional expectations and practices. Indeed, they cannot be fully understood without considering the situational dynamics of combat as experienced by a professional force.

Waterloo Day Contact: Drills and Training

The first contact for 7 Platoon occurred on 18 June 2008, known in the Royal Irish Regiment as 'Waterloo Day' due to the special significance of that battle in regimental history. As 7 Platoon readied themselves to leave the compound they had occupied for a number of days on the Forward Line of Enemy Troops, they picked up radio chatter that indicated an insurgent group was waiting in ambush for them. This information provided extra focus for members of the platoon, and before they left the compound, the author, the commander of the patrol, again reiterated the platoon's formation and reaction to contact drills.[9] On the advice of Royal Marines whom the Royal Irish relieved, platoon tactics had already been altered somewhat from the conventional tactics to mitigate the risk of suicide attacks: the platoon deployed 1 Section one hundred metres off to a flank in the direction of the main enemy threat to shield the rest of the platoon and aimed to form a box if contacted to prevent suicide bombers from infiltrating the platoon and blowing themselves up in its midst. As a platoon untested in combat, it is interesting to examine the crucial role that battle drills played in the platoon's reaction to its first sustained contact and its avoidance of taking casualties in an enemy-initiated ambush where they did not initially hold the tactical advantage and were without air support.

The ambush occurred outside a cluster of Afghan compounds in the heavily irrigated fields of the Green Zone when a group of about six insurgents engaged one section of the platoon with rocket-propelled grenades and small arms from a distance of about two to three hundred metres. Indeed, it was the platoon's tactically bespoke formation—adapted from the Royal Marines—that saved it from taking casualties in the initial and accurate volleys of fire. This ensured only eight men were initially engaged. Each Ranger (private soldier) in 1 Section instinctively went through their personal RTR battle drills in Returning Immediate Fire, Taking Cover, and Returning Effective Fire, as did the rest of the platoon off to their flank. Caught in an open field and amidst the chaos of exploding grenade rounds and bullets clipping all around them, the Section Commander, Corporal Flynn,[10] then sent his contact report to the commander and began to manoeuvre out of the enemy's killing zone. This had been repeatedly practised in the section drill for reaction to contact, to the extent that it had become almost instinctive. At this stage the platoon commander also sent his contact report up to company headquarters, and was relieved that 'at least [he] remembered the first thing [in the platoon battle drill]'.[11] Here, both section and platoon commanders used their respective battle drills in the initial seconds of the ambush to relay crucial information to the chain of command. However, the battle drill not only enabled the critical flow of information, it also provided a check-list for commanders to perform in the first moments of confusion. The drill distracts commanders from the immediacy of the violence and noise around them and cognitively guides them toward the next step of the drill that needs to be performed. The correct execution of the individual battle drill is therefore not only important to the platoon's physical response to combat; crucially, it also has an important psychological dimension, calming individuals and reassuring them that they are acting in the correct way. Following Strachan, when numerous individuals begin to react in a similar manner, psychological confidence in the group grows and collective action becomes easier to coordinate.

It is possible to highlight how the collective battle drills allowed the untested platoon to recover from the initial shock and panic of the ambush and begin to effectively perform as a unit. Firstly, the platoon sergeant told the platoon commander that: 'I got the back door, boss. The backdoor's closed. We're secure...'[12] thereby informing him that he had executed the adapted platoon battle drill and deployed his group at the base of the other sections to face outward. The platoon had now effectively formed a box, all facing outward, with two sides engaging the enemy. This adaption of the platoon battle drill had been repeatedly practised as a drill in base and mitigated the chance of a suicide bomber breaking from cover to detonate in their midst—a favourite insurgent tactic once an ambush had been initiated. Once the platoon had its rear and flanks relatively secured, the platoon could focus on dealing with the enemy threat. Meanwhile, 1 Section, still pinned down by intense enemy fire, had identified the enemy and sent a target location, as per the section battle drill. Corporal Flynn then shouted a fire control order to his section to coordinate their fire: 'Three hundred metres, left edge of compound, two times Taliban, rapid fire.'[13] In issuing the fire control order in the standardized format that his Rangers were expecting, the commander was able to quickly and accurately relay vital information on the enemy location to his section to enable them to collectively suppress the enemy. The standardized instruction 'Rapid Fire' also instructed his section as to the rapid rate of fire is one round every two seconds for riflemen. Understanding and obeying rate of fire instructions are fundamental to individual soldier's and commander's training; this mutual understanding of two simple words had a critical impact on the section's reaction to contact and the conservation of ammunition, and therefore its combat effectiveness. Ritualized words of command were crucial to the section's effective reaction to contact.

It is interesting to note here that the successful execution of the fire control order had a double impact on the platoon's attempts to gain the upper hand in the firefight: not only did it result immediately in an accurate and sustained weight of fire that suppressed the enemy and allowed the other sections to manoeuvre, it also created confidence that the platoon was reacting in the appropriate way, as it had been taught, and it reminded commanders of the next part of the drill to be carried out. As such, the next instruction issued to 1 Section was 'Baseline'; when shouted back to the commander in unison by the section it confirmed to him that his section had heard and understood the instruction that they were to hold where they were and fire another burst. This part of the reaction to contact drill also confirms to each soldier that they are not alone; that they are acting in a choreographed fashion that has been proven to give them the best chance of survival. Indeed, not shouting the reply in the din of battle is an instant indication that the section has taken a casualty and that the casualty extraction drill may need to be implemented.

This particular contact, which lasted approximately ten minutes, ended when the platoon manoeuvred to support 1 Section and suppress the enemy, causing the enemy to take a casualty and break off the attack. However, it was the platoon's bespoke formation and 1 Section's correct performance of the taught battle drills that both saved it from taking casualties and enabled it to initially suppress the enemy, allowing the other sections to manoeuvre. Due to the fact that the platoon reacted to an enemy-initiated ambush who then subsequently broke contact, it did not have time to go through the entire platoon attack drill, but it is clear from this account that the collective performance of battle drills not only ensured the timely and coordinated response to a tactical disadvantage, it also gave the platoon a sense of unity of purpose as it responded to the chaos of an ambush. The choreography of the collective battle drill allowed the soldiers to extract themselves and regain the initiative far faster and more uniformly than if they had reacted individually; a point which is critically important, as initiative is crucial in all military engagements. Finally, the battle drill also acted as a template against which the platoon's actions could be assessed and improved upon, both collectively and individually, after the event. As a result, the successful execution of battle drills also became a source of group pride and confidence within the platoon after the attack: 'We felt good that we hadn't let ourselves or each other down. That we had each others' backs, that we performed our individual tasks that, when combined, formed a cohesive whole that kept us alive. That we hadn't been frozen by fear.'[14] Clearly then, task commitment and collective battle drills were crucial to the inexperienced platoon repelling the ambush without taking casualties, and they remain fundamental to understanding effective British infantry combat performance at the tactical level. Moreover, the constant referral to, and repetitive practice of, these drills during downtime highlights the importance of professionalism to the platoon; the drill not only became a life-saving reality, but its correct and slick execution represented the professionalism of the platoon itself.

Sarwan Qala: Combat, Emotions, and the Profession of Love

In their work on the Israel Defense Forces, Ben-Ari et al. have highlighted the role of training and drills as decisive to the generation of 'swift trust' that allows professional soldiers unfamiliar with each other to interoperate due to common practices.[15] The implication is that cohesion and combat performance is independent of emotional attachment and, indeed, can be effectively, even optimally, executed without emotional cathexis. In fact, despite their emphasis on impersonality, Ben-Ari et al. recognize emotions and emotional bonds in their work, but they understate it in their examination of swift trust. It might be suggested that the contemporary literature on cohesion generally underemphasizes emotions as a corrective to the classical cohesion paradigm in which interpersonal bonds are the almost exclusive focus of interest. Yet it is plainly incorrect to ignore the emotional responses of soldiers in combat; it is, after all, one of the most extreme and terrifying experiences known to humans. Indeed, Collins' arguments on emotional entrainment and the emotional nature of violent interactions are highly informative here. In *Interaction Ritual Chains* Collins argues that almost all human interactions are infused with an emotional energy due to physical and cognitive synchronicity, or entrainment.[16] Building on this theory, in his micro-sociology of violence Collins shows that at the individual interactional level, conducting violence is fundamentally 'about the intertwining of human emotions',[17] as it requires humans to overcome the barrier of confrontational tension and fear associated with the threat of violence by turning 'emotional tensions into emotional energy'.[18] In breaking this barrier—in reaching a certain, highly situational emotional state—one actor is able to commit violence, often at the expense of the other. One of Collins' most important observations on the emotional nature of war is the causal role emotions play in what he terms 'forward panics'. Building on Du Picq's observations about a 'flight to the front' during combat,[19] Collins describes forward panics as a collective emotional reaction to the prolonged build-up of tension in certain violent situations. Much like fleeing, forward panics begin with the breaking of this tension and fear, but they differ from rear panics in that they are followed by an emotional 'hot rush'—often involving rage, frenzy, and elation—that is 'overriding and compelling'.[20] This 'overriding emotional rhythm' is caused by group emotional entrainment in collective action and is particularly evident in combat.[21] Here, soldiers' intense emotional release is directed with the heightened emotional focus of 'tunnel vision' at a particular subject—usually the enemy or civilians—and often results in overkill. Indeed, Collins posits that such is the emotional state of those in the grip of a forward panic, 'they always [have] the look of an atrocity'.[22] While Collins presents much evidence from pre-professional armies, he argues that forward panics have affected professional combat units, citing the Parachute Regiment's shooting of civilians at Derry's Bloody Sunday in 1972 and the similar actions of US Marines in Haditha, Iraq in 2005.[23] As such, forward panics not only represent a physical manifestation of the barrier of emotional tension that often needs to be overcome to kill, they also indicate the importance of group emotional entrainment in combat situations. For Collins then, committing violence itself is typically a highly emotional experience. Moreover, the emotions which are cathected in violence are not random individual or universal genetically cued responses. The collective interactional dynamics of the groups involved in violence generates patterned emotional responses from their participants. However, these patterns are fluid; alternative inter- and intragroup dynamics produce different emotional responses among the members of those groups.

Here, we want to show how emotions are central to any explanation of combat performance, but that these emotions cannot themselves be understood without considering the relationships within and between the combatant groups; crucially, in the current context, that the distinctive solidarity of the professional army evokes distinctive emotional responses. Emotional responses are generated by, and flow along, the contours of established social relations; they are an organic part of interactional dynamics. Moreover, they are precipitated by and articulated through collective understandings of the participants in these processes. There is no doubting the centrality of visceral emotions to professional combat performance—fear, anger, hate, shame, love, and pride are ubiquitous—but its manifestation is always already intertwined with the professional self-understandings of the troops and their relations to each other. As *The Combat Soldier* illustrates,[24] the motivation of professional soldiers and their relations to each other is distinctive, giving rise to equally distinctive emotional responses. Central here are the emotions of shame and pride (or honour) which underpin the relations between soldiers. However, as discussed in Chapter 5, while shame might be a universal human emotion, essential to social life,[25] it takes a specific form among professional soldiers. Unfortunately, many social scientists have typically been insensitive to these differences.

Six days after they had been 'baptised' in their contact on Waterloo Day, 7 Platoon were chosen to accompany C Company, 2 PARA, on an air assault into Sarwan Qala, in the Upper Sangin Valley. This area did not have any coalition or Afghan troops stationed in it and as a result it was assessed as insurgent territory. The nominal idea behind the air assault was to conduct reconnaissance of the 611 road in the area and meet local leaders; yet 2 PARA had lost three soldiers in previous weeks and this was also an attempt to strike at the insurgents' heartland, induce a fight, and get revenge. It succeeded. The battle started at 4 a.m. with incoming fire as the helicopters arrived, signalling a hot landing zone, and ended approximately sixteen hours later, platoons having been engaged in full combat for much of that time. Attack helicopters, fast jets, and artillery strikes were all utilized by the British troops, who were engaged in some of the heaviest fighting British forces had seen in Afghanistan to that date.[26] While it is evident that battle drills were again an important part of the platoon's preparations before battle and its ability to perform during it,[27] it is also clear that during this heavier and prolonged engagement, emotional bonds between soldiers were also very important in maintaining effective performance. Indeed, in one incident, it was the emotional bonds formed over time that enabled the successful completion of the platoon's task.

How this emotional bond, this 'slow-trust' as it were, motivated the platoon to perform in combat becomes apparent when the detail of the Sarwan Qala operation is examined. For approximately six hours 7 Platoon had been pinned on high ground by insurgents firing from the Green Zone below. During this time a 2 PARA Company Sergeant Major attached to the platoon, Warrant Officer Class 2 (WO2) Michael Williams, had been shot dead by an insurgent sharpshooter. As the company was now running out of ammunition and water, and had a body to be evacuated, it became imperative that a helicopter resupply land to sustain the fight, so 7 Platoon were tasked with securing the helicopter landing site (HLS) off to a flank. Due to the abnormally accurate and sustained fire coming from the Green Zone, though, only one section was available: Corporal Flynn's. There were no possible safe routes to the HLS due to the tactical situation and with a helicopter in-bound and the company running out of ammunition and water in the forty degree heat, the task was urgent. The only way to reach the HLS was up a ridge, completely exposing anyone who attempted this to the same sharpshooter who had killed WO2 Williams and had earlier repelled a similar attempt in the same direction. While task commitment remained a key motivator here, battle drills were now relatively redundant. It was abundantly clear to both the platoon commander and Corporal Flynn that to expose oneself on the ridge—given the proven accuracy of the sharpshooter—was akin to suicide. As such, the situation required action beyond the call of normal duty, beyond normal professional standards.

In these circumstances, and according with some of the examples concerning discipline discussed in Chapter 5, it is interesting to note that the platoon commander did not, and ultimately could not, order Corporal Flynn onto the ridge.[28] He simply stated the importance of the task and the realities of the tactical situation they found themselves in: 'Right, Matt, the chopper's inbound in ten. We can't go down the 611 'cause 9 Platoon are being hit. We can't go over H's wall 'cause they are being hit. We know we can't go that way. The only way is up there.'[29] Corporal Flynn looked up the cliff and replied: 'Yeah . . . no worries, boss.' He turned to his men: '1 Section, saddle up, we're going up there.'[30] In the event Corporal Flynn led his section through the withering fire to the HLS without further casualties. Flynn was later awarded a Mentioned in Dispatches for his leadership.

The incident at Sarwan Qala was brief but both extremely serious and illuminating. While battle drills and training are typically critical to the explanation of combat performance and were central to 7 Platoon's actions on Waterloo Day, they cannot so simply explain Flynn's willingness to accept a potentially suicidal mission. He was not ordered to advance up the hill precisely because the dangers were so evident. His response was not an automatically ingrained response to a tactical order therefore; it was a considered decision, not just the automatic execution of a drill. As such, it was the deep professional and emotional bonds between the platoon commander (Patrick Bury) and his Section Commander that provided motivation for successful combat performance in extreme circumstances. Corporal Flynn could have refused to go, citing the threat to his men, but this would have ultimately broken the trust between him and his commander that had developed over time; it would have ended the relationship, both professional and personal. It may also have risked social ostracism from the platoon, and with that, professional shame, providing another powerful motivating force. Lastly, and probably least importantly in his decision to go, it could also have risked later disciplinary action, unlikely as this would have been, given the circumstances.

Yet such a decision would have greatly improved Corporal Flynn's chances of survival.[31]

To understand the incident and Corporal Flynn's willingness to lead a charge up a ridge into enemy fire, it is important to place the action in the historical context of existing relations within the platoon. Originally, the platoon commander and Corporal Flynn had had a relatively poor relationship, with a number of 'arguments' during training and deployment.[32] It is interesting to note that the personal distrust between the two had been caused by Flynn's refusal to accept an order on a training exercise prior to deployment. As a result, the commander had both publicly appealed to Flynn's sense of professionalism and threatened him with being relieved of his command in stating that: 'Corporal Flynn! You are a Section Commander in the British Army. I don't care whether you give a fuck or not. Get your men and get up that hill now or I'll get someone else to do it.'[33] This was a professional dispute in which Corporal Flynn's competence had been called into question. Yet for both parties, and the platoon, this was an extremely serious situation since, as *The Combat Soldier* discusses, trust, cooperation, and cohesion in the professional forces is heavily determined by reference to performance. In refusing an order on a training exercise, Flynn's professionalism had been called into question; he had been shamed by his commander in front of his own troops. The author's criticisms suggested that he no longer trusted Flynn and ultimately cast some doubt as to whether he should even be in the platoon, while Flynn, understandably, felt aggrieved that his credibility has been impugned so publicly. His reputation and standing, at the very least in the eyes of his commander, and possibly in the eyes of the platoon as well, had been significantly questioned and potentially irretrievably damaged.

However, as the Sangin deployment progressed, both parties endured adversity and earned each other's professional respect. There was a decisive moment when relations between them changed. After the platoon had nearly taken casualties in a friendly fire incident and the author was debriefing the Section Commanders on the incident, Corporal Flynn, embarrassed that an argument with his commander in front of the troops had undermined them both, extraordinarily expressed his affection for his commander: 'I love you, boss. I'd do anything for you... I'd take a bullet for you.'[34] There is no doubting the sincerity of this statement, but it is worth considering why the corporal felt impelled to make such an extraordinary claim. At one level, it is possible that Flynn was attempting to mitigate the potential damage to his professional reputation which the very public argument between commanders prior to the blue-on-blue incident—and the pre-deployment dispute—had risked. It constituted repair work in Goffman's sense. He was compensating for an accidental and temporary mistake by the profession of the deepest bonds of personal commitment; he was ready to make the ultimate sacrifice for his commander. It would be easy to take this statement as clear evidence of the persistence of those interpersonal bonds of precisely the kind which are central to the classical cohesion literature. Indeed, a number of scholars and commentators adopt precisely this line. In their work on cohesion during the Iraqi invasion, Leonard Wong et al. argue that soldiers were primarily 'fighting for my buddies':[35] 'This is not simply trusting in the competence, training or commitment to the mission of another soldier, but trusting in someone they regarded as a closer than a friend who was motivated to look out for their welfare.'[36] There is no denying the sincerity of Corporal Flynn's profession of love, nor of the intensity of the sentiment, but it is not at all clear that it can be understood in purely personal terms.

It is not difficult to see that the intense situational pressure which was exerted on the platoon in Helmand played a role in heightening Flynn's emotional dependence on his platoon commander. Precisely because the members of the platoon were so professionally dependent upon each other for survival, their emotional attachments intensified. In Helmand, competent combat performance no longer generated the cool professional appreciation recurrently seen in training, but visceral and intense approbation precisely because soldiers knew their lives depended upon it. There, soldiers no longer simply admired proficient colleagues, they necessarily came to love and honour them because they had protected, defended, and sometimes saved them from death and injury. Moreover, the daily threat of death or being maimed heightened group emotional entrainment. Flynn's profession of love was indivisible from his professional respect for and dependence on his platoon commander, heightened by this operational context. He was effectively implying the superior competence of his commander and subordinating himself to it. He was professing his love not only to repair a personal relationship of friendship between himself and the platoon commander (although one certainly developed), but also to rehabilitate his professional reputation with his commander and the platoon.

This profession had very serious repercussions for Flynn at Sarwan Qala. Having stated his professional and emotional commitment to the platoon commander so definitely, Flynn felt obliged to advance up the ridge even though it put him in great personal danger. To have refused to carry out this task would have demonstrated that his profession of love was empty and that in fact the commander's previous doubts about his professionalism may have been justified. Flynn's reputation had already been questioned somewhat by his unprofessional refusal of an order on exercise and, more collectively, in the argument before the friendly fire incident; he could have been catastrophically shamed if he had refused to secure the HLS, even though it was practically a suicidal mission. No one was forcing Flynn to advance up the ridge, but he was acutely aware of both the necessity of the task at hand and the implications of a refusal, given the context of previous relations with his platoon commander. Chapter 5 notes that, in the professional force, 'the sanction for combat failure is ostracism; it is professional rather than actual death'.[37] Flynn may not have articulated his predicament in this way, but there seems little doubt that this is how he understood his actions that day. His refusal would have preserved his own life, and someone else, or another section, would probably have been found to secure the HLS. However, his professional existence in this platoon would have been over. He would have been definitively shamed. By contrast, his undoubtedly brave actions that day cemented his reputation in the platoon, and because of his subsequent gallantry award, in the wider army, permanently and absolutely. He categorically demonstrated that he could be trusted. His professionalism articulated through a profession of love had been vindicated.

In fact, these emotional commitments extended beyond the corporal's relationship to his platoon commander. Chapter 5 addresses the distinctive informality of discipline in a professional force. Typically, in professional volunteer armies, soldiers are not ordered to conduct missions or tasks against their will. For some missions, the option of refusal is given to them; they will not be subjected to military justice if they refuse. Rather, they will be submitted to the tribunal of their fellow soldiers' good opinion; inadequate performers will be threatened with the sanction of shame and ostracism. This tribunal extends well beyond the primary group. If Corporal Flynn had refused the request to advance up the ridge, his refusal would have become common knowledge across the entire battalion. Flynn's reputation could have been shattered in the Regiment and he would have been rejected even by soldiers who did not know him personally. Thus, the powerful sanction of shame exerted not only by his immediate colleagues but also by the Regiment as a whole provided critical motivation for him.

Indeed, this network of shame extended a good deal further than even the Royal Irish. There is another important complementary explanation for Corporal Flynn's performance. In Sangin in 2008, the Royal Irish—and Ranger Company in particular—were especially sensitive to their reputation and therefore to the threat of professional humiliation. The sanction of professional shame operated at a much wider and more impersonal level. Specifically, regimental politics and the awareness of the potential vulnerability of the Royal Irish to future cuts also informed the performance of Ranger Company at Sarwan Qala.[38] Ranger Company was chosen to deploy with the elite 2 PARA Battlegroup because it had proven itself the best company in 1 Royal Irish during a number of tests set by their Commanding Officer, who subsequently made it clear that members of the company were representing his battalion. At Sarwan Qala, Flynn (and the rest of the company) were deeply sensitive to their reputation in the eyes of the paratroopers from 2 PARA, many of whom they did not know nor would ever meet. Accordingly, while the inducement to perform to the best of the company's ability for reasons of group pride is obvious, it also contained a professional and political element. As the only single battalion regiment in the British army, 1 Royal Irish was protected from potential defence cuts as long as it remained in the high-readiness 16 Air Assault Brigade. Proving capable of combat operations alongside paratroopers was seen as a step toward securing that coveted 16 Brigade position to the institutional advantage of 1 Royal Irish.[39] Interestingly, in terms of the Royal Irish's combat performance, 2 PARA's Commanding Officer chose to deploy Ranger Company on his main effort—the population centre of Sangin town—with his own paratroop companies deployed outside in a loose cordon to block insurgents. Such a move perhaps points to an understanding of both the intrinsically aggressive characteristics of the Parachute Regiment's training and ethos and perhaps the higher acumen of Ranger Company in population-centric counter-insurgency, given that many of its members had grown up on the streets of Northern Ireland. The Royal Irish seemed to have successfully convinced 2 PARA of its combat utility. Thus, regimental political motivations also had an impact on the performance of Ranger Company.

The influence of professional institutional politics on cohesion has been rarely discussed; even in the military memoir, Karl Marlantes' *Matterhorn* remains unique in its examination of the relationship between institutional interests and successful performance in combat infantry units.[40] Illustrating the importance of the institutional element on effective unit performance, it is no coincidence that the major who commanded Ranger Company took over the highly competitive command of 1 Royal Irish, and that the Company Sergeant Major was later promoted to Regimental Sergeant Major upon his return. Thus, it is evident that Flynn's actions in Sarwan Qala—and the emotions which motivated them—were not pristine personal sentiments but situational products of the micro-dynamics of the circumstances in Sarwan Qala on that day, his professional relationship to his platoon commander and the rest of his platoon, and the institutional position of the Ranger Company in Helmand (under command of 2 PARA) and in the British Army, with its own distinctive organizational culture. Paradoxically, Corporal Flynn's love for his commander should not be understood merely as private sentiments which personally motivated him to secure the HLS. This love and the motivation it engendered were generated by the specific expectations and interactional processes which operated in his platoon and the wider company and battalion given its position on this operation and in 16 Air Assault Brigade. At each level, however, the concept of professional shame and honour, and the intense emotions cathected by them, operated with great force, driving Corporal Flynn to accept a dangerous task which he might have refused to perform. His love for his commander and his motivations for acting at

Sarwan Qala can be equally understood as collective responses enjoined in a professional force.

Conclusion

This chapter has used a case study of a British infantry platoon's low-level micro-interactions in combat in Afghanistan to examine the impact of drills, emotions, and regimental politics on cohesion in the professional infantry platoon. In deconstructing two different battlefield experiences of 7 Platoon, it has given some clear examples of how cohesion in a professional British infantry platoon is generated from highly situational and dynamic personal and collective interactions. Clearly, 7 Platoon generated cohesion—whether defined as social bonds or effective collective performance—through a complex array of practices, interactions, and identities, which include drills and training, emotional bonds, and political motivations. However, in examining each of these sources of effective combat performance it also becomes clear exactly how complex and integrated the relationship is between each of these components of cohesion in reality. The execution of collective battle drills not only increases effective performance, it also psychologically increases group confidence in the unit's ability to survive. Yet, in certain situations, battle drills may not be the primary indicator of successful performance; social bonds and emotional commitments can be critical to motivating soldiers to act effectively in combat. However, the nature of these social bonds appears to have fundamentally changed since Shils and Janowitz first identified it; they are now underpinned by pervasive and collective concepts of professional pride, shame, and discipline amongst soldiers. Moreover, unit ethos, traditions, and even internal bureaucratic politics also actively encourage professional performance and infuse the interactions within the platoon with added emotional force. The evidence suggests that even these interpersonal emotions are highly influenced by, and are inseparable from, the impact of professional standards and behaviour. As a result, it is possible to argue that both the traditional literature on social cohesion and the more recent literature on performance-based cohesion are closer to one another than previously argued. At the tactical level of the platoon, the reality of effective combat performance suggests that the relationship between professional and emotional motivations for effective performance is highly fluid, and dependent on the situational dynamics of combat and those involved. This evidence directly correlates with the experience of US infantry platoons.[41] Indeed, in the testing zone of combat, cold professionalism is inextricably intertwined with the intense emotions of love, pride, and shame: combat performance relies upon this fusion.

Notes

1. Casualty statistics available at <http://icasualties.org/OEF/Nationality.aspx>. Accessed 25 March 2015.
2. 'UK General warns of Afghan threat', *BBC News*, 10 August 2006, <http://news.bbc.co.uk/2/hi/uk_news/politics/4779321.stm>. Accessed 25 March 2015.
3. *House of Commons Defence Committee Fourth Report—Operations in Afghanistan, Part 2 Operations in Helmand in 2006*, (London: HMSO, July 2011), <http://www.publications.parliament.uk/pa/cm201012/cmselect/cmdfence/554/55402.htm>. Accessed 25 March 2015.
4. For example, the dismounted French parachute infantry ambushed in Uzbin Valley in Kabul province were found to be poorly equipped and badly organized, according to a NATO report. See Graeme Smith, 'French soldiers unprepared for Taliban ambush: report', *The Globe and Mail*, 19 September 2008, <http://v1.theglobeandmail.com/servlet/story/RTGAM.20080919.wparatroopers20/front/Front/Front/>, accessed 27 January 2014, or Jake Tapper, *The Outpost* (Boston, MA: Little, Brown and Company, 2013) which details the breaching of Combat Outpost Keating in Nuristan province, and Bill Roggio, 'Joint al Qaeda and Taliban Force Behind Nuristan Base Attack', *The Long War Journal*, 14 July 2008, <http://www.longwarjournal.org/archives/2008/07/joint_al_qaeda_and_t.php>. Accessed 27 January 2014, for an American perspective on two attacks that breached American platoon positions causing very heavy casualties.
5. Signals intelligence intercept of Taliban commanders' communications, Nad e-Ali district of Helmand, February 2010, on display in Officers' Mess, 1 Royal Irish Regiment, Ternhill, Shropshire.
6. Anthony King, *The Combat Soldier* (Oxford: Oxford University Press, 2013).
7. Patrick Bury was a Platoon Commander in Ranger Company, The Royal Irish Regiment on Operation Herrick 8. This company was attached to the 2nd Battalion, The Parachute Regiment Battlegroup.
8. King, *The Combat Soldier*, 415–17.
9. Patrick Bury, *Callsign Hades* (London: Simon and Schuster, 2010), 147.
10. The corporal's name has been changed to protect identity.
11. Bury, *Callsign Hades*, 153.
12. Bury, *Callsign Hades* 154.
13. Bury, *Callsign Hades*, 155.
14. Bury, *Callsign Hades*, 158.
15. Uzi Ben-Shalom, Zeev Lehrer and Eyal Ben-Ari, 'Cohesion During Military Operations: A Field Study on Combat Units in the Al-Aqsa Intifada', *Armed Forces and Society* 32(1) (2005): 63–79.
16. Randall Collins, *Interaction Ritual Chains* (Princeton, NJ: Princeton University Press, 2004), 42.
17. Randall Collins, *Violence: a Micro-Sociological Theory* (Princeton, NJ: Princeton University Press, 2008), 4.
18. Collins, *Violence*, 19.
19. Ardant Du Picq, *Battle Studies* (Charleston, SC: Bibliobazaar, 2006).

20. Collins, *Violence*, 92–3.
21. Collins, *Violence*, 85.
22. Collins, *Violence*, 94.
23. King, *The Combat Soldier*, 12.
24. King, *The Combat Soldier*, 12.
25. Thomas Scheff, 'Shame and Conformity: The Deference-Emotion System', *American Sociological Review* 53(3) (1988): 395–406.
26. Bury, *Callsign Hades*, 195; Sam Kiley, *Desperate Glory* (London: Bloomsbury, 2009), 133–61, 219.
27. Bury, *Callsign Hades*. See p. 166 for a description of reminding platoon of drills before battle; pp. 175–7 and 179 for drills being implemented in combat.
28. See King, Chapter 5 of this volume.
29. Bury, *Callsign Hades*, 187.
30. Bury, *Callsign Hades*, 187–8.
31. The enemy's fire was so accurate and heavy that only four men crossed the ridge before the attempt was abandoned.
32. Bury, *Callsign Hades*, 60–1, 133.
33. Bury, *Callsign Hades*, 61.
34. Bury, *Callsign Hades*, 136.
35. Leonard Wong, Thomas Kolditz, Raymond Millen, and Terrence Potter, *Why They Fight: Combat Motivation in the Iraq War* (Carlisle, PA: Strategic Studies Institute, 2003), *Why They Fight*, 9.
36. Wong, *Why They Fight*, 10–11.
37. See King, Chapter 5 of this volume.
38. Anthony King, *The Ethos of the Royal Marines: The Precise Application of Will*, unpublished report commissioned by Commander, Commando Training Centre, 2004; Anthony King, *The Ethos of 16 Air Assault Brigade*, unpublished papers prepared for the commanders of both units.
39. Bury *Callsign Hades*, 294. In fact, the command team were told by the 2 PARA CO at the end of the tour that they were his best company. The Royal Irish subsequently enjoyed a two-year extension to their rotation in 16 Brigade.
40. Karl Marlantes, *Matterhorn* (London: Corvus, 2010).
41. See Elizabeth Rubin, 'Battle Company Is Out There', *The New York Times Magazine*, 24 February, 2008; and Dan Bellavia, *House to House* (London: Simon and Schuster, 2007), 227–56 for many examples of the complex relationship between different motivating factors, training, and successful combat performance.

10

Hierarchy, Status, and Combat Motivation in the French Foreign Legion

Mikaela Sundberg

Introduction

The French Foreign Legion (Légion étrangère) is an elite force within the regular French army, attracting men from all over the world who sign a five-year contract of unlimited liability.[1] The Legion is also a popular cultural icon, portrayed in movies and books, novels, and biographies, but apart from a handful of French studies, the Legion remains sociologically unexplored.[2] Yet this unique force raises a number of interesting issues for long-standing discussions on the role of primary groups for effective military performance, starting with Janowitz and Shils' famous article on the German forces in the Second World War. Janowitz and Shils maintained that the performance of the Wehrmacht was dependent upon close personal bonds between small groups of soldiers. In particular, German soldiers relied upon a common conception of masculinity to encourage solidarity at the front. Janowitz and Shils systematically downplayed the influence of political and ethnic motivation among German troops, an oversight of which scholars like Omer Bartov, Simon Wessely, David Segal, and Stephen Fritz have been critical. For these scholars, Nazi political ideology was critical in forging and motivating the force. Yet, despite their overt differences, Janowitz and Shils and Bartov et al. share a central presumption; the soldiers of the Wehrmacht were bound into tight primary groups united by a dense shared culture consisting of common masculinity, ethnicity, and political orientation. For all these scholars, combat performance relies on cultural homogeneity. Despite its reputation as a fighting force, the Legion, drawing its troops from numerous countries, has never relied upon prior cultural homogeneity to engender primary group solidarity.

Consequently, it represents an interesting, even unique, sociological case study of how an organization might enjoin cohesion when wider cultural bonds, typically invoked in the literature, are absent.

Precisely because of its unusual constitution, the analysis of the Legion forwarded here is intended to contribute to current debates about cohesion. In the past twenty years, scholars have increasingly questioned the necessary connection between these dense bonds in the primary group and subsequent combat performance. For example, King uses the example of a British Commando Manoeuvre Support Group during the Iraq Invasion of 2003 to show how their collective combat performance did not suffer from the existence of 'mutual suspicion and hostility within the group';[3] relations between members of the unit were some way short of those idealized in the classical literature. Given the exceptional linguistic, ethnic, and cultural heterogeneity within the Legion, these types of bonds appear even less likely to develop here. In fact, one may wonder how these soldiers from so many different countries become motivated to fight together at all.

This chapter draws on organizational sociology to discuss this issue in relation to power in the Legion as organization and institution respectively. In sociological writing, organizations and institution are defined in multiple ways. I use Ahrne and Brunsson's definitions where the principal difference between an organization and an institution is that an organization is an order based on explicit decisions, whereas institutions are taken-for-granted, imagined orders.[4] Organizations are formal and explicit; institutions informal and implicit social orders. In an organization, hierarchical decision-making determines who gets to decide what others do, whereas 'power' in an institution often operates through more personal and contingent forms of status, connected to conceptions of legitimate authority based on particular qualities. This chapter explores the way in which formal organizational and informal institutional power or orders and status operate in the Legion. It begins with a historical background to the Legion, including its present arrangement and composition. I proceed with a brief methodological note before examining hierarchy in the Legion, including presenting the formal rank system and the conception of orders. I then turn to the question of status, what it is based upon, and how it relates to prestige of different Legion regiments, but also discuss how status clashes with hierarchy. The chapter ends with a concluding remark regarding the hierarchical culture of the Legion.

Historical Background

Throughout history it has been common for polities to use foreign troops in their armies, but a gradual 'nationalization' of the European armed forces is evident from the mid-fifteenth century.[5] Universal conscription in France dates from the end of the eighteenth century, based on the conviction that voluntary conscription alone could not provide France with the numbers of troops required and some form of compulsion was therefore needed.[6] During the nineteenth and twentieth centuries, conscription became an important ideological tool, evoking an obligation that the nation could demand of its citizens.[7] In the last decades of the twentieth century this process began to be reversed and, as part of this general trend, France decided to abolish universal conscription in 1996 with effect from 2000. The Legion, which had always been an all-volunteer professional force, was unaffected by these developments.

When the Foreign Legion was installed in 1831, it institutionalized the previous tradition of having foreigners serving as French troops. In particular, King Louis Philippe founded the Legion to handle the unemployment crisis by 'usefully employing' the mass of refugees on French territory.[8] Regiments were, however, situated in Algeria until Algeria's independence from France in 1962. At present, the Legion consists of ten regiments and a small recruitment regiment in Paris. The *1er régiment étranger* (1 RE) in Aubagne deals with central administration, recruitment, security, and hosts the *Commandement de la Légion étrangère*. Basic training, as well as many special courses, takes place at the *4ème régiment étranger* (4 RE) in Castelnaudary. There is one cavalry regiment, two engineering regiments, and one parachute-infantry regiment, *2ème regiment étranger de parachutists* (2 REP), stationed on Corsica. One infantry regiment is located in Nîmes in metropolitan France, with a smaller one situated in French Guyana. There is also a detachment in Mayotte and one regiment in Abu Dhabi, United Emirates.

Importantly, the Legion is not a separate force but an integral part of the French Army, currently representing about 6 per cent of its total size, and therefore commanded by French officers, who have graduated from the military academy *École Spéciale Militaire de Saint-Cyr*. It has traditionally been considered a great honour for officers to serve in the Legion, granted only to the best graduates. At present, the Legion consists of about 5545 legionnaires and 1741 noncommissioned officers (NCOs), commanded by 413 officers (2013). The so-called *officiers de titre étranger* (OTE) is a special category of officers, usually recruited among NCOs with at least fifteen years of service through a special nomination, or after they have taken an exam to become officers. There are never more than about forty OTEs in the Legion at a time, and, whereas regular officers might be deployed anywhere in the French Army, OTEs were, until recently, guaranteed to stay in the Legion.

In terms of ethnic diversity, the Legion is not only unique among contemporary armed forces, but its heterogeneity is probably greater than ever considering that over 140 nationalities are represented in it. Using the Legion's own classification, the national/ethnic proportions are the following: 10 per cent France, 14 per cent Western Europe and North America, 24 per cent Central Europe and Balkan, 18 per cent Slavic world, 11 per cent Asia, 4 per cent Arab world, 9 per cent sub-Saharan Africa, and 10 per cent Latin America (2010). This composition of nationalities is strikingly different from earlier periods. German speakers represented about 30 per cent during the first thirty-five years of the Legion, but peaked just before 1930 to around the 1950s, when about 50 to 60 per cent of the legionnaires were German.[9] There has been a steady decline of Germans since the 1960s, which is the first decade during which non-Europeans began to enrol, at the same time as Anglophones became a notable linguistic group.[10] Anglophone representation peaked in the early 1980s, when they represented about 20 per cent of the legionnaire population. This is commonly regarded as an effect of British soldiers enrolling after having fought in the Falklands War. It is also an example of how the Legion's recruitment mirrors international politics. Currently, 43 per cent of new recruits are Slavs or from the Balkans, commonly assumed to derive from high unemployment rates in these countries and a surplus of military staff in former Soviet Union states. These figures indicate that structural factors have explanatory value regarding at least where legionnaires come from, against the myth of the Legion as being a hide-out for criminals.[11]

Method

This chapter is based on material collected through an ethnographic study of everyday life in the Legion, primarily at regiments, where most interviews and observations have been carried out. I have interviewed eleven officers, three of which are OTE, thirteen NCOs, twenty-seven enlisted, and nine former legionnaires.[12] Most interviews have been conducted in French, but a few in English and Swedish. Thirty-seven interviews were recorded; field notes were made for the rest.

I have conducted observations and participant observations at the education regiment (4 RE), the cavalry regiment (*1er régiment étranger de cavalerie,* 1 REC), the parachute regiment (2 REP), and the main administrative regiment (1 RE).[13] At all regiments except for the last, each visit lasted for about one week, and I visited one of the combat regiments twice. In addition to conducting interviews, I observed activities such as control of guard duties, office work, shooting exercises, language classes etc., and on a few occasions, participated during the regular morning exercise. My ten shorter visits to the main administrative regiment were mostly related to meetings regarding the other regimental visits and interviews, but also twice involved participation in the major annual celebration of the Battle of Camerone, during which regiments are open to the public for two days. At every regiment, an important part of participant observation took place during the regular two-hour lunch breaks and after-work beer at company clubs. These occasions involved many informal conversations with officers and legionnaires of various nationalities and formal ranks. I have also spent time with a few legionnaires on their leisure time outside the regiments. Many of the presented quotes have been translated from French or Swedish into English. Their length is partly intended to offer the reader some more of the flavour of the environment. In preparing for field work, I also studied a number of autobiographies by former legionnaires. Some of these are cited as supporting evidence as well.

Hierarchy

Janowitz and Shils emphasized the paternal role which officers and noncommissioned officers played in the Wehrmacht, fulfilling the personality needs of the soldiers. In this way, they fostered the personal interdependences of the primary group. Given the strict discipline of the Wehrmacht, Janowitz and Shils perhaps underemphasized the formal authority which German officers exercised over their troops. In the Legion, formal hierarchy and discipline is central to the organization and has to feature as a major factor in any explanation of its cohesion and combat performance. The formal rank system is the basis for hierarchy.[14] The enlisted are divided into legionnaire, corporal, and senior corporal (*caporal chef*). One becomes a legionnaire after having received the famous white cap, the *képi blanc*, during a ceremony about five weeks into basic training. Legionnaires can be promoted to corporal after about three years and to senior corporal after seven or eight years of service. It usually takes between three and at most six years before a corporal can be promoted to sergeant, the lowest noncommissioned officer rank. This involves significant differences, beyond the rise in formal rank per se. During the first five years of service, enlisted legionnaires live in barracks at the regiment, sharing rooms with one to (more often) five others and they are, for example, not allowed to buy a car. These limitations cease on becoming a NCO, or with more than five years of service, and NCOs live in separate buildings with one-room apartments just beside the regiment, or organize their own housing.

The officer ranks of importance at the company level are lieutenant and captain. Each company is commanded by a captain, who is rotated every second year. As part of *cadre*, NCOs interact with the captain in planning the activities of the platoons and during the morning coffee and after-work beer in the company club. Enlisted personnel see the commanding captain every day at assemblies, but have very little personal interaction with him. In terms of interaction, the lives of enlisted personnel revolve around people on the platoon level. An officer-enlisted gap is common within the armed forces, primarily due to a class-based demarcation line between the officers, of noble and professional backgrounds, and the enlisted, from lower socio-economic classes.[15] In the Legion, most officers and legionnaires are divided by culture, language, and not least nationality.[16] Most importantly, officers are not legionnaires, and this boundary between legionnaires and officers is expressed explicitly in many different ways, for example by emphasizing that they have never worn the *képi blanc*.[17] The expression *'on aime pas les officiers'* ('one doesn't like the officers') is heard occasionally, and there are even expressions such as *'fils d'officer'* (son of an officer) used in a derogative manner. This does not mean that every officer is personally hated—there are many examples of the opposite—but officers are *by definition disliked as officers by legionnaires as legionnaires*.

Orders: Who's Right

The hierarchal order of decision-making determines who decides what others do in an organization.[18] An *order* is the communicative directive from a superior, telling a subordinate exactly what to do. While the notion of an order is a core aspect in every military hierarchy, the conception and practice of an order are not the same,[19] especially if we include considerations of *how* orders are expected to be given and obeyed on the interactional level. For example, King reports that in the British Royal Marines, there used to be more distance between officers and enlisted in the 1970s and 1980s, whereas officers nowadays are expected to take advice from enlisted marines and NCOs.[20] Another example is the Italian Alpine units *Alpini*, where it is claimed that it is 'important to talk and not give orders'.[21]

It is quite otherwise in the Legion. The Legion is one of the most rigid and authoritarian Western military organizations. Since the Legion has been completely professional for most of its history, this is a remarkable and counter-intuitive fact. In Chapter 5 in this volume, Anthony King suggests that disciplinary systems operate differentially in citizen and professional forces. Precisely because professional soldiers are motivated by their career and have been subjected to intense regimes of training, traditional military discipline tends to be relatively light in most all-volunteer Western forces today. In particular, convictions for cowardice have all but disappeared. In the light of this general move among professional forces away from martial discipline, the Legion's authoritarianism might be seen as surprising, but it is important to recognize the organizational role which its adherence to hierarchy and its investment in the order serves. The utter priority of the order in the Legion is central to the generation and maintenance of cohesion and, ultimately, to its combat performance.

Central to the Legion's culture is the sanctity of the order; even if it is wrong. In the words of one former senior NCO: 'In the Legion, you say the officers command, the non-commissioned officers give the orders and the legionnaires execute.' Orders in the Legion are given in French. Only about 20 per cent of the legionnaires are Francophone when they join the Legion.[22] Consequently, learning French, not least military terms, is for most legionnaires a crucial part of the four-month-long basic training.[23] In addition, each Francophone is also responsible for helping one or, in more recent years, two non-Francophones to learn French. In passing, several officers have mentioned that the decreasing rate of Francophones is problematic. There is an explicit organizational effort to separate legionnaires with the same native language throughout different platoons to facilitate French learning, but also to avoid the creation of *mafias*; network-like groups of legionnaires based on nationality and language, which could potentially cause problems since they provide a conflicting base for loyalty.

Not surprisingly, a sergeant at 4 RE said it is easier for the Francophones at the beginning of training because they understand orders, but he also stated that they make more *bananes* and *conneries*—mistakes and stupidities—than the others. To exemplify this, he said that when the French are told to pack a list of things, they pack a bit more; when they are supposed to clean something, they clean a bit more. Thus, trying to do things better, or more than they are told to, generates 'mistakes'. In fact, orders should always be executed precisely. When I asked one OTE-captain (2 REP) about making mistakes due to difficulties in understanding French, he answered:

> During the first year here, there's lots of imitation and if the guy one imitates takes the wrong path, one could take the wrong path too, you understand? But well, in general, no. You imitate. You understand what happens. (. . .) So it's like 'you take that and you put it here'. 'The beret, you take it like this, put on like this'. It's a lot of imitation, lots of... know-how. After that, it is also up to the person himself to act in such a way as to try to understand what happens. But for me the legionnaire, he's a man who doesn't reflect, that's the force we are talking about, you see? Tomorrow, if the commanding officer says 'Faros, tomorrow the company will assemble at this place'. Captain Faros says, 'Good, my commander'. I don't reflect. I can't say, 'Yes, but well, tomorrow 10 o'clock... ' No. If the commander says 8 o'clock, everyone is there [at 8 o'clock], because the commander said so. You understand? You don't ask questions. Why? How? Why? As captain I *could* ask the commander a few questions, but it's rare. I should listen until the end, without any state of mind. Because if I had a state of mind in my head, I start to think and that means I'm on the way to disobeying already.

This is an illuminating statement. First, the captain (Faros) claims that in the beginning, following orders are not so much about understanding the verbal statements as imitating what the others are doing. Interestingly, he suggests that language is not very important for orders and for learning how to act more generally. Second, he claims that the legionnaire's supposedly lack of reflection is a strength. At least among superiors in the Legion, it is commonly stated that legionnaires are particularly obedient as soldiers and that this is one of the Legion's strengths compared with *la regulière*, the regular French Army. This is a source of pride, in line with appreciation of obedience in the military in general and during combat in particular. Finally, the quote indicates a difference among ranks in terms of how acceptable it is to ask questions for clarifications. It might be acceptable for a captain, but he should still listen without 'starting to think'; a process regarded as a step towards disobedience in and of itself. In this sense the understanding of obedience goes beyond visible, overt actions (e.g. hesitation, questions) into the very thought process. We could perhaps speak of a notion of all-encompassing obedience. However, this interpretation raises the questions to what extent this conception of, or perhaps rhetoric of, obedience is related to the high rank of the informant as captain, albeit OTE. It should also be considered in relation to the fact that orders from commander to the captain are not as precise as from, for example, a sergeant to legionnaire. Regarding the execution of orders, the enlisted are certainly aware that they are not supposed to reflect or 'talk back'. For example, one corporal (4 RE) tried to explain to me what the Legion is like: 'You are not allowed to question and if a superior says 1+1 is 3, this is the way it is.' He also said ironically that one of his 'favourite' expressions in the Legion is *arrête de te réfléchir*—stop thinking—said when orders are not executed immediately or as a response to questions. Although obedience is required in extremis, these examples indicate that an extreme order culture, where the superior is always 'right', may have serious shortcomings, especially when the superior is obviously 'wrong'.

What is more, in the Legion, the superior is always right, when the superior's word is against that of a subordinate. A story from a corporal (1 REC) reveals this very clearly. This corporal was considered a few minutes late to a group assembly and asked to do push-ups as a punishment, but refused as he claimed that the superior had changed his mind about what time to be there. When the incident was discussed with the commanding captain, the captain said he believed that the corporal (a legionnaire at the time) was right, but that the 'the system has to win'; consequently, he had to trust first the lieutenant, then the NCO, then the corporal, and, lastly, the legionnaire. A different corporal (1 REC) gave an example of another case of conflict involving a legionnaire and a lieutenant, where the latter 'tries to put all blame on the legionnaire that it's his fault' by repeatedly revising the legionnaire's report on what happened so it would not appear as his fault. According to the corporal, 'Everyone knows it's [the lieutenant's] fault but he's an officer so like what can you do? I mean... you never win that case.... Everyone knows the system is like that but there's nothing we can do about it, so it's always like that.' This quote is a concrete example of the perspective that according to this system, an officer is always right.

Importantly, the implications of hierarchy in the lives of legionnaires extend further than simply following individual orders. It means being a pawn in the organizational game all the time—on an hour-to-hour and day-to-day basis, as well as on a monthly and yearly basis. The key notion is *disponibilité*, availability.[24] Besides regular morning exercise and in spite of general orders (*note de service*) being presented the day before, the individual does generally not know for sure what he will be asked to do in the afternoon, or the next morning, and definitely not the day after tomorrow.[25] Vacant positions are filled through assignment, offering very little room for negotiation, and when it comes to courses, special duties such as *Vigipirate* (public security work) or long-term positions, a superior sometimes asks for the opinion of the subordinate, but this does not mean that there is a choice. To give a typical example from my study, one corporal (2 REP) was about to go with a part of his company to do *Vigipirate* in Paris. He did not want to go, but was selected. Someone just as available was volunteering, but was not selected. 'It's like that' ('*C'est comme ça*'), the corporal wrapped up, shrugging, as if this was common, accepted practice. As Ewan McGorman puts it in his autobiography when writing that recently arrived legionnaires at the parachute regiment are asked which company they would like to be sent to: 'They won't probably be considerate about your choices... I'm not even sure why they bother asking.'[26] This appears as typical *practice* with regard to many decisions about legionnaires, but also a typical *perspective* among legionnaires.

The authoritarian culture even informs the apparently voluntary process of signing on and extending service with the Legion. NCOs and officers frequently refer to the fact that the Legion is a voluntary organization when talking about legionnaires who complain, invoking the first commandment of the *Code d'honneur du légionnaire: Légionnaire tu es un volontaire servant la France avec honneur et fidélité* (Legionnaire, you are a volunteer serving France with honour and fidelity). Concretely, terms of service are negotiated voluntarily through signing a contract. After the initial five-year contract, soldiers can agree to serve a further one to seven years, though recently contractual extensions have tended to be for shorter periods of time (one to three years). (This may go on; there are members with up to about thirty years of service.) On the surface, this appears as a moment where the power relation switches to the benefit of the legionnaire, as there is a possibility to negotiate one's service by threatening not to sign a contract. Yet even here the hierarchy operates.

Promotion to NCO or participation in some special courses requires signing a new contact for three or four years of service, in addition to those years that are left of the current contract. This is legitimate practice and systematically used.

The attempt to try and force soldiers to stay in the Legion reflects a more fundamental problem the Legion confronts which could be related to the Legion's authoritarian culture. One reason for using the contract system to hold onto soldiers is that the Legion in fact struggles to keep its personnel. Most legionnaires do not sign a new contract after having completed their first five years and the Legion has about 300 deserters per year. One hundred and fifty of them have spent less than a year in the Legion, corresponding to almost 19 per cent of new recruits per year. 'We have a real battle in trying to keep our personnel as long time as possible', as one officer in charge of human resource management at one regiment put it.

In the treatise *Exit Voice and Loyalty*, Albert Hirschmann discusses under what organizational conditions exit prevails over 'voice', and vice versa.[27] In the Legion it is clear that exit—by not signing a new contract or deserting—is the dominant reaction mode to dissatisfaction, partly due to the small prospects for the effective use of voice. Even if 'voice' would be feasible in principle, language difficulties diminish possibilities in practice. For example, the French one learns through daily routines at a Legion regiment is not the French required to grasp the clauses of a contract. Members usually remain loyal to, as in staying in, organizations because of interest in the resources that an organization provides, but these may be of different kinds.[28] Besides the obvious role of material resources such as a salary, there may also be symbolic rewards such as status and respect.

The authoritarian culture of the Legion is a peculiar phenomenon in Western professional forces, which are typically open to degree of discussion and debate. Yet orders serve a crucial purpose in the Legion. Precisely because of the diverse recruitment of the Legion, most of whose soldiers are not even Francophone, it is impossible to rely on the shared cultural backgrounds of soldiers to generate cohesion which has featured so heavily in sociological explanations of the armed forces from Janowitz and Shils onwards. In fact, several legionnaires also mention that the Legion needs discipline, precisely because members have such varied backgrounds. The only way in which an ethnically and linguistically diverse organization is able to unite its troops and demand unlimited liability from them is through unquestioned hierarchy and discipline in which the system and the order are sacred. Certainly, there may be times when a particular order is wrong, negligent, or absurd, but to question one order is to weaken the single centripetal force which holds this organization together.

Yet organizational costs are to be born here. Retention rates are low and desertion rates are high. The Legion cannot hold onto its troops, and precisely because they lose so many troops each year and rely on sustaining manpower through intakes of new recruits from diverse cultural and ethnic backgrounds, it is imperative that the Legion's strict hierarchy is sustained. The authoritarianism of the Legion is self-referential, therefore. There is little doubt that the extreme discipline of the Legion generates high levels of instrumental cohesion and a famous robustness. Yet it may also limit the level of combat performance, because it seems to impede the development of a cadre of experienced junior officers and NCOs who can apply their knowledge in creative ways to new problems.

Status as Adjustment to Hierarchy

The Legion is a hierarchical organization. Indeed, it is perhaps the most hierarchical military organization among professional Western forces. Yet it would be wrong to ignore other institutional factors which operate within the organization to influence relations between the legionnaires and who can be ordered to do what. Status, as a form or informal source of power, plays a critical role here. Considering the strength of hierarchy in the Legion as an organization, what is the significance of status within the institution? And what is the basis of it? The two most important factors in the Legion are length of service and experience of military operations (*missions*) and we will primarily discuss them in relation to the tensions they pose relative to hierarchy.

The position of so-called *fut-fut* corporal illustrates such a tension between formal organizational rank and authority and status. The *fut-fut* corporal is a type of corporal who has bypassed normal promotion procedures, having been specially selected immediately after basic training. After their four months of basic training at 4 RE, they receive special training for four additional months, before going to the two-month corporal school. Usually, legionnaires have to serve two or three years before being sent there. The special *fut-fut* corporal then spends another eight months as a training corporal for volunteer recruits, before finally being assigned to a different regiment. I met one of these *fut-fut* corporals, who was rather disappointed because, according to him, corporals like him are not appreciated at the regiments and when they realize that after having already started their courses, some ask to leave, or desert.[29] In principle, the benefit of becoming a *fut-fut* corporal is the freedom to choose a combat regiment after qualification, as opposed to being only able to express a preference. Yet the *fut-fut* corporal suggested that one 'can't' choose the parachute regiment, 2 REP, because it is too 'hard'; the other corporals there would harass the *fut-fut* corporal who had earned his rank in what is deemed the wrong way. In spite of his rank, they would force him to clean, a demeaning duty reserved for junior soldiers in the Legion. Indeed, one legionnaire (2 REP) said it was not 'too good' to be the youngest in the section for ten months as 'You have to go by *matricule*, the army number, from youngest to oldest, so when there's work to do the youngest get the worst job and so on, and then the oldest, they don't do anything.' Effectively a *fut-fut* corporal in 2 REP would be reduced to the status of a new recruit into the regiment. 'What would the legionnaires say then?' the previously cited *fut-fut* corporal asked rhetorically, indicating that the legionnaires would not respect him, since he would have to do a job which is 'below' his rank. In this way, the indignities forced upon the *fut-fut* corporal exposed the gap between formal hierarchy and status; he was in the anomalous position of having rank but no status.

Considering the strength of hierarchy discussed above, it is not evident to what extent the clash of status and hierarchal position with regard to the so-called *fut-fut* corporal is noticeable in relation to both peer-corporals and subordinate legionnaires. Importantly, at some point after the end of the working day, NCOs and officers leave the regiment, whereas most enlisted men have to stay. The total institution-like conditions create the possibility for informal status orders to arise.[30] These are likely to disturb the formal rank hierarchy among legionnaires and corporals rather than, for example, among senior corporals and NCOs. Thus, the consequences of clashes between status and hierarchy are potentially smaller further up the ranks, when hierarchy supersedes any status. One senior NCO described how he had a tough time as a young corporal. Since he was in a company with few legionnaires, he had to do the cleaning despite his rank. However, by chance, he was sent to sergeant school (at 4 RE) shortly after this period, thus advancing much faster than customary. He also rapidly earned a higher rank in relation to his former superiors, the senior corporals. When I asked if it was not difficult to command when he returned, he simply said 'They had no choice.' He added that 'of course, they could turn a bit obstinate' if given 'a bad order', but since disobedience could lead to sanctions, this was unlikely to happen. Thus, the NCO implicitly admits that his orders could be 'bad', but that he expected (silent) obedience anyway.

Another, military basis for status is operational experience. The informal appreciation of people with these experiences is manifest in various instances and forms.[31] For example, during one ceremony (1 REC), the commanding officer of the regiment gave a short speech of thanks to two senior NCOs who had more than fifteen years of service and who were leaving the regiment. The president for the NCO described some of their careers as legionnaires, including the regiments in which they had served and the military missions in which they had participated. The sergeant standing next to me was impressed by one of the presentations as he nodded slowly, commenting: 'Eight, nine missions, that's not bad', adding that one usually has (only) two or three.

In line with the suggestion that missions are the basis for higher status, it follows that a lack of operational experience is correlated with lower status. For example, one legionnaire (2 REP) talked about how good a fellow corporal in his platoon was, repeating that the corporal had experienced a mission (to Afghanistan), in contrast with himself: 'I am medic, but I have not done a mission or something, there's less respect.' The quote also indicates the relevance of specialty in relation to status. In line with the assertion that experience of missions is the key to higher status, distance from combat is related to the status of a specialty.[32] The secretary is an emblematic case. On the one hand, one NCO (2 REI) recounted his career and, by emphasizing, 'I wanted to go to war!', he expressed how dissatisfied he was when being assigned an administrative position to start with. On the other hand, when a senior corporal (1 RE) told me he was a secretary, he immediately added that he liked it, as if he had to defend his position. This is in line with common knowledge that 'everyone' prefers combat companies to support companies, but this probably also reflects an ideology regarding what everyone is *supposed* to prefer.

It is interesting to consider the authority of officers in relation to status. Formally, all officers, no matter how junior, have primacy; their orders must be obeyed. As noted above, captains are influential but distant, whereas lieutenants are slightly more present in the everyday life of the platoons, collaborating closely with NCOs. The following quote from a legionnaire (2 REP) is typical in describing lieutenants as ignorant and lacking experience: 'You get a completely new lieutenant that comes from a French school somewhere, who becomes boss, he doesn't know *anything* in the beginning.' Thus, a junior lieutenant typically only has authority, not status. It is therefore expected that lieutenants should learn from NCOs, who have longer experience of life in the Legion.[33] Their status trumps the officer's rank. For example, one former legionnaire compared a platoon sergeant with a lieutenant he had during his time serving at 2 REP. The sergeant had jumped into combat with the Legion, had been in the free fall platoon; 'everybody looked up to him' as a 'God', whereas the lieutenant, a fresh graduate from a military academy, was a 'wimp'. The recognition of status is important. It qualifies and moderates the extreme hierarchy which undoubtedly operates in the Legion and which demands that plainly incorrect orders issued by inadequate officers are followed. Because of the status order, there is some negotiation. Yet, even here, the latitude for flexibility is very narrow. A high-status sergeant might, indeed, be able to exert influence, but he himself is part of the hierarchy and holds significant rank. He is part of the military order. The status hierarchy involves no radical displacement of the hierarchy.

The Prestige of 2 REP

In line with the high status of missions, working at the training regiment (4 RE) and administrative headquarter (1 RE) is considered low status. As non-combat regiments, they are usually excluded from overseas operations. This is also reflected in prejudice about the members who work at these regiments. For example, 1 RE is conceived of as full of 'bureaucrats' and a 'deposit' for those injured who cannot participate in combat. Without a doubt, the parachute regiment (2 REP) on Corsica is considered the most elite among all Legion regiments; and the 'Repmen' as the best legionnaires. This prestige is not only manifested by legionnaires, who serve or have served at this regiment, consistently confirming that 'it's the best', after having professed their affection for it, but also by instructors at the education regiment (4 RE) who will ask high-performing volunteer recruits if they want to go to 2 REP. The prestige of 2 REP can also be related to the status marker discussed above. For example, 2 REP is part of a particular group of French regiments concerned by '*Alerte Guépard*'—a constant state of alert for overseas deployment—and companies prepared for said alert rotate within the regiments. For this reason, legionnaires at 2 REP are more conditioned than other legionnaires to head off overseas, at a moment's notice.

Let us connect the discussion of the status of 2 REP with the earlier discussion on hierarchy and obedience. In general, rumours about 'Repmen' indicate that they are in better physical shape and more disciplined (at work) than legionnaires at other regiments, but interestingly they are also obedient to the point of becoming '*gâteux*'. The literal meaning of the word is senile, but in the Legion it is an expression that signifies conditioning to obey to the point that one can no longer think for oneself. For example, a corporal at a different regiment said legionnaires from 2 REP distinguish themselves when sent to courses with other legionnaires by their dedication to orders. He gave the apocryphal example of a soldier from 2 REP, who had been jokingly told by a superior to jump out of a window, and who did so. 2 REP is known as the strictest of all regiments with regard to discipline and hierarchy and, among NCOs and officers within the Legion, it is a common view that the Legion works well because of its discipline and emphasis on obedience and this is especially the case at 2 REP. For example, one OTE at 2 REP emphasized the importance of obedience, especially during combat, and that this was the reason behind why the Legion was 'best' in Afghanistan, whereas soldiers in the regular French Army did 'a bit as they wanted to'. Another captain at 2 REP also claimed that the companies from 2 REP deployed in Afghanistan did very well, and proudly recounted how US General McChrystal visited the field himself to see how 2 REP implemented the restrained method he advocated, notably by only shooting on identified targets. Indeed, an article in *Le Figaro* recorded how the 'perfect fire discipline' and 'the solid training of the NCOs' when 2 REP successful cleared the Tagab Valley.[34]

However, the organizational requirement for extreme adherence to orders also involves significant disadvantages. On operations, the narrow adherence to orders which characterizes the Legion may also be problematic. The shortcomings of such a strictly imposed hierarchy become clearly visible in Pascal Vennesson's analysis in Chapter 11 of this volume of how members of the French Army obeyed an illegal order in the Ivory Coast. Some years ago, an illegal response to perceived disobedience at 2 REP shared certain similarities with the event Vennesson's describes. During an exercise, a sergeant shot a legionnaire in the genitals because he interpreted his subordinate's disrespect for the firing directives as disobedience, whereas the subordinate claimed that he had not understood the directives correctly.[35] According to articles in Corsican newspapers, the superior was sanctioned with forty days' regimental imprisonment, but the Legion did not want the story to come out. Instructively, the legionnaire was himself dismissed for having complained about his superior.[36] The occurrence of such responses to perceived disobedience during operations remains unknown, but it certainly leads to doubts about the exclusive benefits of the hierarchical culture analysed above.

Conclusion

This chapter has examined cohesion in the French Foreign Legion, focusing on everyday life at the regiments. The Legion is particularly useful in contemporary discussions of cohesion because it is anomalous in a number of ways. It is a professional force which nevertheless relies primarily on discipline and hierarchy in order to generate cohesion and ultimately combat performance precisely because its recruitment is so diverse. It is, therefore, quite unlike other professional forces. The emphasis on unthinking obedience has some obvious organizational advantages for the Legion; indeed, it is widely admired in military and civilian circles as a robust, aggressive, and trustworthy force. At the same time this dedication to orders has also potentially rendered it less flexible than some of its less authoritarian peers; this may be a disadvantage in an era of complex hybrid operations. The question of whether the organizational culture of the Legion will evolve more towards the type of professionalism among other Western armed forces in the contemporary era is unknown.

Notes

1. This chapter contains excerpts from Mikaela Sundberg, *A Sociology of the Total Organization: Atomistic Unity in the French Foreign Legion* (Farnham: Ashgate/ Gower, 2015).
2. See Marie Larroumet, *Mythe et images de la Légion étrangère* (Paris: L'Harmattan, 2004); Ana Pouvreau, *Le système Légion: Un modèle d'intégration des jeunes étrangers* (Sceaux: L'esprit du livre editions, 2008); Mélanie Téxier, 'Identités et Langues à la Légion étrangère', *Res Militaris* 1(3) (2011).
3. Anthony King, 'The Word of Command: Communication and Cohesion in the Military', *Armed Forces & Society* 32 (2006): 494–5.
4. See Göran Ahrne and Nils Brunsson, 'Organization Outside Organizations: The Significance of Partial Organization', *Organization* 18 (2011): 83–104. Organizations and institutions can be defined according to basis for affiliation, power, expectations, control/knowledge, and consequences. In this chapter I only explore the power dimension.
5. See Lars Mjøset and Stephen Van Holde, 'Killing for the State, Dying for the Nation: An Introductory Essay on the Life Cycle of Conscription into Europe's Armed Forces', in Lars Mjøset and Stephen van Holde, *The Comparative Study of Conscription in the Armed Forces. Comparative Social Research,* vol. 20 (Emerald Group 2002, Greenwich, CT: JAI), 10. Distinguishing 'foreigners' from natives requires some classification system and since single-nationality armed forces are preconditioned upon the formation of nation-states, it follows that they are relatively new.
6. Alan Forrest, 'Conscription as Ideology: Revolutionary France and the Nation in Arms', in Mjøset and van Holde, *The Comparative Study of Conscription in the Armed Forces*, 96.
7. Forrest, 'Conscription as Ideology', 97.
8. Jean Hallo, *Monsieur Légionnaire* (Panazol: Lavazuelle, 2007), 17–18.
9. Hallo, *Monsieur Légionnaire,* 152.
10. Hallo, *Monsieur Légionnaire,* 152–3.
11. According to official presentations of the present situation, the profiles of candidates when they join indicate that they come from 'complicated situations' and want to 'get out', but it is difficult to know what constitutes these 'complicated situations' that the candidate wants to escape; it certainly cannot be assumed to be criminality as opposed to mass unemployment, for instance. Historically, foreign volunteers have volunteered for other armies to escape from deteriorating living conditions, to seek adventure, or to aspire to a professional military career (see e.g. Lars Mjøset and Stephen Van Holde, 'Killing for the State', 15–16.). It is likely that these factors account for many legionnaires too. If so, these various interests and subjective preferences do of course not simplify the task to make them fight together.
12. The nationalities of the interviewees, including OTEs and veterans *anciens* but excluding the eight regular officers, are the following: Albanian, American, Austrian, Brazilian, Belorussian, Canadian, Chinese, Croatian, English, Finnish,

French (14), Hungarian, Icelandic, Indian, Italian (2), Latvian (2), Nepalese, Portuguese (2), Romanian (2), Russian, Swedish (9), Swiss, and Ukrainian (2). The length of interviews varies greatly. Most of the interviews at regiments lasted for about thirty to sixty minutes. A few interviews with officers and NCOs lasted around one-and-a-half hours. Some interview guides were especially tailored in relation to the organizational position of the interviewee (e.g. human resource officer, chancellor), but most of them were relatively open, focusing on different aspects of the everyday life at the regiment, including working duties, experiences of rules and punishments, atmosphere, relation to superiors and other legionnaires of different and the same ranks. In this chapter, citations only include information about formal rank.

13. During data collection 1 REC was situated in Orange, but moved to Carpignan in 2014.
14. To aid understanding, I list the ranks of legionnaires in the French Foreign Legion, with existing equivalents in the British Army: Enlisted Ranks: *Légionnaire* (private), *Caporal* (corporal), *Caporal-chef*. Noncommissioned officers: *Sergent* (sergeant), *Sergent-chef* (staff sergeant), *Adjudant* (warrant officer 1), *Adjudant-chef* (warrant officer 2). I refer to *caporal chef* as senior corporal and for simplifying purposes, all NCOs above staff sergeants as senior NCOs.
15. Historically, commanders from noble families have been common. See, for example, John Levi Martin, *Social Structures* (Princeton, NJ: Princeton University Press, 2009), 261.
16. It is in fact possible that the division is comparatively smaller regarding class, or at least more varied. For example, the education level among those enlisted in the Legion is higher compared with the regular French Army and, according to rumour, many legionnaires from Eastern Europe have university degrees and come to the Legion to avoid unemployment and/or obtain a residence permit in France which would enable them to find a job corresponding to their qualifications, after having completed their service in the Legion.
17. See Chiara Ruffa, Chapter 12 of this volume.
18. Ahrne and Brunsson, 'Organization Outside Organizations', 86, 94.
19. Compare, for example, Anthony King, 'The Word of Command', 496–7.
20. Anthony King, *The Ethos of the Royal Marines. The Precise Application of Will.* Unpublished report commissioned by Commander, Commando Training Centre, 2004, 11.
21. See Ruffa, Chapter 12 of this volume.
22. Most francophones are French, but some come from the Maghreb and former French colonies and in a very few cases Belgium and Switzerland. Very few of the non-French interviewees and none of those I have chatted to in the regiments said they spoke any French before enrolling. When, for the purpose of validating this claim, I occasionally used French to address candidates waiting in front of 1 RE or those volunteer recruits waiting to be sent to 4 RE from 1 RE, they have never understood what I have said. This is, of course, only anecdotal evidence, but it supports the view of widespread French-ignorance among recruits, at the same time as there are variations among different nationalities and groups.

23. For a detailed analysis of French teaching in the Legion, see Texier, 'Identités et Langues'.
24. See also Henri Weill, *Légionnaires* (Saint-Malo: Pascal Galodé, 2011), 74.
25. On the regimental level, the planning machinery is detailed, and the bureau of operation and instruction and human resource personnel plan extensively months and/or years ahead, but on the level of the individual legionnaire, which this discussion focuses on, everything can change at any time.
26. Ewan McGorman, *Life in the French Foreign Legion: How to Join and What to Expect When You Get There* (Ashland, OR: Hellgate Press, 2000), 64.
27. Albert Hirschman, *Exit Voice and Loyalty. Responses to Decline in Firms, Organizations and States* (Cambridge, MA: Harvard University Press, 1970).
28. Göran Ahrne, *Social Organizations. Interaction Inside Outside and Between Organizations* (London: Sage, 1994), 12.
29. See also McGorman, *Life in the French Foreign Legion*, 24.
30. Compare Levi Martin, *Social structures*, 146.
31. These differ analytically from formalized recognitions such as decorations and medals.
32. See also McGorman, *Life in the French Foreign Legion*, 150; and Jaime Salazar, *Legion of the Lost* (New York: Berkeley Caliber, 2006), 126.
33. See also Weill, *Légionnaires*, 64.
34. Adrien Jaulmes, 'Afghanistan: Les légionnaires du 2e REP pacifient la vallée de Tagab', *Le Figaro*, 26 April 2010 <http://www.lefigaro.fr/international/2010/04/25/01003-20100425ARTFIG00172-le-2e-rep-pacifie-la-vallee-de-tagab-.php>. Accessed 6 February 2014.
35. See, for example, Jean-Paul Cappuri 'Tribunal correctionnel de Bastia Le sergent-instructeur du 2e REP condamné à deux ans ferme', *Corse Matin*, 12 May 2010 <http://www.corsematin.com/article/corse/tribunal-correctionnel-de-bastia-le-sergent-instructeur-du-2e-rep-condamne-a-deux-ans-.3362.html>. Accessed 9 May 2013.
36. See Jean-Paul Cappuri, 'Une vie brisée pour une autorité bafouée', *Corse Matin*, 17 March 2011 <http://www.corsematin.com/article/corse/une-vie-brisee-pour-une-autorite-bafouee.307694.html>. Accessed 9 May 2013.

11

Cohesion and Misconduct

The French Army and the Mahé Affair

Pascal Vennesson

Introduction

In May 2005, during the 'Licorne' operation launched by France in September 2002 at the beginning of the Ivorian civil war, the force commander General Henri Poncet gave an ambiguous oral order to Colonel Eric Burgaud to kill Firmin Mahé, an Ivorian man accused of being the leader of a dangerous group of 'road blockers' active in the area of operation (General Poncet subsequently denied giving such an order).[1] Mahé was at that time wounded and in the custody of a patrol. Colonel Burgaud transmitted this illegal order to Guy Raugel, a senior noncommissioned officer (NCO), who, with the help of two privates, executed Mahé and later filed false reports dissimulating the actual circumstances of his death. Eight months later, the events were publicly acknowledged by the Minister of Defence Michelle Alliot-Marie and the military leadership, leading to the suspension of those involved, and to the temporary imprisonment for several months of the NCO and the privates but not of the officers who gave the order. The trial of the five army personnel involved—Colonel Eric Burgaud, Adjudant-chef Guy Raugel, Brigadier-chef Johannes Schnier, and Brigadier Lianrifou Ben Youssouf—took place from 27 November until 7 December 2012 before the *cour d'assise* (Assize Court) of Paris. It was the first time that military personnel were put on trial before such a jurisdiction of common law, since they were previously put on trial before the military tribunal of Paris, which was closed in 2011.[2] They were found guilty of murder but received lenient sentences with suspended jail terms.

Why and how does cohesion, a key factor that helps soldiers fight together and risk themselves for their comrades, lead them to unethical and unlawful behaviour? The goal of this chapter is to examine the relations between military cohesion and misconduct. Specifically, I seek to address two gaps in the cohesion literature. First, in discussions of cohesion, the role of discipline and command structures that connect national political goals and micro-practice is often missing. I aim to show empirically how command affects military performance—both collective action and cohesion—during the course of an operation. Second, deviant cohesion commonly refers to situations in which subgroup solidarities overturn organizational goals. By contrast, I explore a situation in which deviant cohesion supports organizational goals at the tactical and operational levels but generate a misconduct with important consequences at the strategic and political levels. I do not claim that the 'Mahé affair' reflects cohesion- and misconduct-related challenges that are specific to the French army. For example, in 1993 soldiers from 'Two Commando' of the Canadian Airborne Regiment were involved in the illegal beating and torturing of a Somali youth. In 2003, British soldiers subjected Baha Mousa, an Iraqi civilian, to serious, gratuitous violence resulting in his death, and between 2004 and 2008 there have been four trials by court martial for alleged misconduct committed by British soldiers in Iraq. Other proceedings in different jurisdictions involved Russian soldiers in Chechnya; American military personnel for alleged civilian deaths and cover-ups in Iraq and Afghanistan, including the abuse of prisoners at Abu Ghraib in 2003; the killings of twenty-four Iraqi men, women, and children by Marines in 2005 in Haditha; the murder of an Iraqi family in Yusufiya in 2006; and the Maywand District killings in which a group of army soldiers murdered at least three Afghan civilians in 2010; as well as Canadian forces in Afghanistan; the Israel Defense Forces against Palestinian civilians; and several allegations of misconduct against UN peacekeepers.[3]

The chapter proceeds in the following three steps. In the first section, I locate the 'Mahé affair' in the evolving theoretical discussions on cohesion and misconduct and I specify the contribution of this case study. In the second section, I present the operational context and analyse the killing of Firmin Mahé and the subsequent unfolding of the 'Mahé affair'. In the third section I further explore the most salient issues of this case study, highlighting the effects of cohesion on misconduct.

Why Study Cohesion in the Context of Military Misconduct?

Military cohesion, understood as 'the bonding together of members of an organization/unit in such a way as to sustain their will and commitment to each other, their unit, and the mission', has long been recognized as crucial and began to be identified as such, systematically studied and theorized in the second half of the nineteenth century.[4] Military institutions in the accomplishment of their missions depend on cohesion, particularly a shared commitment to the unit's task-related goals. However, the almost axiomatic reference to cohesion in studies of combat motivation and military effectiveness has been increasingly put into question, which led to a richer and more differentiated research agenda.[5] I seek to contribute to this evolving assessment of the significance of cohesion in three ways.

First, in their landmark article 'Cohesion and disintegration in the Wehrmacht', Edward Shils and Morris Janowitz explored the paternal role played by Wehrmacht officers in generating cohesion.[6] However, they downplayed the fact that officers could order men to fight; and punish them if they did not. Subsequently, command structures have been often quite neglected in discussions of cohesion, which is either attributed to the dynamics of the primary group or accounted for through the question of political motivation of soldiers. The middle level of command which connects national political goals to micro-practice is often missing.[7] Moreover, I suggest a shift in the boundaries of the existing debates about cohesion which usually focus on small units at the tactical and at the operational level. I claim that military cohesion of the small unit can affect an entire service, the military as an institution, the Ministry of Defence, and ultimately the entire set of relations between armed forces and society. This shift in the scale and scope of the effects of cohesion and of the debate about cohesion shows that small-unit cohesion issues are embedded in, and interact with, a wider set of actors and strategic and sociopolitical logics that cannot be neglected for long. Specifically, I explore the ways in which the military personnel involved evoked cohesion and its consequences, including the tensions between interpersonal trust and legality of orders. I underline the unintended role of cohesion in deviant behaviour, including secrecy and the hiding of deviant behaviour. I also analyse the ways in which both during the events and, in a spectacular way, during the trial itself, the cohesion issue at the small infra-unit level was, in different ways, connected to the broader issue of the cohesion within the army and the military as a whole, including the relations between officers at the highest levels of the hierarchy and the rank-and-file.

Moreover, I explore the second gap in the study of military cohesion, that concerning the inadequate attention placed on how cohesion actually happens in action. For example, Janowitz and Shils do not provide an examination of the dynamics of cohesion during an actual military operation.[8] Significant cases of military misconduct are useful in that regard because, when they are investigated, they place in the public domain a wealth of information that makes visible actors and logics that are usually buried deeply within the institution and not accessible to outsiders.[9] Indeed, one of the key purposes and consequences of cohesion is to maximize within-group exchanges and restrict these exchanges with the 'outside'. The 'Mahé affair' helped to uncover a wealth of fine grain empirical material that is usually out of reach for outside analysts of military organizations, including information about the personalities of the main actors involved, the ambivalent role of face-to-face, interpersonal relations and the underlying views of those involved regarding obedience to orders, loyalty, legality, and cohesion. The logic of the trial also ensures that the different actors involved publicly lay out their views and conceptions of what cohesion is and should be, including the trade-offs involved. Finally, some segments of the broader public, here in France and in Ivory Coast, reacted to these claims and counterclaims about cohesion. The role and interpretations of cohesion, the honour and status of military personnel were discussed, assessed, and publicized. Moreover, it is important to examine military cohesion through the prism of the operational context of peace and stability operations which—while problematic and challenging in their own ways—is not characterized only, or even mainly, by combat in high-intensity warfare. I ask what happens to military cohesion when the frontline is blurred and hard to identify, and when the main task of the soldiers deployed is to police the frontline of other belligerents in order to keep them apart. While the character of the mission and the operational context in Ivory Coast in 2005 was, to some degree, different from counter-insurgency operations in Iraq or Afghanistan, I suggest that there is a sufficient enough overlap between these situations that the underlying logics might be relevant beyond this case study.

Finally, I address the question of deviant cohesion and military misconduct, that is, a breach of military discipline—here the ordering and execution of an illegal order—which has an effect on an individual (or property) not belonging to the armed forces of the perpetrator.[10] Deviant cohesion usually refers to the breakdown of command as subgroup solidarities overturn organizational goals. But in this case, I explore a situation in which deviant cohesion supports organizational goals at the tactical and operational level but contradicts the self-identity promoted by the Army as a whole and has negative strategic and political consequences. Approaching the problem of cohesion from a case of military misconduct leading to a major trial is fruitful because it helps to focus on the trade-offs involved in military cohesion and explore some of its limits and unintended consequences. Cohesion is widely acknowledged as a powerful force and a key factor in military effectiveness. Yet, under some circumstances, it can become militarily, strategically, and politically problematic. I build upon Donna Winslow's classic study of the Canadian Airborne Regiment, which provided an insightful appraisal of a situation where high cohesion led to conformity and favoured military misconducts leading to major military and political consequences.[11]

The 'Mahé Affair'

The Licorne Operation and Joint Tactical Group 2 in Ivory Coast, February–May 2005

The Licorne (Unicorn) operation was launched by France in September 2002 at the onset of the political and military crisis in Ivory Coast. It was ongoing in 2013, albeit on a much smaller scale after the resolution of the crisis in 2011. Its main mission was to separate the belligerents and create the conditions for a process of national reconciliation. The 'Force Licorne' was joined by military contingents from the Economic Community of West African States (ECOWAS) and in April 2004, the United Nations Opération des Nations Unies en Côte d'Ivoire (ONUCI) took over from the contingents of the ECOWAS while the Force Licorne remained in place under French command to support UN forces.

Between May 2004 and June 2005, the 'Licorne force' was under the command of General Henri Poncet, while his second in command was General Renaud de Malaussène. In February 2005, Colonel Eric Burgaud, commander of the 13th mountain troops battalion (13ème bataillon de chasseurs alpins), became the commander of Joint Tactical Group 2 (Groupement tactique interarmes 2) deployed in the western part of Ivory Coast with his command post in the city of Man. A second Joint Tactical Group covered the north-eastern part of the country and a third the south. From February to June 2005, Colonel Burgaud commanded about 800 military personnel and was in charge of a vast territory.[12]

In that area of operation, one of the key security problem at the time were the 'road blockers' (*coupeurs de route*), armed bandits concentrating on axes of communications and preying on cars, taxis, and transport trucks, committing robberies, rapes, murders, and even attacks on entire villages. Since different type of goods, such as gold, diamond, cocoa, coffee, and wood, transit on roads, the banditry often focus their activities on and around these axes of communication. While 'road blockers' were not a new phenomenon in Ivory Coast, the weakening of the state and of public order during the civil war led to an intensification of their action. They particularly targeted the so-called 'zone of confidence', a supposedly demilitarized buffer area 1200 km long and in places 100 km wide stretching from the Liberian border in the west, to the frontier with Ghana in the east which, until 2008, separated the government-held south from rebel-held north. Despite the presence in the country of about 6000 UN peacekeepers and 4000 French troops, insecurity remained high, especially in the west. In May 2005, the head of the UN human-rights programme in Ivory Coast Simon Munzu noted:

> It has to be recognized that the United Nations does not have an administrative or policing or for that matter general security mandate in this zone. It is a very peculiar situation, where even though there is no local administration, nevertheless the United Nations, although it has a certain presence there, does not have a mandate to administer the territory. There is therefore an administrative and judicial void in the area.[13]

UN CIVPOL officers deployed to the ONUCI mission were too few in numbers, unarmed, and had no power of law enforcement. Consequently, they were unable or unwilling to fulfill a meaningful role.

This situation of lawlessness generated an important insecurity limiting the freedom of circulation of goods and people, essential for the survival of civilian populations, who were helpless and taken hostage in the so-called 'zone of confidence'. Moreover, in the western part of the country the situation was exacerbated because cocoa was cultivated by northern ethnic groups who sided with the rebellion. Most of the actions of the 'road blockers' who claimed to belong to the presidential camp targeted both the valuable commodity and those who cultivate it. There were, in part, used by southern authorities in violent actions and disruption against the north. During the first and the second semesters of 2005 the situation became even more alarming as the northern rebellion threatened to break the ceasefire and enter the zone of confidence to establish security, since international forces were unable to do so. Such a move would have instigated the resumption of direct hostilities between north and south, a direct challenge to the mission of the Licorne force.

As a consequence, to avoid a further breakdown of the situation, police mission and maintenance of law and order in this area became a first order necessity for the commander of Joint Tactical Group 2. After the attack by armed partisan of President Gbagbo on a UN post in Logoualé in February 2005, the violence against civilian populations increased in April and May. The Joint Tactical Group identified two groups of 'road blockers', one of them led by Firmin Mahé, which caused numerous crimes, with twenty-four deaths, four rapes, and eight wounded as a result of their actions.[14] Between the end of April and mid-May 2005, a number of road blockers and bandits were arrested by mixed patrols ONUCI/Licorne Force and brought to the police authorities of the north or of the south depending on their ethnic origins. However, they were quickly freed on procedural grounds.

Arrest and Killing of Firmin Mahé

On 13 May 2005, an informant embarked in the military vehicle of a patrol identified Firmin Mahé and his wife. While he was intercepted, he escaped in the bush. Without orders, the patrol fired warning shots, then interdiction shots, but Mahé escaped. Contrary to the procedure, no fire report was sent to the Joint Group headquarters.[15] A helicopter equipped with thermal camera to identify individuals hidden in the vegetation was sent over the zone but could not find Mahé. In late afternoon, a patrol which remained in the area finally found Firmin Mahé grievously wounded by gunshot in the leg. Under the orders of Joint Group headquarters, he was brought back half-conscious at the Peloton de reconnaissance et d'intervention antichar (PRIAC) in Bangolo. General Henri Poncet then contacted Colonel Burgaud and, after asking about the health of the prisoner, gave him the following verbal order (as reconstituted by Colonel Burgaud but subsequently denied by General Poncet): 'you bring him back to Man, he does not escape, you drive slowly, you understand what I mean' ('Vous le remontez à Man, il ne s'échappe pas, vous roulez doucement, vous comprenez ce que je veux dire').[16] On his website and during his testimony during his trial, Colonel Burgaud explained that he understood this oral order to mean killing Firmin Mahé. General Poncet's order mentioned bringing the prisoner to Man (70 km from Bangolo) but by adding, 'drive slowly' and 'you understand what I mean' Colonel Burgaud claimed that the actual meaning of the order was that Firmin Mahé should not arrive in Man alive. He repeated the same order, to the letter, to the adjudant-chef Guy Raugel, who was in charge to bring back Mahé to the camp of the Joint Tactical Group in Man. Raugel confirmed that he received this order and asked Burgaud to confirm that Mahé should not arrive alive in Man. He later confessed to having suffocated Firmin Mahé, wounded and handcuffed, with a plastic bag during this transportation.[17]

It took a few days for Colonel Burgaud to clarify what happened that day as Adjudant-chef Raugel initially reported that Mahé had died of his wounds. The morning patrol also explained that the initial fire exchange was based on legitimate defence when, in fact, Mahé did not fire. The soldiers involved initially claimed to have fired warning shots only, but later recognized that they had actually executed an intimidation or neutralization fire, which explained Mahé's leg wounds. Colonel Burgaud reported all the information to General Poncet, who decided to keep it to himself and not report back to the chain of command in Paris.

The 'Mahé Affair'

From the end of April until October 2005 silence was kept about the real circumstances of the death of Firmin Mahé. However, four months after the murder, Lieutenant-Colonel Gilles Neviaski, head of the operation centre and Colonel Burgaud's second in command, decided to inform a fellow officer about the circumstances of the killing of Firmin Mahé. This officer, who had not been present on the ground, alerted the hierarchy. At the end of

November 2005, the crew of the vehicle in which Mahé was killed, Colonel Burgaud, and General Poncet were placed in temporary police custody (*garde à vue*). After this period, the judge decided to put into temporary detention (*détention provisoire*) the two military personnel who admitted having participated directly in the murder of Firmin Mahé (one Adjudant-chef Raugel for seven months, the other Brigadier Schneir for five months). The other three—General Poncet, Colonel Burgaud, and the driver of the vehicle—remained indicted but were not put into temporary detention.

At the end of the inquiry, in July 2010, General Poncet saw his case dismissed, as the judge found no ground for prosecution. He denied that he gave the order to kill Firmin Mahé and there was no material proof to implicate him. The testimonies of General de Malaussène, Colonel Burgaud, as well as different commanding officers who claimed that he had given similar orders in other circumstances were dismissed.[18] Adjudant-chef Raugel, the platoon leader, was sent to the Cour d'Assise for murder, as was Colonel Burgaud for complicity in murder and falsification of documents, and one private for complicity in murder, as well as the driver for omission to prevent the realization of a crime.

In sum, the main cause of misconduct was a leadership and command failure related to an unlawful order probably given by General Poncet (although he denied giving the order) to Colonel Burgaud (who recognized transmitting an illegal order) to Adjudant-chef Raugel. Burgaud explained that his superior Poncet gave the order orally and that it indicated clearly what the force commander wanted. In addition, Burgaud added the justifications to himself that this action served some military purpose and that the situation in which he found himself altered his ingrained 'moral code'.

Military Cohesion and Misconduct in the 'Mahé Affair'

Mission, Task Cohesion, and Misconduct

Military misconduct can take many forms and conform to multiple logics. It is important to specify the character of the misconduct that occurred in the case of the killing of Firmin Mahé and its distinctive link with cohesion. First, in that particular case, the 'bad apple' explanation—that the misconduct can be explained by seeing the alleged culprits as 'bad apples', or individuals with a propensity to certain form of criminal behaviour, which, once removed, would allow the military to continue to operate without further case of wrongdoing—is not satisfactory. All the individuals involved, far from being 'bad apples', were some of the best product of the Army's selection system: three highly ranked Saint-Cyr graduates—two generals and one colonel—serving with distinction in some of the most prestigious Army units, the

Troupes de Marine, spécialité parachutiste (Marine Corps, airborne), and Chasseurs Alpins (Mountain troops). The same was true for the NCOs and the privates. Moreover, the psychological and psychiatric expertise conducted as a routine matter for the trial did not identify any of the factors commonly identified in the 'bad apples' explanation. Moreover, the 'Mahé affair' cannot be seen as the result of a lack of interest or awareness of the ethical issues. On the contrary, it took place in a context of much greater reference to ethics in army training.[19] This points to more systemic issues, related, I argue, to the broader notion of cohesion and the ways in which it is perceived and used.

The ill-treatment, and in the most egregious cases the intentional killing, of those detained by the armed forces in some form of military custody is one of the most common forms of military misconduct.[20] In these situations, the actors involved believe that their misconduct was serving some military purpose, which is notably perceptible in the ways in which they frame what they had done at the time and retrospectively. In the case of the 'Mahé affair', the misconduct was not a random ill-treatment or massacre, but it was directly related to perceived operational needs. Some local civilians were taking part in activities, such as frequent acts of robberies and terror directed at the civilian population in the area of operations, which, if not directly hostile to the force, were undermining its mission. This not only generated a sense of frustration and helplessness, since those who committed these illicit acts benefited from a de facto impunity, but the absence of law and order also undermined the credibility of the missions, especially since protecting and ensuring the security of the local population was at stake. Those involved were convinced that by ordering, or executing, the killing of Mahé, they were carrying out a military function and hence that their actions was 'morally justifiable'.[21] Adjudant-chef Raugel explained, for example, that by killing Mahé he thought that he contributed to the protection of the population.[22] He claimed that in the year following Mahé's death and the subsequent weakening of this group of bandits there were no further victims in the area.[23] In the end, the Court was sympathetic to the soldiers placed in serious difficulty by the situation they faced.

It is also important to note that the group involved had a high task cohesion, as it was composed of members who shared the common goal and who were motivated to coordinate their efforts as a team to achieve that goal. However, that task conflicted with the broader goals and expressed values of the Army as a whole. Peacekeeping and peace enforcement operations require a demanding situational judgement capacity at the tactical level, as police-like constabulary functions and combat engagement can be both necessary. As a result, efficacy often depends on ground-level initiative, unrestrained by complex rules and bureaucratic procedures, in creating and capitalizing upon opportunities. These situations lead the personnel involved to rely less on formal authority and more on personal rapport, fostering a more informal approach to leadership. This relative autonomy can also become a double-edge sword, however, as it can create a permissive environment favouring misconduct.

Fractured Cohesion, Group Loyalty, and Obedience to Orders

Cohesion, group loyalty (and perceived disloyalty), and obedience to orders and their consequences—together with critical assessments of honour and shame, comradeship linked to collective, professional practices—played different roles and were examined and discussed in multiple ways throughout the events.[24] This significance of the small group was on display. The close and personal relations between soldiers were an important dimension of the process and of the subsequent ways in which the protagonists portrayed their action and their decisions. However, the affair also showed the limits and the tensions involved in these small-group dynamics.

Cohesion, understood as the collective effectiveness of military groups in combat, played an important role in the decision and in the implementation of the order to kill Firmin Mahé, as well as in the cover-up. With the direct support of Colonel Burgaud, Adjudant-chef Raugel and his platoon had previously acquired a significant autonomy, and did not systematically report back to its immediate hierarchy, Captain Le Segretain du Patis and his second in command Captain Ricard. An important characteristic of the small group is that it identifies itself by its difference from others. The judge (*juge d'instruction*) identified a number of serious tensions within the Army,[25] and some of the actors involved noted that the chain of command was broken.[26] But Raugel explained that in the name of military effectiveness, it was normal for him to receive his orders directly from Colonel Burgaud.[27] Raugel had put together a network of local informants who did not know each other to double-check the intelligence received. At times, one of these informants was dressed in a French military uniform and included in the patrols to provide information.

The issue of cohesion and group loyalty took a wider dimension once the affair was revealed. The cover-up, based on a perceived need to avoid embarrassment and remain loyal to the group and the institution, was an important aspect of the 'Mahé affair' but by no means unique.[28] The individuals concerned were convinced that none of their colleagues would report what they had seen or heard or would support them, out of either unit or, more broadly, military loyalty. While cohesion brings significant operational benefits to the unit, or to the military concerned, it also has a problematic side.[29] According to Adjudant-chef Raugel: 'what made us bad soldiers is not that we accomplished our mission, it is just that this affair came out'.[30] Burgaud recognized that his relations with his second in command who revealed what happened were soured. He saw him as incompetent and not trustworthy.[31] He considered that this officer revealed the 'Mahé affair' because he was motivated by resentment and vengeance. Burgaud had given him such a negative professional assessment that he would be unable to become battalion commander.

However, as he admitted during the trial (recognizing 'I have been a coward, at times I have been unworthy'), Colonel Burgaud wavered in his attitude vis-à-vis the event. It is only fairly late that he recognized having transmitted the illegal order. He recognized that he did not have the moral courage to confront General Poncet.[32] Initially, like General Poncet, he denied giving the order and claimed that Guy Raugel alone took the initiative of suffocating Firmin Mahé. He claimed that what he meant when transmitting the order of General Poncet was that Mahé should die from his wound.[33] However, in February 2006, during the inquiry, the judge obtained the report of the Army doctor on the death of Mahé which mentioned explicitly that Colonel Burgaud had explained to him that he ordered that Mahé should not arrive alive at the post.[34] In fact, he was in full agreement with the order of General Poncet and felt that it was an appropriate decision given the situation.

From the perspective of cohesion and loyalty one of the most striking aspects of the 'Mahé affair' was the fact that General Poncet consistently refused, including in his testimony during the trial, to acknowledge that he gave the order to kill Firmin Mahé. One of the most spectacular tension of the November–December 2005 trial took place when the former force commander denied giving the alleged order at all. He attributed the behaviour of Colonel Burgaud and the other protagonists to a failure of moral judgement. He also argued that they became too personally involved in the mission and too close to the population and their suffering. Colonel Burgaud responded immediately: 'I did not know that I was put under the command of a psychiatrist. I thought he was a leader.' And he added: 'I take responsibility for transmitting this order. Not this man.' The version of Colonel Burgaud was supported by General De Malaussène, who considered that General Poncet had undoubtedly given, if not an actual order, at least an 'impulse', and criticized his style of command as prone to flashes of temper during external operations and even crime-inducing behaviour. Malaussène also explained that, as the one who gave the initial order, his refusal to take responsibility was indicative of a lack of courage. Poncet's attitude—which some attributed to an implicit agreement with political leaders that, in exchange for not being put on trial he would keep silent on other controversial aspects of the French involvement in Ivory Coast—was highly criticized, notably within the army.[35] The issue of cohesion, lack of trust, and disloyalty was also deeply felt by both the Chief of the Army and the Chief of the Joint Staff, who both felt betrayed and misled by General Poncet.

In addition to the layers of cohesion rooted within the Army, the issue of cohesion also involved the broader role of the political leadership. For example, the website created by the friends of Eric Burgaud (<http:www.eric-burgaud.fr>) is entitled: 'what support of France for her soldiers?' (*Quel soutien de la France pour ses militaires?*), which is just one of the connections between cohesion in the operation and the broader cohesion issue between the armed forces and the nation. Colonel Burgaud criticized Michèle Alliot-Marie, the defence minister, for not sufficiently supporting those involved in the affair. He noted, for example, her haste to immediately open an internal administrative inquiry and suspend the generals Poncet and Malaussène, Colonel Burgaud, and Adjudant-chef Raugel.[36] He also noted that before the conclusions of this internal inquiry, the Ministry of Defence publicized the facts, including the names of the military personnel involved. He claimed that these military personnel were thrown to the media without any capacity to defend themselves.[37] In early November 2005, the procedure became judicial, with the opening of an inquiry for murder and falsification of documents, since no report mentioned the real causes of Firmin Mahé's death. In his narrative, Colonel Burgaud also noted pointedly that the inquiry was attributed to the police criminal brigade despite the fact that the Gendarmerie has some prerogative to conduct inquiries in cases of crimes and offences concerning military personnel. He claimed that the personnel of the Gendarmerie were allowed to consult classified documents, while the police were not allowed to do so.

Conclusion

Michael Walzer notes the tension between military cohesion, on the one hand, and the notion of disobeying illegal orders. 'Theirs is the rough solidarity of men who face a common enemy and endure a common discipline.... To disobey is to breach that elemental accord, to claim a moral separateness (or a moral superiority), to challenge one's fellows, perhaps even to intensify the dangers they face.'[38] As he explains: 'the path that I have marked out as the right one is often a very lonely path'.[39] The 'Mahé affair' shows that the problem of military cohesion is not confined to the small group. In that case, it quickly reverberated and reached higher levels both within the Army and beyond. The logics of cohesion cannot be isolated from an increasingly vast set of concentric circles. This case, as well as the numerous reactions that it generated in blogs, comments, and so on, also shows a widespread misunderstanding or an outright rejection of the notion of an illegal order, particularly its relevance in a tense operational context.

Notes

1. An earlier version of this chapter was presented at the conference *Frontline: Combat and Cohesion in Iraq and Afghanistan*, University of Oxford, All Souls College, Changing Character of War Programme, 18–19 March 2013 and at the 2013 International Biennial Conference of the Inter-University Seminar on Armed Forces and Society, Chicago, 25–7 October 2013. I thank the participants, especially Anthony King and Michael Noone, for their comments and suggestions. I gratefully acknowledge the intellectual and material support of the S. Rajaratnam School of International Studies, Nanyang Technological University. This chapter is dedicated to the memory of Bastien Irondelle.
2. A 'Cour d'assise' (Assize Court) is a criminal trial court composed of a jury trial of six jurors and a panel of three active judges. It has original and appellate limited jurisdiction to hear cases involving defendants accused of major felonies or indictable offences, notably any criminal act punishable by over ten years of prison, including murder and rape. The very existence of such a trial with condemnations (although lenient) is part of the indirect effects of the existence of the International Criminal Court. France being party to the Rome Statute, if such a trial had not been conducted, the Court might have invoked its jurisdiction and required handing over of the soldiers concerned for trial.
3. See notably: Donna Winslow, *The Canadian Airborne Regiment in Somalia: A Socio-Cultural Inquiry* (Ottawa: Canadian Government Publishing, 1997); Peter Rowe, 'Military Misconduct during International Armed Operations: "Bad Apples" or Systemic Failure?', *Journal of Conflict and Security Law* 13(2) (2008); Jim Frederick, *Black Hearts. One Platoon's Descent into Madness in Iraq's Triangle of Death* (New York: Harmony Books, 2010); A.T. Williams, *A Very British Killing. The Death of Baha Mousa* (London: Jonathan Cape, 2012).
4. John H. Johns et al., *Cohesion in the US Military* (Washington, DC: National Defense University Press, 1984), 9. On the classics in the study of military cohesion: Charles Ardant du Picq, *Battle Studies. Ancient and Modern Battle* (1st edn. 1880. trans. 1920) in *Roots of Strategy. Book 2* (Mechanicsburg, PA: Stackpole Books, 1987); Morris Janowitz and E. Shils, 'Cohesion and Disintegration in the Wehrmacht in World War II', in Morris Janowitz (ed.), *Military Conflict* (London: Sage, 1975), 177–220; Samuel A. Stouffer et al., *The American Soldier* (Princeton, NJ: Princeton University Press, 1949); S.L.A. Marshall, *Men against Fire. The Problem of Battle Command in Future War* (New York: William Morrow, 1947). See also: Simon Wessely, 'Twentieth-Century Theories on Combat Motivation and Breakdown', *Journal of Contemporary History* 41(2) (April 2006): 269–86. On leadership and the exercise of command: Yves Cohen, *Le siècle des chefs. Une histoire transnationale du commandement et de l'autorité (1890–1940)* (Paris: Editions Amsterdam, 2013).
5. Robert J. MacCoun and William M. Hix, 'Unit Cohesion and Military Performance', in *Sexual Orientation and U.S. Military Personnel Policy: Options and Assessment* (Santa Monica, CA: RAND, 1993); Hew Strachan, 'The Soldier's Experience in Two World Wars: Some Historiographical Comparisons', in Paul Addison and Angus Calder (eds.), *Time to Kill. The Soldier's Experience of War in the West* (London: Pimlico, 1997),

369–78; Hew Strachan, 'Training, Morale and Modern War', *Journal of Contemporary History* 41(2) (2006): 211–77; David R. Segal and Meyer Kestnbaum, 'Professional Closure in the Military Labor Market: A Critique of Pure Cohesion', in Don M. Snider and Gayle L. Watkins (eds.), *The Future of the Army Profession* (Boston, MA: McGraw-Hill, 2002), 441–58; Anthony King, *The Combat Soldier. Infantry Tactics and Cohesion in the Twentieth and Twenty-First Centuries* (Oxford: Oxford University Press, 2013). On the ambivalence of comradeship: Gerald F. Linderman, *The World Within War. America's Combat Experience in World War II* (Cambridge, MA: Harvard University Press, 1997), 263–99. On the significance of patriotic and ideological convictions not just small-group loyalty: James M. McPherson, *For Cause and Comrades. Why Men Fought In The Civil War* (Oxford: Oxford University Press, 1997), 85–9, 90–103.

6. Morris Janowitz and E. Shils, 'Cohesion and Disintegration in the Wehrmacht in World War II', in Morris Janowitz (ed.), *Military Conflict* (London: Sage, 1975), 177–220.
7. Anthony King, Chapter 5 of this volume; Mikaela Sundberg, Chapter 10 of this volume.
8. Roger W. Little, 'Buddy Relations and Combat Soldier Performance', in Morris Janowitz (ed.), *The New Military* (New York: Russell Sage Foundation, 1964), 195–224.
9. Including the websites and blogs of some of the key protagonists: Colonel Burgaud: <https://www.eric-burgaud.fr/valeurs/vie-familiale.html>; Adjudant-chef Raugel: <http://adjudantchefraugel.over-blog.com/article-2992978.html>. Accessed 2 February 2015.
10. Rowe, 'Military Misconduct during International Armed Operations': 166.
11. Winslow, *The Canadian Airborne Regiment in Somalia*. See also: Isabel V. Hull, *Absolute Destruction. Military Culture and the Practices of War in Imperial Germany* (Ithaca, NY: Cornell University Press, 2005).
12. <https://www.eric-burgaud.fr/valeurs/vie-familiale.html>. Accessed 25 December 2012.
13. 'Violence Continues in Ivory Coast's Zone of Confidence', 24 April 2005—<http://worldnewssite.com/News/2005/April/2005-04-24-26.html>. Accessed 13 March 2013 <http://www.voanews.com/content/a-13-2005-04-24-voa26-66381122/547704.html>. Accessed 11 February 2015; James Copnall, 'Ivory Coast's zone of no-confidence,' BBC News, 18 May 2005 <http://news.bbc.co.uk/2/hi/africa/4556335.stm>. Accessed 13 March 2013. See also: Médecins sans frontières, *'People are left defenceless'—Testimonies from Ivory Coast: An Overview 2003–2006*, Operational Center of Brussels (2006), 9.
14. <https://www.eric-burgaud.fr/firmin-mahe/exactions-coupeurs-de-route.html>. Accessed 25 December 2012.
15. <https://www.eric-burgaud.fr/firmin-mahe/arrestation-et-mort.html>. Accessed 25 December 2012; 'Entretien avec Guy Raugel: "A ce moment-là, la fin justifiait les moyens"', *Le Dauphiné libéré*, 27 November 2012 <http://www.ledauphine.com/faits-divers/2012/11/27/guy-raugel-a-ce-moment-la-la-fin-justifiait-les-moyens>. Accessed 25 December 2012.
16. <https://www.eric-burgaud.fr/firmin-mahe/arrestation-et-mort.html>. Accessed 25 December 2012.

17. 'Entretien avec Guy Raugel', *Le Dauphiné libéré*, 27 November 2012.
18. <https://www.eric-burgaud.fr/affaire-mahe/volet-judiciaire.html>. Accessed 25 December 2012.
19. Jean-René Bachelet, *Pour une éthique du métier des armes. Vaincre la violence* (Paris: Vuibert-Espace éthique, 2006); Jean-René Bachelet, 'La formation des militaires à l'éthique dans le métier des armes', *Droits Fondamentaux* 6 (janvier-décembre 2006): 1–10; Benoît Royal (Colonel), *L'éthique du soldat français: la conviction d'humanité* (Paris: Economica-Collection 'Stratégies et doctrines', 2008); Bastien Irondelle, 'L'armée française et l'éthique dans les conflits de post-Guerre Froide', *Critique internationale* 41 (2008/4): 119–36.
20. Peter Rowe, 'Military Misconduct during International Armed Operations': 169. See also: Jeff McMahan, *Killing in War* (Oxford: Oxford University Press, 2009), 203–35 (on civilian immunity and civilian liability); Stephen Coleman, *Military Ethics. An Introduction with Case Studies* (Oxford: Oxford University Press, 2012), 51–4.
21. Rowe, 'Military Misconduct during International Armed Operations', 171.
22. 'Entretien avec Guy Raugel', *Le Dauphiné libéré*, 27 November 2012.
23. 'Entretien avec Guy Raugel'.
24. Wm. Darryl Henderson, *Cohesion. The Human Element in Combat* (Washington, DC: National Defense University Press, 1985); Christopher R. Browning, *Ordinary Men. Reserve Police Battalion 101 and the Final Solution in Poland* (New York: Harper Perennial, 1993); Donna Winslow, *The Canadian Airborne Regiment in Somalia*; Anthony King, 'The Word of Command. Communication and Cohesion in the Military,' *Armed Forces & Society* 32(4) (July 2006): 493–512.
25. Jean-Dominique Merchet, 'Affaire Mahé: la défiance régnait chez les officiers…', Marianne Blog, 27 November 2012 <http://www.marianne.net/blogsecretdefense/Affaire-Mahe-la-defiance-regnait-chez-les-officiers_a852.html>. Accessed 25 December 2012.
26. 'Affaire Mahé : "Vous m'avez compris"', *Le Monde*, 4 December 2012 <http://www.lemonde.fr/international/article/2012/12/04/affaire-mahe-vous-m-avez-compris_1799621_3210.html>. Accessed 2 February 2015.
27. 'Affaire Mahé : "Vous m'avez compris"', *Le Monde*.
28. On cohesion and secrecy in the cover-up of the accidental death of Pat Tillman: Mary Tillman (with Narda Zacchino), *Boots on the Ground By Dusk. My Tribute to Pat Tillman* (New York: Modern Times, 2008); Jon Krakauer, *Where Men Win Glory. The Odyssey of Pat Tillman* (New York: Doubleday, 2009).
29. Peter Rowe, 'Military Misconduct during International Armed Operations: 'Bad Apples' or Systemic Failure?' *Journal of Conflict and Security Law*, 13(2) (2008): 187.
30. 'Affaire Mahé : "Vous m'avez compris"', *Le Monde*.
31. Merchet, 'Affaire Mahé : la défiance régnait chez les officiers…', Marianne Blog.
32. 'L'indignité est maintenant dans le camp du général Poncet', 29 November 2012.
33. Stéphane Colineau and William Assayag, 'Affaire Mahé : un colonel déstabilisé', *20 Minutes*, 27 April 2006.
34. Colineau and Assayag, 'Affaire Mahé : un colonel déstabilisé'.

35. Jean-Dominique Merchet, 'L'armée en mots troubles,' *Libération*, 5 January 2006 <http://www.liberation.fr/grand-angle/010134446-l-armee-en-mots-troubles>. Accessed 2 February 2015.
36. <https://www.eric-burgaud.fr/affaire-mahe/volet-judiciaire.html>.
37. <https://www.eric-burgaud.fr/affaire-mahe/volet-judiciaire.html>.
38. Michael Walzer, 'Two Kinds of Military Responsibility', in *Arguing About War* (New Haven, CT: Yale University Press, 2004), 315–16.
39. Walzer, 'Two Kinds of Military Responsibility', 315.

12

Cohesion, Political Motivation, and Military Performance in the Italian Alpini

Chiara Ruffa

Introduction

As many contributors have already noted, in their article on the Wehrmacht, Morris Janowitz and Edward Shils famously attributed the 'extraordinary tenacity' of German soldiers to the social cohesiveness of its primary groups.[1] Yet a number of scholars have noted that while close interpersonal bonds have often encouraged soldiers on the ground, there is no necessary correlation between primary group solidarity and combat performance. On the contrary, dense social commitments in the primary group can just as easily facilitate combat refusals, deviance, and, ultimately, mutiny: 'The most serious problem in relying on primary group cohesion to prevent disintegration is the well-documented fact that bonds between members of primary groups can develop into a basis of opposition to the larger organization.'[2] In this case, primary groups manifest 'resistance to organization goals'. The French and Australian mutinies in the First World War and, of course, the conduct of black American soldiers in Vietnam have been widely taken as notorious examples of the problem of deviant cohesion.[3] As suggested by both Charles Moskos and Stephen Wesbrook, political motivation, even if only in the form of a 'latent ideology', has been a recurrent method to avoid deviant cohesion. 'The prevention of military disintegration in modern military organizations first depends on the soldiers' moral involvement with their primary groups and either their unit or their national socio-political system, but ideally both.'[4]

In Chapter 5, Anthony King explores the way in which new forms of motivation have emerged in the professional army, reducing the dangers of deviant cohesion and displacing the requirement for political motivation. Indeed, Stephen Wesbrook observed that highly trained professional armies can sometimes only rely on unit affiliation to ensure allegiance to higher organizational goals: *esprit de corps* provides sufficient motivation in and of itself to ensure combat performance. In *The Combat Soldier*, King seeks to show how a professional ethos is detectable across most Western forces in the twenty-first century.[5] Yet, as he himself admits, the increased levels of training, which normally accompany professionalism, increase only the chances of effective combat performance; 'even among the most professional forces, it is a possibility not an inevitability'. There are likely to be differential levels of performance across professional forces and, indeed, King's work hints at this. Interestingly, variation in levels of professionalism and performance could derive directly from specific forms of unit cohesion. In particular institutional circumstances, the bonds of cohesion may generate greater levels of political autonomy and lower levels of allegiance to higher organizational goals. Hence, deviant cohesion is possible in the professional army too. Here, the professional force becomes a status group in Weber's sense of the term. In this context, soldiers are not necessarily characterized by their expertise or by a sense of responsibility towards society, as Huntington defined them. On the contrary, they simply 'monopolize' opportunities and resources, which their privileged status as soldiers gives them. In this case, the internal cohesion of the professional military unit does not generate higher levels of combat performance but actually enables these forces to avoid unpalatable forms of military service, while enjoying the privileges of state patronage.

In the contemporary era, the Italian Army's Alpini Regiment provides one of the most striking and interesting examples of deviant professional cohesion. The Alpini is a highly specialized and professional force dedicated to mountain warfare, which, due to its own regimental culture and institutional and political position, has remained highly internally cohesive while performing consistently poorly on conventional military operations, despite its political autonomy and its popularity among the public.[6] The analysis in this chapter suggests that exceptionally high levels of internal social cohesion coupled with high levels of political autonomy produced low levels of political motivation and trust in political leaders and state institutions. While these elements prevented the Regiment from being particularly effective in military operations, strong cohesion and desire for public support have made the Alpini very effective in contemporary peacekeeping and humanitarian missions.

The Alpini [literally Alpine/Mountaineer] is an infantry specialism in the Italian Army. The term 'Alpini' refers both to the specialism (singular) and to the soldiers belonging to it, the Alpini (plural). Created in 1872, the Alpini were developed to support and sustain a new concept of defence for the newly unified Italian state. Since their creation, recruitment has been based on the concept of territorial defence and the idea that soldiers have to build the support of 'their people' rather than merely obey the institutions that directly commanded them. During the First and Second World Wars, the Alpini were at times openly critical of constituted authority, such as the army and the government, and they constantly worked for increasing consensus within the Italian civil society. With the disbandment of the Italian Army after the surrender of 1943, most of the Alpini units, in contrast to the behaviour of other army units, joined the Italian resistance against the Fascist dictatorship and the Nazi occupation.[7] Consequently, at the end of the Second World War, the Alpini avoided much of the delegitimization suffered by the rest of the Italian armed forces. They partly distinguished themselves by their role in the resistance, as well as relying on their established image, and were more quickly rehabilitated in operational roles. They also worked hard to regain their lost public legitimacy.

Italian Military Performance

The Italian 'reputation for military incompetence was securely established as early as the sixteenth century and is referred to by writers as diverse as Rabelais, Machiavelli, Erasmus and Montaigne'.[8] Since the creation of a unified Kingdom of Italy in 1861, Italian military history is a long litany of defeats. In 1888, a foreign military attaché wrote about the Italian Army: 'recent history does not shine with great military exploits, or brilliant victory on the part of Italian arms, in fact rather the reverse'.[9] The worse experiences were probably the First and Second World Wars. The First World War resulted in 650,000 dead and more than a million wounded, although modest gains were achieved: Alto Adige, Istria, and very small territories in Africa.[10] In 1940, Mussolini 'ignored the state of the Italian armed forces, which were not ready to enter the conflict' as he sought to exploit the opportunity presented by Hitler's apparently decisive victories.[11] During the war, the performance of Mussolini's army was, excepting the cases of the *Ariete* and *Folgore* divisions in North Africa, disastrous on every front, culminating in its humiliating disbandment in 1943.

Historians have identified a number of factors to explain the disastrous Italian military performance in all recent wars: tactical ineptitude, ineffective leadership, low morale, poor training, and national recruitment.[12] Some scholars have explored the effects of low morale on 'constant and pervasive' Italian military incompetence.[13] Others have rather explained Italian military ineffectiveness by a lack of dedication and commitment 'to the proclaimed aims of the armed forces among the generals, officers and the civilian administrators and politicians involved in military matters'.[14] Deep political divisions in Italian society are also often invoked to explain failure. Among the most common sources of those divisions are profound differences across regions, the relative power of the Papacy, and chronic weakness of the central state during the nineteenth century. Since the Pope was also the head of state immediately after the unification, Italians were divided between being loyal to the Pope and to the newly founded Italian state, 'with the result that they never developed a strong dedication to either'.[15]

The Alpini reflect the characteristically poor military performance of Italian forces and exemplify a particular Italian trait noted by Christine Davies. She identifies the patrimonial structure of Italian society consisting of tight local networks of allegiance and interdependence as the central cause for military ineffectiveness: 'Italians growing up in such a society develop a capacity for loyalty to a small group of persons known to them personally but not for loyalty'. Accordingly, Italian soldiers have often been highly innovative in small specialist military operations, such as the use of mini-submarines in the Second World War or, more recently, their quick adaptation to improvised explosive devices during operations in Afghanistan.[16] The Alpini seems to be the epitome of this phenomenon; they are a military specialism based on territorial recruitment and capitalizing on primary-group bonding. Indeed, Davies cites the Alpini as precisely the kind of localized, face-to-face grouping in which Italian society excels. The Alpini are an interesting case because, against conventional expectations, their exceptional levels of internal cohesion have not translated into combat effectiveness. Instead, cohesion and political autonomy have encouraged low political motivation, close association with civil society, and, subsequently, great results in recent non-conventional military, humanitarian operations. The remainder of this chapter explores the preconditions at play in constructing the Alpini's high levels of cohesion and political autonomy and how these conditions impact on culture and operational performance.

The Origins of Cohesion and Political Autonomy

The creation and development of the Alpini is directly relevant to the regiment's cohesiveness today. The Alpini were conceived as a military specialism with a specific strategic role in mind. In the late nineteenth century, the newly created Italian state needed new, effective units that could defend its sovereign national territory. Until the mid-nineteenth century, the dominant defence concept in Italy had been to stop potential aggressors on the plains (in the Po valley) after the enemy had crossed the Alps. But, in 1871, in a pamphlet, a captain of the Italian Royal Army, Giuseppe Perrucchetti, proposed the creation of a specialist corps that could defend the Italian territory in the Alps themselves. This new concept of territorial defence required soldiers with unique characteristics. Ideally, the defence of the Alpine borders would be best assured by people coming from these regions. To achieve this, Perrucchetti proposed territorial recruitment. Thus, the Alpini specialism required a recruitment system distinct from the national enrolment of the newly formed Italian Army, substantially devised to minimize the risk of unit rebellion and social tensions. National recruitment was very cumbersome, particularly in the case of crises, however. If standard national recruitment had been applied to the Alpini, recruits from the Alpine regions would have trained on the plains like everyone else, acquiring no mountain skills. Furthermore, since only a small section of the Alpine population would have been drafted into Alpini units, they would have been made up of people coming from all over Italy, without pre-existing ties to the local communities. Hence, applying the national recruitment system to the Alpini was deemed inappropriate.

Perrucchetti proposed the introduction of regional recruitment and divided the Alpine territory into military districts, each with its own defensive units. His ideas emerged from observing the mountainous areas:

> Whoever has wandered around for long through the mountains shall have realized how among these, that remarkable book called terrain is extremely difficult to read; for reading it correctly it is necessary to get used to it: the practical knowledge of the terrain that has to be defended can be acquired only by maneuvering over it and considering all the foreseeable options [lit. *ipotesi* hypotheses].[17]

On this logic, a profound knowledge of territory was much more important than conventional military skills. Alpini had to be able to endure the dangers and difficulties of mountain life, much more than they needed to be disciplined soldiers. They had to work well together in small dynamic teams rather than in traditional infantry platoons.

While the new concept of defence was well received, political elites were sceptical about the idea of territorial recruitment, as it posed, according to them, concerns of public order. Perrucchetti and the Minister of War managed to overcome this scepticism by showing that the risk in the specific case of the Alpine population was minimal: 'The alpine valleys were all profoundly conservative, catholic and monarchist, absolutely incapable of rebelling', demonstrated by the extremely low percentage of people who failed to report for military service in comparison to other Italian regions.[18] Perrucchetti managed to convince the Italian Army Chief of Staff of both the new defence concept and the plan for territorial recruitment. In 1872, the Italian parliament finally approved the creation of the Alpini specialism.

Political Motivation and Cohesion in Alpini History

When Italy became involved in colonial warfare towards the end of the nineteenth century, there were discussions as to whether the Alpini would be used outside the national territory. Instead, they were mainly used to reinforce defensive fortification across the Alps. In 1911, things changed for the first time: the Alpini were deployed in Libya during the colonial campaign: 'They wanted to use them because they could adapt more easily to the difficult terrain.'[19] In the First and Second World Wars, Alpini regiments were widely used, both in defensive Alpine functions and in conventional combat roles in other areas, including the operations on the Italian northeastern front in the First World War and the northwestern front in the Second World War, together with contributions to military campaigns in the Balkans, North Africa, and Ukraine during the Second World War. During this time, the Alpini were widely employed because of relatively high levels of cohesion and discipline. The Second World War and its aftermath, however, constitute a decisive collective—and political—juncture for the specialism. The political autonomy of the Alpini engendered institutional 'un-bonding'; the Alpini tried to differentiate themselves from the rest of the army and started to become more critical towards the rest of the military and political leadership. In this critical shift, traditional autonomy translated into lack of political motivation due to the Alpini's disconnect from the dominant frame for political reasons.

This rupture became pronounced after the fall of Mussolini, the declaration of the armistice on 8 September 1943, and the disbanding of the Italian Army, which left the Army without orders. In general, some Army units kept fighting alongside the Germans; others surrendered to either the Germans or the Allies; while other units switched side and started fighting alongside the Allies. While this period of Italian history is still one of the most politically contested, it is uncontroversial that most Alpini units joined either the resistance, integrating in the existing command structures, or the Allies, reconstituted as two battalions, called 'Piemonte' and 'L'Aquila'.[20] There was nothing particularly heroic in this process, according to many Alpini who joined the resistance. During the resistance, some Alpini had simply ascended into the mountains 'to do something', unaware of precisely what was asked of them.[21] Others wanted to oppose the Fascist dictatorship in some ways. In the account of an Alpino who joined the resistance, these motivations become quite clear:

> There is nothing heroic in what we did. We did what we had to do. We are Alpini. On September 8, I realised quickly that something was wrong. I gathered my Alpini, told them to prepare their backpacks. Over night we left the town and we

> hid in forests close to the Alps. There were 50 of us. I told them they could go home if they wished to do. None left me. After a while we contacted the resistance groups and we joined the resistance movement.[22]

Even those oppositional movements traditionally more critical of the military (such as members of the Italian Communist Party) welcomed the Alpini resistors. In many cases, they allowed the Alpini to maintain their uniforms and military status. 'A lot of the partisans were officers from the Alpini specialty. They [the civilian partisans] immediately gave us command responsibility: they knew they could trust us and they knew we loved our people.'[23] During the resistance war, partisans, of which many were Alpini, came down from the mountains and helped the local populations to harvest and do other things: 'it was the most obvious thing to do'.[24]

At the end of the Second World War, 'the role of the armed forces in Italy was drastically reduced'.[25] Several factors accounted for it: the prior strong connections of the armed forces with the Fascist dictatorships, the total collapse of the Royal Army after the fall of Mussolini, the legacy of a lost war, the hostility of the main political parties towards nationalism and militarism, and, above all, the rejection of war as means of settling international disputes, explicitly written into the Constitution of 1948. The events following the fall of the Mussolini regime severed the armed forces from civil society because the public felt betrayed by its own Army.[26] The Italian Army faced a legitimacy problem. While the Alpini shared part of that delegitimization, the population acknowledged and was grateful for the contribution of the Alpini during the resistance war. The Alpini were often seen as a separate group that shared less responsibility, in contrast to the rest of the armed forces. The Alpini have continued to draw upon a tradition of independence and resistance to justify their legitimacy and garner favour with the public. Thus, they were much more present and visible during the postwar period than other units.

Although the Italian armed forces were dismantled, the two Alpini battalions who had joined the Allies were reconstituted in 1945 and progressively more were re-established until the stable establishment of five Alpini brigades (Julia, Tridentina, Taurinense, Orobica, Cadore) in 1953. During this period of transformation, the Alpini maintained and if anything emphasized their cultural traits, which also served functional purposes. Since the early 1950s, the Alpini were the only specialism of the Italian Army visible in the domestic context, where the military was severely frowned upon. For instance, the day after the ceremony reconstituting the Brigata Alpina Taurinense in 1952, the main Italian newspaper *Corriere della Sera* published an image of Italian Alpini climbing the Matterhorn on the front page. No other reconstituted brigade of the Italian Army received any major media coverage. The very fact that the Alpini were seen as distinctly non-military, engaged primarily in mountaineering activities and only nominally dedicated to external territorial defence, rendered them less problematic in a widely anti-militaristic context. Even today, the mountains remain central to the Alpini. In terms of fighting ethos, the Alpini still do not define themselves as warriors: 'We like to have our feet well anchored to the ground, we are a bit less exaggerated, a bit humbler than other specialisms within the military.'[27]

The Alpini has also benefited from its political position. Throughout the Cold War period and until very recently, political elites in Italy minimized the role of the army, which was associated with Fascism. 'Only 27.7% of the Italian public opinion thinks that the armed forces are an indispensable instrument of foreign policy.'[28] Furthermore, a number of events during the 1960s and 1970s eroded the already limited trust towards the armed forces. High-ranking military personnel were more or less directly involved in a series of subversive plots by right-wing groups and intelligence units, such as the Masonic Lodge P2 in 1981 and the Gladio affair in 1990. During the Ustica affair, in which an Italian commercial airplane was hit and destroyed in mysterious circumstances, the Italian Air Force refused to release information to clarify the event. All of these events exacerbated an overall situation of general distrust. Notwithstanding the unfavourable structural situation, Alpini brigades tried to increase their legitimacy by intervening in the cases of natural disasters, such as the 1951 flash flood in Polesine or the 1976 earthquake in Friuli. They were also readily deployable in the framework of the NATO defence contingent following Italy's accession to NATO in 1949.

In the early 1990s, Italian policymakers started to re-use the military as an instrument of foreign policy. Alpini contingents were extensively used in out-of-area operations at this time. Interestingly however, they continued to portray their interventions as peacekeeping and humanitarian operations even when they were not.[29] They did not describe themselves as being involved with combat: 'no soldier is ordered into combat: our most important role is what we can give to our community—national and international'.[30] Even in Afghanistan, Alpini troops prioritized humanitarian aid and reconstruction over patrolling and combat.[31]

During the past decade, a period of professionalization and defence transformation, much has changed for the Alpini. First and foremost, their demographics have altered, with most volunteers now coming from the South. Being an Alpino is no longer a matter of birthplace; anyone can volunteer. Symbols have evolved to reflect these patterns. For example, the motto for the annual Alpini reunion used to be: 'from one side to the other of the mountains: Alpini!'. Now, it has been modified into a more inclusive 'from the plains to the mountains: Alpini!'. However, the Alpini specialism continues to recreate a narrative that celebrates social cohesion and autonomy. There is still a strong feeling that the mountains bring the Alpini together, notwithstanding their now different origins: 'Once you were born as an Alpino, now you can become one of them.'[32] 'Now we also have Sicilian Alpini, Alpini from Lazio and from other regions in Italy.'[33] A commander of the Alpini artillery puts it plainly: 'We are lucky: we have the mountains, mountains are our big, big fortune. The mountain is what makes us invincible.'[34] Every Alpine soldier has to cultivate 'the value of being an Alpino [lit. of the Alpinity, *dell'alpinita'*] because mountains do not forgive, they unite us and it is wonderful to live together. Those who do not like the mountains cannot understand that we all love something that imposes big sacrifices on us but makes us strong and cohesive.'[35] As these statements clarify, even within an increasingly diverse force, Alpine values and mountaineering practices have encouraged high levels of solidarity and institutional autonomy. The very basis of internal social cohesion is ultimately a non-military function: it is about climbing and skiing rather than fighting.

Civil society has, to date, encouraged the non-military image and functional priorities of the Alpini, a fact that reinforces the Alpini's distinctive cohesion. Alpini are regularly invited to take part in celebrations and official ceremonies. As a result of their respected status, they were the most popular regiment for recruits to enrol in during the conscription era. Their role in the highly antimilitaristic Italian society remains striking. Even today, at weddings in rural areas, if the groom has done his military service with the Alpini, tradition requires at least one group picture in which he wears the Alpini wide-brimmed hat, accompanied by his bride, veterans, and members of the associations. Annual Alpini national reunions gather about 200,000 Alpini in various Italian cities. Alpini are also involved in several community activities, such as organized marches and providing security on ski slopes. The distinctive Alpini culture has sustained and cemented the interplay between political autonomy and cohesion, and informs the actions of Alpini soldiers, even in contemporary operations.

Alpini Ethos in Action

Regional recruitment meant that the Alpini culture became based on three core elements: closeness to the local population, a predominantly defensive role, and exceptionally strong peer-group solidarity.[36] The Alpini's 'military superiority...derives directly from the regional recruitment system, which was the decisive element to create a strong cohesion where the links among men were authentic because pre-existing and the fracture with the external world much less clear'.[37] These three elements engendered extremely high levels of cohesion, coupled with high levels of political autonomy from the rest of the Army and the government. However, the political autonomy and cohesion that define the Alpini have also impacted on their operational performance. In the remainder of this chapter, I will exemplify how the ethos of unity and political autonomy influenced Alpini soldiers' behaviour, even during the recent NATO mission in Afghanistan.

First, since its inception, the state, Army, and the Alpini themselves have stressed the 'natural' and strong territorial connection between the specialism and the local communities in the mountains. 'These were people who had hung out in the same stables, who had had the same girlfriends, who had gone to the same schools, people who spoke the same dialect.'[38] As such, the Alpini cemented allegiances and the sense of belonging to communities to which they were already closely tied: 'We were tight to one another like brothers.'[39] As a result, they sometimes developed distinct opinions, implicitly criticizing national institutions, and usually aligning with what 'their' people thought. This was coupled with considerable institutional autonomy from both the Army and the state, preventing the Alpini from developing any strong political motivation for engaging in war fighting.

Second, since the creation of the first regiments, the legendary social cohesion of the Alpini has been closely related to their tactical role. Mountain warfare requires intense in-group solidarity and cooperation; physical survival depends in the first place on collective mountain-craft and coordination skills, and does not necessitate tactical military acumen. Intimately related to the harsh environment in which the Alpini must operate, a strong but distinctive ethos of self-sacrifice has developed, recorded in the diaries and memoirs by Alpini soldiers serving on various fronts.[40] While the importance of self-sacrifice is recurrent more generally in soldier narratives, for Alpini soldiers the core element was to help each other survive against the elements, not primarily by killing the enemy. In a war diary, an Alpini who fought in Russia during the Second World War called the tie among Alpini a 'sacred [holy] human pact among each other' rather than against the enemy.[41] This ethos is apparent in the specialism's rituals and symbols.

Since the arm's origin, several newspaper articles and figurative images celebrated the elements that define the Alpini: namely closeness to local communities, defensive posture, and solidarity. The recurring images were Alpini on a crag, one hand held to the side and the other holding up the rifle, the face looking at the horizon. However, the motto 'from here you do not pass', 'vigilantes', and 'lookout of the summit' refer to a defensive rather than offensive function.[42] Oliva notes that Alpini troops were always represented in the mountains, 'standing on the highest summit'.[43] In iconography, another recurring element was the natural Alpine environment, which was perceived as wild and unexplored by a significant portion of Italians. In what is considered his masterpiece, Italian writer Edmondo De Amicis wrote in 1886 about the Alpini: 'you could see them coming forward, hundreds of long straight feathers, which grew taller than the public: they were the Alpine troops, the defenders of Italy's doors; all tall, rosy and strong with wide-brimmed hat and bright green uniform, the same color of the grass of their mountain'.[44] In these representations, the Alpini evoke bonds to the local communities and stress their relative autonomy and strong bonds.

The defensive role of the Alpini is not only celebrated in public, civilian representations but also in their own regimental self-understandings. Their daily routines are structured around devotions to symbols, which they cultivate much more than other specialties of the Italian Army. Flag rituals punctuate daily routine in Alpini military bases, both in Italy and in Afghanistan. Significantly, even symbols, which seemed overtly martial, have ultimately been demilitarized. Six core symbols are particular relevant in this respect as they represent and enjoin the Alpini's distinctive identity. First, shortly after the creation of the various Alpini regiments, a unique green uniform was introduced, signifying the 'landscape of the mountain valleys'.[45] Second, the symbol of the Alpini became the black feather on a wide-brimmed hat of the same colour (*sic Cappello alla calabrese*). The Alpini hat is completely different from the more military basque beret used by the rest of the Army. Like other specialisms, both officers and men earn the right to wear this special headdress only once they have passed training and selection, at which point the hat is ceremonially decorated with a black feather. At that moment, a soldier becomes an Alpino for the first time. The black feather of the Alpini headdress might be mistakenly taken as a warrior symbol, especially since it is awarded after successful training. In fact, the feather signifies the natural environment of the mountains and the individual character required to survive there. Moreover, the Alpini hat is a markedly non-military item; it is the headdress of a mountain guide or shepherd rather than a soldier and contrasts notably with the distinctively oversized blue berets of the French *Chasseurs Alpins*, an elite force which was created at a similar time to the Alpini to defend France from Italian invasion. The *Chasseurs'* beret is adapted for mountain conditions: it is very large so that a flap can be made from the excess material to protect the face from the sun, wind, and rain. Of course, in the mountains today the French *Chasseurs* employ modern, waterproof hoods and hats to protect them from the conditions, and their distinctive beret has become purely symbolic. Nevertheless, closely related to the standard infantry beret and adorned with the bugle insignia of the *Chasseurs*, it plainly remains a ceremonial piece of military attire. By contrast, the Alpini's hat with its black feather and jaunty brim civilianizes the Alpino.

Similarly, the black flag of the Alpini, which may appear to be a warrior symbol, actually signifies the black plumage of the mountain crow. The black flag symbolized defensive values appropriate in the mountain environment.

Furthermore, the Alpini adopted a mule as its mascot in 1972 and used mules during training and on operations until 1993.[46] Mules are still used as mascots and are displayed in ceremonies and celebrations. 'The mule shared so much with us all and sacrificed so much. And it is such a powerful symbol of the Alpini: so humble and modest and yet so powerful.'[47] Again, the mule is not a particularly militaristic symbol. It suggests independence, toughness, and endurance, and above all an ability to survive and move freely in the mountains. It is not a warlike animal, however.

Fifth, the Alpini prayer underlines the defensive role of the unit and the attachment to the motherland:

> On the bare rock, on the perennial glaciers, on every crag of the Alps where providence has put as bulwark of our districts, we, purified by the duty dangerously implemented we elevate to you, Oh Lord, who protects our mothers, our spouses, our children our distant brothers (. . .) God all powerful make our feet [lit. *piede*, foot] stand safe on the vertiginous peaks, on the steep face, beyond the insidious crevasse, make strong our weapons against whomever menaces our Homeland, our Flag our thousand-year Christian civilization.[48]

Finally, the Alpini's protecting saint suggests relatively low value for respect of authority in general. Saint Maurice is the patron saint of the Alpini. The cult of this saint is particularly widespread in Northern Italy, sanctified because he refused to obey an order asking him to persecute Christians and to sacrifice and pray to idols. Although he is usually represented as a soldier with armour and a sword, he carries a palm frond, prioritizing his non-military role.[49]

These traditions and symbols remain central to the professional Alpini of the twenty-first century. They constitute an imagined history for this regiment celebrating its social unity, institutional autonomy, and civil roots. Crucially, as with the Alpini of the past, these symbols do not primarily denote military service in protection of the Italian state. Rather they signify the dense social cohesion of ski-mountaineers dedicated to the protection of Italy's territorial borders and, above all, those communities living in those northern regions. These ritualized practices have informed the Alpini's operational performance over time and continue to do so.

For instance, during the NATO mission in Afghanistan, Alpini soldiers' behaviour was wholly consistent with the values and beliefs outlined in this chapter. For instance, soldiers from Alpini units conducted eight to twelve sorties from their base per day, which sometimes included genuine combat patrols but, far more often, involved contact patrols, visits to the local authorities, and humanitarian projects; French units of comparable size conducted twenty to thirty patrols a day, even though both forces were deployed under very similar circumstances. In contrast to their French counterparts, Alpini soldiers emphasized civil–military coordination activities over patrolling.[50] In the words of an Alpino in Afghanistan: 'we are not here to fight, we are here to defend and help the people'.[51] During the period of observation, the Alpini implemented more than 150 development and humanitarian projects, while French soldiers implemented less than thirty. They even explicitly connected the Afghan locals with the local communities back in Italy, which they were used to helping and supporting. As an Alpino remarked: 'we are here to help the people, the Afghans, it is like helping our people, the people from the mountains'.[52] This is typical of the Alpini. In comparison with other units of the Italian Army, the Alpini seemed to care a lot more about ties with the local community. For instance, in stark contrast to the Alpini, despite being deployed in a much more permissive context in Lebanon, the Ariete, Italian mechanized infantry troops, did not seek to establish much of a connection with the locals.

It could be argued that the Alpini's focus on assisting the local population represented the ideal counter-insurgency strategy for Afghanistan, which aimed not at defeating insurgents but rather at securing the population. Anecdotal evidence suggests that while the Alpini conducted some beneficial development work in Afghanistan, their military performance was widely regarded as inadequate by their NATO allies, since they made fewer attempts to take control of important areas from the Taliban.[53] Their low patrolling rates and reluctance to fight were taken as indicative of this unwillingness to exert control over their designated areas of responsibility. However, their empathy and ability to interact with the local population was much appreciated by the local population.

Conclusion

This chapter has argued that the distinctive history, institutional position, and role of the Alpini has given rise to a highly developed corporate ethos and organizational solidarity which frames the way this arm conducts military operations to this day. As such, the Alpini constitute an interesting example for the analysis of cohesion. They are a highly cohesive regiment, which nevertheless has performed consistently poorly and reluctantly on military operations. This paradox can partly be explained by reference to their traditional function. As mountain soldiers, they worked in small groups in an environment which prioritized skiing and climbing skills and mere survival over fighting. Accordingly, dense bonds and close team work are evident among the Alpini, but they do not display the tactical capabilities of traditional infantry. The mountain is their symbol and their role. At the same time, the special political conditions of Italian civil–military relations and the political traditions of the Alpini themselves also have to be recognized. The British Royal Marines, the French *Chasseur Alpins*, and the German

Gebirgsjäger are also specialist mountain troops with a similar role to the Alpini but whose military effectiveness is well known. The political and institutional independence of the Alpini, partly as a result of their tradition of political resistance, has been critical here. Not only has it allowed the Alpini to abjure conventional military service in the name of the state—and effectively to resist performing combat missions—but this tradition of political independence is central both to the regiment's identity and to the way it unifies its troops. The very cohesiveness of the Alpini as an organization, based on political autonomy, impedes its effectiveness as a military force. In the Alpini, professionalism does not prioritize combat missions, which the regiment is singularly poorly motivated to perform anyway, partly because of its independence from political direction. The persistence of a fragmented state and the low trust in formal institutions explains plausibly why even the most strongly cohesive specialist corps in the Italian Army, the Alpini, struggled to perform effectively in war from the First World War to the present. At the same time, and for the same institutional reasons, Alpini soldiers have, however, a proven ability in peace operations, where their acumen for generating social support and taking difficult decisions on the ground due to their long-standing political independence has been widely noted. As such, the Alpini as a professional force are an unusual and even unique case, but they illustrate some of the presumptions and problems of the cohesion debate extremely well.

Notes

1. This chapter is dedicated to Giovanni Reineri '*Capitano Gianni*' (his enduring nickname) (1917–2013), a great man, resistant, and Alpino, who introduced me with patience and accuracy to the experiences of the Alpini in 1943–45 during several interview sessions in spring/summer 2013.
2. Stephen Wesbrook, 'The Potential for Military Disintegration', in Sam Sarkesian (ed.) *Combat Effectiveness and Cohesion* (London: Sage, 1980), 257.
3. Charles Moskos, 'The American Combat Soldier in Vietnam', *Journal of Social Issues* 31(4) (1975): 34.
4. Moskos, 'The American Combat Soldier in Vietnam', 274.
5. Anthony King, *The Combat Soldier* (Oxford: Oxford University Press, 2013), Chapter 7.
6. King, *The Combat Soldier*, Chapter 7; Stanislav Andreski, 'Causes of the Low Morale of the Italian Armed Forces in the Two World Wars', *Journal of Strategic Studies* 5(2) (1982): 257–65; Alexander Lopasic, 'Italian Military Performance in the Second World War: Some Considerations', *Journal of Strategic Studies* 5(2) (1982): 270–5; Christie Davies, 'Itali sunt imbelles', *Journal of Strategic Studies* 5(2) (1982): 266–9; John Gooch, 'Italy during the First World War', in Allan R. Millett and Williamson Murray (eds.) *Military Effectiveness*, vol. 1: *The First World War* (Cambridge:

Cambridge University Press, 2010), Ch. 5; John Gooch, 'Italian Military Competence', *Journal of Strategic Studies* 5(2) (1982): 257–65.
7. Gianni Oliva, *Storia degli Alpini* (Milano: Mondadori, 2001), 213–30.
8. Davies, 'Itali sunt imbelles', 266.
9. Gooch, 'Italian Military Competence', 258.
10. Lopasic, 'Italian Military Performance in the Second World War', 272.
11. Lopasic, 'Italian Military Performance in the Second World War', 270.
12. MacGregor Knox, 'The Italian Armed Forces: 1940-3', in Allan R. Millett and Williamson Murray (eds.) *Military Effectiveness* (Boston, MA: Unwin Hyman, 1988), 162.
13. Andreski, 'Causes of the Low Morale of the Italian Armed Forces in the Two World Wars', 276.
14. Andreski, 'Causes of the Low Morale of the Italian Armed Forces in the Two World Wars', 276.
15. Andreski, 'Causes of the Low Morale of the Italian Armed Forces in the Two World Wars', 276.
16. Chiara Ruffa, *Imagining War and Keeping Peace? Military cultures and peace operation effectiveness* (Florence: European University Institute doctoral dissertation, 2010).
17. Oliva, *Storia degli Alpini*, 21.
18. Giovanni Massobrio and Giorgio Rochat, *Breve Storia Dell'esercito Italiano Dal 1861 Al 1943* (Torino: Einaudi, 1978), 95.
19. Author interview with Capitano Gianni, July 2013 and January 2013.
20. Laura Ceci, *La Resistenza dei Militari* (Roma: Biblink, 2006).
21. Nazzareno Peano, 'Salimmo in montagna per fare qualcosa', *Rivista dell'Istituto Storico della Resistenza e della Societa' Contemporanea*, 81 (2012): 207–21.
22. Author interview with Alpini officer and former member of the resistance, January 2013.
23. Author interview with Capitano Gianni, July 2013 and January 2013.
24. Author interview with Capitano Gianni, July 2013 and January 2013.
25. Paola Barrera, 'Crisis in the Military: Rethinking Conscription and the Military Code', in Robert Y. Leonardi, Raffaella Nanetti and Piergiorgio Corbetta (eds.) *Italian Politics: A Review* (London: Pinter, 1988); Stefano Battilossi, 'Armed Forces', in G. Moliterno (ed.) *Encyclopedia of Italian Political Culture* (New York: Routledge, 2000), 29.
26. Leopoldo Nuti, *L'esercito italiano nel secondo dopo guerra, 1945–1950: La sua ricostruzione e l'assistenza militare alleata* (Roma: Ufficio Storico, Stato Maggiore dell'Esercito), 15. And also for similar arguments: Chiara Ruffa, *Imagining War and Keeping Peace?*; Piero Ignazi, Giampiero Giacomello and Fabrizio Coticchia, *Italian Military Operations Abroad- Just Don't Call It War* (London: Palgrave MacMillan, 2012).
27. Author interview with Alpini colonel, Taurinense Brigade, January 2013.
28. Giovanni Massobrio and Giorgio Rochat, *Breve Storia Dell'esercito Italiano. Dal 1861 Al 1943* (Torino: Einaudi, 1978), 273.
29. Chiara Ruffa, 'Imagining War and Keeping Peace? Military Cultures and Forces Employment in International Operations' (Uppsala: Uppsala University Archives,

2012), paper under review. For an analysis on the impact of the Alpini approach on the level of violence: Chiara Ruffa, *Imagining War and Keeping Peace?*.

30. Author interview with Alpini officers in Afghanistan.
31. Chiara Ruffa, 'Imagining War and Keeping Peace?'.
32. Conversation with Mario Renna, January 2013. See also, Mario Renna, *Ring Road. Sei mesi con gli Alpini in Afghanistan* (Milano: Mursia, 2011).
33. Author interview with Alpini colonel, Taurinense Brigade, January 2013.
34. Author interview with Alpini colonel, Taurinense Brigade, January 2013.
35. Author interview with Alpini colonel, Taurinense Brigade, January 2013.
36. For an interesting perspective on why military culture matters for military cohesion see: Charles Kirke, 'Group Cohesion, Culture, and Practice', *Armed Forces & Society* 35(4) 2009: 745–53.
37. Kirke, 'Group Cohesion, Culture, and Practice', 745.
38. Author interview with Capitano Gianni, July 2013.
39. Author interview with Capitano Gianni, July 2013.
40. Giulio Bedeschi, *Il Segreto degli Alpini* (Mursia: Milano 2004), 17; Nuto Revelli, *Mai Tardi. Diario di un Alpino in Russia* (Cuneo: Panfilo Editore, 1943), 30.
41. Revelli, *Mai Tardi*, 13.
42. Oliva, *Storia Degli Alpini*, 50.
43. Oliva, *Storia Degli Alpini*, 51.
44. Edmondo De Amicis, *Cuore* (Torino: Einaudi, 1974), 41.
45. Gianni Oliva, *Storia degli Alpini dal 1872 a oggi* (Milano: Mondadori, 2001), 35.
46. To be precise, mules have been used by Piedmontese mountain troops from 1831. Nowadays, the debate about killing the last Alpini mule triggered an uproar: '"Eutanasia" per l'ultimo mulo alpino', *La Tribuna*, 11 September 2013 <http://tribunatreviso.gelocal.it/cronaca/2013/09/11/news/eutanasia-per-l-ultimo-mulo-alpino-1.7725900>. Accessed 23 October 2014.
47. Giulio Bedeschi, *Il Segreto degli Alpini* (Mursia: Milano 2004).
48. Francesco Somale, 'Le Compagnie Alpine', *Rivista Militare* 4 (1878): 193.
49. Rosa Giorgi, *I Santi* (Milano: Electa, 2002), 270.
50. Ruffa, *Imagining War and Keeping Peace?*.
51. Author interview with Alpini noncommissioned officer (NCO), NATO mission in Afghanistan, 2011.
52. Author interview with Alpini NCO, NATO mission in Afghanistan, 2009.
53. Author interview with Alpini NCOs, officers, and non-Italian troops, NATO mission in Afghanistan, 2009–12; 'French troops died after Italy stopped bribing local Taliban, Times reports', *France 24*, 16 October 2009 <www.france24.com/en/20091015-french-troops-died-after-italy-stopped-bribing-local-taliban-times-reports/>. Accessed 22 May 2014.

Part V
The Home Front

13

Of FOBs and Families

Leonard Wong and Stephen Gerras

Introduction

The role of cohesion—or the emotional bonds between soldiers—in contributing to combat motivation has been well documented in the literature, as Anthony King discusses in Chapter 1.[1] While the research that began in the First World War and continues through today has steadily confirmed a relationship between cohesion and combat motivation, comparatively little research has explored the variables that may moderate that relationship. One such moderating variable is the family; personal domestic relations often mediate and inform the relations between soldiers. Indeed, Roger Little demonstrates that during the Korean War, the reputation of soldiers with their families did not play an insignificant role in combat motivation; they did not want to return home as cowards or failures.[2] This chapter explores the impact of the family on combat soldier cohesion. We accomplish this by first presenting a brief review of the combat cohesion literature. We then describe a relatively recent development in the Western (in this case, American) way of waging war—the use of forward operating bases (FOBs)—that results in two outcomes that may influence the development of cohesion. First, FOBs minimize many of the traditional causes of apprehension and stress. In the past, combat cohesion has usually been viewed as a key counterweight to the anxiety and fear found in war. The use of FOBs may be modifying that relationship. Second, FOBs often provide the technology to permit the introduction of the family as a more salient variable in the development of cohesion. This too may be affecting the cohesion found in modern warfare. We conclude the chapter with an examination of how cohesion as demonstrated by today's professional Western forces may differ from their citizen forebears in the twentieth century due to the influence of the family.

The Oasis

Most wars have had rear areas, base camps, or firebases from which soldiers could come and go to wage war. FOBs are different from past places to regenerate combat power because the disparity between the environment inside and outside the wire is much more pronounced than what was experienced with forward bases in past conflicts. FOBs are, as one Army general put it, 'little oases in the middle of a dangerous and confusing world'.[3] FOBs are also situated literally on the battlefield. They are where combat soldiers actually live, not just visit for a few days of rest and recuperation.[4]

FOBs have developed into places where many of the stresses, frustrations, and discomforts commonly thought endemic in the combat soldier's life are mitigated. On a FOB, soldiers are able to renew their fighting spirit and also find a psychological escape from the rigours of battle. More importantly, while rear areas in the past have afforded soldiers amenities and safety, forward deployed combat soldiers on FOBs have been able to be in daily, or even minute-to-minute, contact with their families.

War has typically been an environment of physical and emotional stress. In his study of American enlisted men in Vietnam, Charles Moskos described the physical factors influencing the behaviour of soldiers in combat: 'There are the routine physical stresses of combat existence: the weight of the pack, tasteless food, diarrhea, lack of water, leeches, mosquitos, rain, torrid heat, mud, and loss of sleep.'[5]

Peter Kindsvatter, in his analysis of the American combat soldiers from both world wars, Korea, and Vietnam, noted that in addition to the physical pressures of combat, there is a pervasive emotional strain on soldiers in a combat environment:

> Even as the soldier contended with a harsh, seemingly evil, physical environment, he also found that war stressed him emotionally. He was never completely safe from harm, except when far to the rear, perhaps attending a military school or in a hospital, and even then he was hounded by the thought of having to return to the fighting. Thus the combat soldier was never completely free from fear and anxiety.[6]

Most combat veterans from the Second World War through the war in Afghanistan would be able to relate to Moskos's and Kindsvatter's description of the combat situation. From the Battle of the Bulge to the Battle of Wanat, soldiers have understood combat to be tiring, draining, and emotionally exhausting.

FOBs have changed the battlefield environment in many ways. The most obvious is providing a safe haven from danger. Off the FOB, troops are hypervigilant as they conduct missions or follow up on intelligence reports of insurgent activity. On the FOB, soldiers clear their weapons, requirements for helmets and body armour are relaxed, and while the threat of a rocket, mortar attack, or even direct attack is always present,[7] it is greatly minimized. The result is that a primary cause of soldier fear and stress—omnipresent physical danger—is largely absent from most FOBs.

FOBs give soldiers the rare opportunity to be engaged in brutal combat one minute and yet return seconds later to a place where the threat of harm is exponentially diminished. The psychological benefit of not feeling threatened by the enemy is significant. A convoy commander interviewed for this study whose mission was delivering supplies, described the psychological release of arriving at each FOB, 'There is a sense of relief. You pull in at the next gate and it's "Aaaahhhhhh... Relax!"'[8] Another soldier commented on what life would be like without the safety found in FOBs, 'Imagine you are under constant threat.... Your own tent wouldn't relax you. You couldn't relax when you go to take a hot shower. You couldn't relax having a meal... I can see how that would be way harder on people.'

Another persistent factor affecting soldier stress during combat is the lack of sleep. Studies have repeatedly shown that when conducting continuous combat operations, soldiers should ideally get six to eight hours of sleep every twenty-four hours.[9] A cumulative lack of sleep leads to soldier fatigue, disorientation, and will eventually diminish the ability to resist fear. Second World War surveys administered in the Italian theatre shortly before the end of the campaign (when the fighting had quieted and conditions were 'probably as favorable as ever exist in combat') showed that despite a lull in combat, nearly one-third of the men averaged only four hours or less of sleep each day.[10] Similarly, researchers chronicling the Vietnam War point out that most infantrymen in Vietnam could expect an average of about four hours of sleep a night in combat.[11]

Several aspects of FOBs influence the ability of soldiers to receive adequate amounts of sleep. First, as described earlier, insulating soldiers from the physical danger outside the wire eliminates the requirement for them to be constantly concerned about the security for themselves and their unit. Second, the conditions on FOBs generally allow soldiers to escape the harsh environment of Southwest Asia or the Middle East in order to rest. Air conditioning and comfortable beds are critical luxuries in providing rest for soldiers. Finally, because of the safety afforded by the FOB, commanders can set up a combat rhythm that includes allowing time for sleep. As one soldier noted, 'My first sergeant and commander—they are adamant about giving the guys their personal time. We get five hours of personal time every night and six hours of sleep every night. ... That takes the fear away from me.'

Cumulative hunger or thirst, like the lack of sleep, adds to combat stress and a decreased psychological readiness. In one of the earliest studies on fear in battle, John Dollard interviewed veterans of the Abraham Lincoln Brigade who fought in the Spanish Civil War. He concluded that 'hunger and thirst, excessive heat and cold, prolonged fatigue—all tend to force men out of battle'.[12] Despite the recognition of the link between hunger and a soldier's ability to deal with fear, over half the combat soldiers in a Second World War survey reported that they did not get as much to eat as they needed.[13]

Soldiers on FOBs would be hard pressed to claim they could not get enough to eat. Even after fifty-seven eateries in Afghanistan such as Burger King, Pizza Hut, and Subway were eliminated to cut back on logistical support, soldiers could still go to the dining facility 'where fried calamari rings, fish fingers and burgers are free for soldiers and available in abundance'.[14] As one infantryman dryly noted, 'If my worst complaint is that there aren't any tacos in the salad bar, I've got it pretty good.'

In his analysis of American soldiers in past wars, Kindsvatter noted that, 'Once the troops had attended to the basics of cleaning up, eating, and resting, they began looking for diversions. Any entertainment was greatly appreciated.'[15] Examples of amusements included group singing; listening to radio broadcasts, especially music; and watching movies. A Second World War soldier commented that the movies 'took our minds off the war and transferred us into a fantasy world for a couple of hours'.[16]

Distractions on the FOB allow soldiers to also mentally leave the war zone—although in newer ways. As one soldier pointed out, 'Everybody has their way of unwinding. A lot of the younger guys have Xboxes they play with. . . . I got my little DVD player I watch every once in a while.' A team leader stressed the importance of recreation, 'We work out every day, which takes away the stress. They play Playstation together, pool together, ping pong. . . . That is what they need.'

Diversions on the FOB appear to minimize idleness, which can have a detrimental effect on psychological readiness. Dollard pointed out in his study of fear in battle that while some fear is beneficial to prevent complacency, a 'good way to combat useless fear is to think about something else'.[17] Stress and fear germinate in the minds of soldiers who have nothing to do but ruminate on upcoming missions and think about time away from home. As one infantryman noted, 'Staying busy, keeping your mind active with other things. . . . When we are not going outside the gate, I try to keep busy with something.' Another soldier gave similar counsel, 'If you stop and think about what you are doing and being away, then that's when it gets to you. If you don't stop and think about it, then it's not as difficult.' Distractions on the FOB also temporarily remove (at least mentally) soldiers from the combat environment. In the words of one soldier, 'When I come [to the FOB] I don't feel like I'm in Iraq.'[18] A combat engineer suggested, 'We need to trick our minds that we're somewhere else. Otherwise, we wouldn't be able to go on, knowing today might be our last.'[19]

The analogy of the FOB as an oasis is appropriate because the FOB has become a haven for safety, rest, and refreshment in an unforgiving environment. The FOB, however, can be more than just a refuge or a sanctuary. While it can be a place that serves to isolate soldiers from the debilitating aspects of the battlefield, it also can be a place that connects soldiers with the families in the home front.

Connecting with Home

Of course, communicating with families has always been a critical factor in the morale of combat soldiers. Mail call traditionally has been a highlight (and occasionally a low point) of a soldier's time in combat. In addition to postal mail being used to communicate with home, other methods have been introduced through the years. V, or Victory, mail was introduced in the Second World War to cut down on the valuable shipping space required for postal mail. Even with the patriotic lure of V-mail, however, most people continued to use regular first class mail, which was much more private and less regulated.[20]

In 1948, the Military Affiliate Radio System (MARS) was established to assist in the training of amateur radio operators. A side benefit was that deployed soldiers could phone home free of charge.[21] Because the calls were relayed over amateur radio networks, the calls were limited in availability and quality, were vulnerable to eavesdropping (including the two radio operators relaying the call), and subjected callers to using proper radio procedures (e.g., 'Hi mom, this is your son, over'). The cumulative effect of the disadvantages discouraged most soldiers from using the system.[22]

While telephone calls were possible but extremely rare during the Vietnam War,[23] the potential for commercial telephones for more than a once-a-tour phenomenon became apparent during the invasion of Grenada when an officer called home in order to coordinate fire support.[24] The telephone as a means to communicate with the home front slowly increased during military operations in Panama,[25] the Gulf War, Somalia,[26] the Balkans, and the wars in Iraq and Afghanistan. Despite the growing availability of commercial telephones, however, expense and availability continued to be limiting factors.

During the Gulf War, and especially during operations in Bosnia and Kosovo, email emerged as a key means of communicating for deployed soldiers.[27] As the use of email surged in the civilian arena, soldiers were quick to adopt it as a means of communicating from the war zone. In the Iraq War, soldiers initially siphoned off bandwidth from military networks for access to the internet and email capability.[28] As FOBs became established, internet access became cheap (or free), ubiquitous, and expected. Soldiers routinely began using Skype, Google Talk, AIM, Facebook, or texting to communicate home.

Against this backdrop of electronic interconnectedness, soldiers interviewed for this study were asked how often they communicated from the FOB with their families. Researchers were surprised to receive replies such as: 'I talk to her every day when I am here,' or 'I would say at least twice a day.' Amazingly, many soldiers reported that they were not just 'keeping in touch' with their families with emails, but were instead communicating real time several times a day.

Soldiers develop high expectations concerning life on the FOB. Security, plenty of food and drink, a place to rest, and morale-building activities are soon taken for granted by combat veterans. Instantaneous communications with home join the list of amenities expected by soldiers in combat. As one section leader observed: 'We just had a census over in the company. The commander did his census and my guys' biggest problem was the internet. They want internet. We have three MWR [Morale, Welfare, and Recreation] computers, but they would like some internet in their rooms. I said if that is your biggest concern, then we are doing okay.'

Distant, but not Disconnected

FOBs remove much of the anxiety and emotional stress of combat experienced by previous generations of soldiers. When examining the role of the FOB as a place where soldiers can stay connected to their families, however, the situation becomes more complex. As pointed out earlier, the amount of contact that soldiers on FOBs had with their families was surprising. With many soldiers reporting that they talked to their families 'every day, all night', issues emerge that were not as prevalent with other forms of communication such as letters or email.

With phone calls, texting, and instant messaging (IM) from the front becoming almost routine, the concept of the family for deployed soldiers shifts from the abstract to the concrete. The family is no longer a fond memory tucked in the back of a soldier's mind resurrected only prior to turning in for the night. Instead, for many soldiers, the family becomes part of their battle rhythm—a tangible factor affecting their daily routine and perspectives.

It is important to note that this phenomenon is found mostly in junior enlisted soldiers. Interviews in the field reflected that officers and more senior noncommissioned officers (NCOs) are less likely to communicate as frequently with home and thus tend to view the family in a more abstract perspective. Three reasons help to explain why. First, officers and senior NCOs are older and therefore tend to have relationships with spouses, significant others, and children that have had more years to mature. There is less of a perceived need for frequent communication as relationships are more stable and family members become more independent. Second, senior NCOs and senior officers are less comfortable with the emerging technologies. For older NCOs and officers, the traditional letters, email, and occasional phone call remain the standard means of communication. Finally, and most importantly, NCOs and officers do not have as much discretionary time or mental energy to contact home due to leadership responsibilities. While junior enlisted soldiers may have periods of personal time after missions, NCOs and officers continue to be engaged with mission planning or checking on the troops.

Soldiers frequently in contact with their families from the FOB reported that communications moved from fleeting occasions of abstract assurances that everything was fine to more practical interactions.[29] One soldier describing the daily exchanges remarked that the conversations were 'just day to day stuff . . . basically like things going on with the kids and stuff like that'. This is not surprising, as research in the social aspects of communication media suggests that an increase in the frequency of electronic communications leads to more informal and intimate interactions.[30]

Soldiers were asked what types of topics were discussed with families. As expected, soldiers quickly pointed out they seldom mentioned any details of life at the front. The initial reason given for guarding their conversations was concern for operational security. When pushed for a deeper explanation, however, soldiers expressed a desire to shield their families from any information that could cause concern or worry. For example, in one interview, a soldier had these responses to a researcher's questions:

Interviewer: Do you tell other people about what you do here?
Soldier: No.
Interviewer: Fiancée? Anything?
Soldier: No I try to keep them out of it. So that way she's not worried. She knows I go on missions.
Interviewer: And you don't give details?
Soldier: No, I told her about the IED [improvised explosive device], which I shouldn't have done.
Interviewer: What made you feel afterwards that you shouldn't have done that?
Soldier: She got all worried. I really shouldn't tell her anything.

Another soldier commented similarly, 'I try not to tell her a whole lot except that I am doing OK and just the normal stuff. . . . I just don't want her to worry as much as she does.' Soldiers on FOBs, much as soldiers from previous generations, learn not to burden families with the stresses of the combat zone. The primary motivation in restricting the amount of detailed information conveyed about day-to-day happenings in the war zone is to prevent families from worrying. Soldiers believe that it is better to spare the details than to provoke apprehension in the home front.

Despite the paucity of problems being communicated homeward, most soldiers surprisingly reported that family problems at home were being communicated to the FOB. While soldiers were careful that no potentially anxiety-producing information was relayed home, there was an unusual expectation—and a desire—to hear about the problems at home. An interview illustrated this interesting phenomenon:

> Interviewer: Now, does she tell you problems from home?
> Soldier: Oh I hear all of them. Well, I can't say I hear *all* of them. I feel I have a stressful situation but I realize that she has a stressful situation and she does everything she can. It just gets away from her sometimes and she needs to vent [to someone].
> Interviewer: And that's you?
> Soldier: That's me, no matter how far away I am.

When asked why concerns and anxieties seem to flow in only one direction, another soldier observed:

> I like to think of her job—with four kids at home—as much worse than my job here. I try to make it as easy as possible on her. I don't vent about my job. She asks, 'How's work?' and I say, 'Work is work and I am here' and that's all it's going to be.... She knows I can handle that burden a lot more than she can hear about body parts blowing up in front of a position down the street. She doesn't need to hear about that or she won't sleep well or she's going to worry. Versus me worrying about the kids, you know, 'You got to email your son; he's kind of out of control right now.'

Another soldier put it simply, 'I want to be able to listen to her and listen to her worries and I don't want her to worry about what's going on over here.'

Why would soldiers facing the stresses of war want to take on additional concerns from the home front? Researchers have long noted the detrimental effects of family problems on combat troops. For example, Michael Doubler wrote of Second World War soldiers:

> Mental anxiety was perhaps the greatest cause of low morale. The army learned that war fostered worries that could drain a soldier's working and fighting efficiency. Troops thousands of miles from home, living and fighting under heart-breaking and nerve-wracking conditions, worried about immediate problems and the *welfare of loved ones at home*. Common concerns among officers and enlisted men included being gone from the United States for long periods, *family matters at home*, and the overall progress of the war.[31]

To help keep family problems from becoming distractions to deployed soldiers, the US Army has traditionally relied on informal and formal organizations such as Family Readiness Groups (FRGs) to support and assist families. FRGs operate under the assumption that providing support to families allows soldiers to 'concentrate on the mission at hand, and have the emotional readiness to carry out that mission'.[32] FRGs continue to assist in vital problem-solving support for families, but communication technology has allowed soldiers to retain more of the role of providing *emotional* support to the family. Deploying soldiers understand that while they cannot physically contribute to the family, the informal and intimate nature of phone calls, texts, and instant messages can allow them to share the emotional burdens of parenting and marriage. One infantryman who called home on a daily basis explained why he called so often: 'Being the man of the family, your job is to support them and protect them. Well I can't protect them right now because I am not with them.... I know I am the one in the combat zone, but I just worry about them—making sure they are all right constantly, stuff like that.' The availability of affordable synchronous communications results in many deployed soldiers attempting to retain much of their roles as husbands and fathers. While they might be physically distant, they are not disconnected. Despite being thousands of miles away in a combat zone, soldiers have learned that technology and the protective qualities of the FOB enable them to still help redirect a wayward child or encourage a disheartened spouse.

The most obvious outcome of this increased connectivity is an increase in morale. Soldiers, just as their civilian peers, crave relationships and staying connected. Being involved in the lives of their families—including their problems—is preferable to being isolated from the home front. As one soldier commented on the alternative, 'One letter every month from your wife or fiancée? That would be hard. That would be very hard because I would be worried about her and what's going on back home.'

Besides boosting morale, increased interaction with home may have some less obvious, but extremely significant effects. In his examination of combat trauma during Vietnam, psychiatrist Jonathan Shay describes the 'berserk state'—a condition characterized by rage, callousness, and unrestrained anger. According to Shay, soldiers descend into the berserk state as they become socially disconnected and 'cut off from all human community'. Often driven by revenge, soldiers in a berserk state 'cannot see the distinction between civilian and combatant' and are prone to committing atrocities.[33] In a harsh counter-insurgency environment where the insurgents are not easily identified and the populace may be sympathetic or even assisting the enemy, the potential for soldiers to move toward the berserk state is increased. One potential implication of remaining emotionally connected with home via technology on the FOB, however, is the calming effect on soldiers.

With an instant messaging session, text, or a phone call home, moral and social restraints on soldiers can be restored. The social isolation normally found in a combat zone is diminished by the soldier's interaction with a familiar voice from home as the soldiers are reminded of their humanity. A sergeant observed that lower ranking soldiers were often at the receiving end of orders and bore the psychological brunt of counter-insurgency operations. He pointed out that it was therefore important for soldiers to hear 'a friendly voice'. Especially, he added, because 'I am not a friendly voice. I am not their friend. I am their leader.'

Another potential outcome of the increased connections with home concerns the reintegration of the soldier back into the family upon redeployment. Upon returning home, soldiers typically have wondered if their families still need them or if their children will recognize them. With deployed soldiers increasingly involved through technology in the family's everyday rhythm of life, the redeployment transition is less troublesome. When soldiers were asked about the purpose of their communications, they replied with responses such as 'We just talk about our daily routine and the kids', or 'I try to call home and talk to her and let my kids hear my voice, especially my little one.' In other words, many soldiers were taking deliberate steps to remain involved in their families despite their deployed status.

Ironically, it is actually possible that many young soldiers spend more time interacting with their spouses from the FOB than when not deployed. Soldiers reported that they had established routines of conducting phone calls or instant messaging sessions very early in the morning (e.g., 4 a.m.) and again late at night (e.g., midnight) because of mission cycles and the time difference in the United States. Consistently setting aside specific times of the day for the sole purpose of interacting with one's spouse is a habit that would probably benefit the marriages of all soldiers—not just those deployed. Interestingly, a captain related that one of his concerns was that some of his troops were noticeably tired from waking up very early to communicate with home. Nevertheless, the availability of synchronous communications has great potential in reducing the debilitating effects of a deployment on family relationships. A significant subsequent implication is that soldiers and families could be more apt to tolerate future deployments given this trend toward increased connectivity while deployed.

One might think that the increase in communication could result in compromised security due to sending sensitive information over unsecured lines. Two factors, however, suggest that phone conversations, texting, and instant messaging sessions may be less of a concern. First, nearly all soldiers interviewed were careful to state that they were cognizant about operational security concerns when using technology. In other words, soldiers tended to be heavily sensitized to the hazards of using open internet media. Second, the information transmitted over phone, texting, and IM was usually less fact-laden than letters, emails, or discussion forum postings. As discussed earlier, soldiers tend to use synchronous communications more to discuss family matters, not operational details. Still, even telling a family member that a Skype session has to be discontinued because the soldier is going 'out the wire' is a security breach.

A more likely misuse of instantaneous communications is sidestepping the chain of command during an incident involving unit casualties. It is standard procedure for the chain of command to hit the 'kill switch' on outgoing communications immediately after casualties are known. For official channels (e.g., publicly shared morale computers), the kill switch could be literal. For unofficial channels (e.g., private cell phones), however, the kill switch relies on following the chain of command's directive to enforce a communication blackout.

The Greedy Institutions of the Family and the Military

A less obvious, but potentially more troubling issue concerns the competing demands of the family and the military. Communication advances on the FOB give the family access into the combat environment—a world traditionally dominated by the military's hold on soldier attention and energy. Soldiers have a finite amount of energy and emotional bonds, yet both the family and the military are what military sociologist Mady Wechlsler Segal calls 'greedy institutions'. Greedy institutions depend for their survival on the commitment of their members and as a result demand considerable amounts of their members' loyalty, time, and energy.[34] The concept of the greedy institution has attracted attention in military studies, but in fact, it may substantially underestimate the extent of the transformation which is currently occurring in the domestic–military nexus. The FOB may not be simply a new site of familial–professional interaction where competing claims on service-personnel are enacted and negotiated but may represent a profoundly new institutional arrangement in which the nature of the family and its relationship to the military has been reformed.

At the most obvious level, because technology on the FOB has shifted the family from an abstract to concrete concept, the competition for soldier time and commitment is renewed. A very real possibility emerges that soldiers might be spending time thinking about family concerns when they should be focusing on the mission or nurturing the bonds with their comrades. In other words, increased connectedness with family might adversely affect aspects of task (the mission) or social (the primary group) cohesion. During interviews, soldiers were questioned about such potential problems.

Soldiers responded that while pressing family issues require increased attention, it was possible to remain focused on the mission. One soldier talked about turning off thoughts of his family while out on a mission, 'It's pretty tough, sir. It is. You just shut it off. You make it seem that whatever you're doing at that time is the only thing that matters in this world.' Another soldier commented, 'Once I roll out of the gate, my wife does not cross my mind because I talked with her before, and I can't do my job properly if I think about her. That is the deal.' Another soldier talked of the attention demanded by both his family and his unit, 'I had to learn not to focus on my family while I am out on my mission but focus on my crew, which is my immediate family here.' One soldier spoke of temporarily blocking his family from his mind when going out the wire, 'I am mission focused. I am all about coming back alive and then get my lovey-dovey face back on.'

Some soldiers, however, candidly acknowledged the possibility that family concerns could be a distraction. For example, one soldier reported, 'When I talk to my mom, and my mom will yell at me for not doing the right thing or for not being there—which I can't do anything about—I'll get [distracted] in the head for a day or two.' Nevertheless, a large majority of the interviewed soldiers contended that they were able to balance the attention required by their families and the mission. Nearly all interviewed soldiers asserted that when the situation demanded, the mission received their total focus. It may be that soldiers were over-confident in their ability to divide their attention and are underestimating the effects of the family on their focus. Or it might be that this generation weaned on the frenetic pace of *Sesame Street* and MTV and accustomed to surfing the internet while watching television are truly adept multitaskers capable of compartmentalizing the many demands on their attention.

Greedy institutions demand energy and time, but they also demand a person's total commitment and loyalty. Loyalty to one's comrades is a key aspect of combat effectiveness. The institution of the military—especially at the small-unit level—requires that its members place loyalty above self-preservation. In combat, loyalty to fellow soldiers and the mission must overcome the fear of death. This is the essence of combat cohesion. As Israeli military psychologist Shattai Noy stated, 'The main conflict of a soldier in the battlefield is survival versus duty and loyalty.... In combat, this threat is pervasive and difficult.'[35]

Ben Shalit, Chief Psychologist of the Israel Defense Forces (IDF) during the Yom Kippur War in 1973, interviewed soldiers after battles and asked what soldiers found to be the most frightening or stressful aspect in combat. The most common frightening aspect of battle, as indicated by 40.4 per cent of the junior enlisted soldiers surveyed, was the fear of letting comrades down. Loss of limb or injury was the second most common fear, comprising 26.6 per cent of responses. The fear of death was ranked third, as noted by 20.7 per cent of the respondents.[36]

Combat effectiveness relies upon soldiers placing responsibility for their fellow soldiers above their own personal safety. In the extreme conditions of combat, the military becomes the ultimate greedy institution by demanding that soldiers fear letting their comrades down more than the possibility of their own death. With the greedy institution of the family entering the scene via increased connectedness, however, an interesting question emerges—does the family compete for deployed soldiers' loyalties? In other words, as the family becomes a more concrete concept in the minds of young soldiers, does it affect their obligations to the unit?

Samuel Stouffer addressed this issue with Second World War soldiers. He wrote:

> Closely related to this struggle [of overcoming concern for personal safety] is the problem of the tension between the requirements of military duty and the individual's felt obligations to a social circle at home. Many men felt a deep obligation to a wife and children, or to a dependent mother and father. The society at large indicated that their highest obligation was to fight for their country. But it would be surprising if such men did not feel a strain between these obligations. Should they take risks which they might avoid but which seemed necessary for most effective accomplishment of their military mission?[37]

Stouffer proposed that the competing demands of the mission and the family should be manifested by greater worry and fear in the battle situation. In a survey of combat soldiers in Italy, he confirmed that married men were more likely to say that they worried a lot about their chances of becoming a casualty when they were in combat.[38] Yet the family was a motivating factor for soldiers on the frontline precisely because they were concerned that any inadequacies they displayed in combat would redound upon the reputation of their kin at home. Roger Little noted the use of similar moral leverage during the Korean War. Little noted that 'events which reinforced the soldier's relations to some meaningful element of the larger society—especially his family—correspondingly strengthened his relation to his organization'.[39] The family was actively employed to encourage honourable performance from US soldiers. The US Army recognized that the letters were vital in maintaining this connection and pressure: 'The conception of his role as the citizen of a community or a member of a family was influenced by the letters written to him by persons whose evaluations of him were very important.'[40]

Soldiers interviewed for this study were asked the same question posed by Ben Shalit—'What was the most frightening aspect of battle?' The responses were interesting. Because the improvised explosive device (IED) was the most prominent threat at the time of the study, soldiers were generally fatalistic about what they feared. They often related that there was no reason for any fear since the main threat was, in their minds, a random occurrence. Sniper attacks and IEDs were dangerous, but it was not worthwhile to be frightened because there was nothing they could do about it anyway. One soldier put it this way, 'It is not going to happen unless the good Lord intends. If he punches your number, there is nothing you can do.'

Very few soldiers commented that they were concerned about the possibility of their own death or injury. Most mentioned little about the fear of letting their comrades down. After some reflection, however, a surprising majority of soldiers commented on how the *consequences* of their death or wounds frightened them. One infantryman gave a typical response, 'I could care less if I get shot, sir... My kids growing up with no dad—*that* scares the [life] out of me.' Another soldier noted, 'I try not to think about it. If it happens, it happens. You know, I sometimes think about what my family is going to do if something happens to me—to where I am incapacitated or killed, but I try not to put too much thought into that.'

An infantryman who commented that he had no specific fears in battle stated: 'If anything, I worry. I worry about the well-being of my soldiers. I worry about me not coming home. I don't worry about me personally, I worry for my kids. I have taken such great pride in being a father to my children.'

He went on to add:

> I just want to be there for [my wife]. She just wants me to be there for her. So I worry about not being able to come home to her. It doesn't matter if I came home in a different condition as long as I come home for her. So those are my worries—never fear.... Since I have been in the Army, sir, I just haven't been scared of anything.

As with Shalit's findings with the IDF in the early 1970s, self-preservation for the twenty-first-century soldier appears not to be a major source of fear or stress. Unlike Shalit's findings and more in line with Stouffer, however, today's soldiers do show concern for their own safety *when considering the impact on their families*. It appears that because the family has emerged as a concrete concept in the world of today's troops, the institution of the family is competing for soldiers' attention, as evidenced by observations such as, 'I worry more about how my wife would react to something happening to me more so than I would worry about me personally getting hurt.' Concerns voiced by soldiers revolve not around their own personal injury, but instead on the effect an injury would have on their ability to continue being the family provider and supporter. Soldiers listed concerns such as, 'What will happen to my family? How are my wife and kids going to handle it?' and 'How will it affect my wife? Will they be taken care of?'

Ironically, while FOBs have removed many traditional distracters and sources of stress for deployed soldiers, allowing the greedy institution of the family access through improved technology may have put demands on soldiers at levels not experienced in previous conflicts. Soldiers on FOBs interact daily with their families, resulting in stronger emotional family bonds and in turn raising their family's welfare as an aspect of battle that frightens them the most.

Of course, other factors may have also been contributing to the increased salience of the family in the minds of soldiers other than just increased interaction with the home front. First, today's force is more married than in the past. Currently, 41 per cent of the enlisted force is married, while the junior enlisted soldiers studied by Stouffer in the Second World War reported a 36 per cent marriage rate.[41] Additionally, today's soldiers may be more motivated to cite family concerns as a source of fear in battle based on their own family experiences. They come from an American cohort increasingly accustomed to growing up in single-parent homes. Currently, 27 per cent of American youth come from single-parent families—three times the amount in 1960.[42] Soldiers may be stating their desire to avoid that situation. Finally, soldiers may be more apt to be concerned about their role in supporting their families simply because of changing views on the responsibilities of parents for children. Figure 13.1 provides analysis from a recent Pew Research Center report showing that younger people are consistently more likely than older people to say that it is 'extremely important' for a father to provide more than just income for his family. (Similar results were found for roles for mothers). Twenty-first-century soldiers see their responsibility as parents as more than being just the breadwinner. More young people perceive instilling values, providing emotional support, and enforcing discipline as critically important aspects in their parental role.

It is important to recognize that the importance which soldiers attach to their domestic relations is not merely a personal matter. Rather it points to some profound institutional changes in Western society. Families are, of course, ubiquitous in all human societies, and the nuclear family has become institutionalized in Western Christian society for centuries. Accordingly, it is easy to overlook its changing significance and the alterations of relations within the family because the structure of the nuclear family (so long as parents remain together) has remained the same. Yet, as a result of globalization, established class, gender, ethnic, and racial hierarchies have been substantially revised in Europe and North America. At the same time and especially since the 1970s, the state has retreated from many areas of activity which it regulated in the middle decades of the twentieth century, and withdrawn or privatized some social welfare provisions which were available at that time. Changing patterns of employment have also demanded more

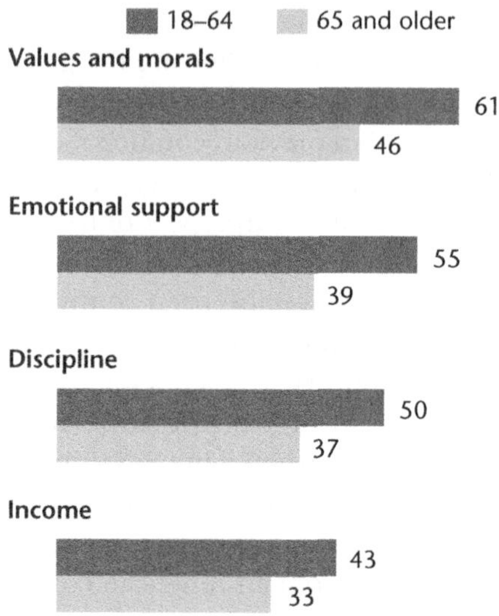

Figure 13.1. Father's role by age. Percentage of each age who say it is 'extremely important' for a father to provide this for his children.[43]

Source: "The New American Father" Pew Research Center, Washington, DC (June, 2013) http://www.pewsocialtrends.org/2013/06/14/the-new-american-father/

Notes: Sample size of 18–64 year old age group is 685; for 65 and older n = 294. An additional 25 respondents refused to give their age.

geographical mobility while offering less security to the labour force, displacing the workplace and its unions or the professions and their associations, clubs, and communities as a prime site of social solidarity. Of course, female emancipation has also become an important factor here. As women have participated more fully in the workforce and become more equal in the domestic sphere, male participation in external social organizations has declined as their contribution to childcare and domestic duties has increased. The family has turned inwards and relations between parents and children have been intensified to a degree which would have been unusual in the 1950s and unheard of before the First World War. The consumer market has played a role in this elevation of the family because firms have increasingly recognized the potential of the domestic sphere as a market and have sought to develop products for safe familial consumption.

In his widely read book *Bowling Alone*, Robert Putnam seeks to analyse the decline of American civic life in the last three decades of the twentieth century, and his work is directly relevant to the rising importance of the family. For Putnam, one of the most important collective goods which any society possesses is what he terms 'social capital'. Social capital refers to the dense bonds of reciprocity between citizens which encourages them to give and receive help from each other independently of the market. A strong society, according to Putnam, is one whose members are willing to band together in communities and associations providing not only essential services, like social and health care for each other, but also a vibrant public life in which individuals interact and communicate with each other.[44] For Putnam, the United States has been going through a collapse of civic engagement or of social capital since the 1970s. Although Putnam does not discuss the issue, the rising importance of the family as a prime site of social solidarity might be understood as compensation for these communal organizations: 'Large groups with local chapters, long histories, multiple objectives and diverse constituencies are being replaced by more evanescent, single-purpose organizations, smaller groups.'[45] One of these smaller groups might be the family, which is a central source of social (and economic) capital.

The emergence of the family as a prime site of social solidarity has necessarily altered the domestic–military nexus. Yet the effects have been paradoxical. At one level, because wives (or husbands) now work, families are often more detached from the armed forces as an institution; they are not completely dependent upon the military with their social life indivisible from regimental duty as they were in the past. At the same time, because of the cultural transformations, soldiers are bound into intensified domestic relations with their parents, siblings, and children which invade once purely professional domains. Soldiers increasingly understand themselves and define their service in domestic terms. This has been particularly obvious in new practices of commemoration which have emerged in Iraq and Afghanistan, as Anthony King demonstrates in his work on memorial practices in Britain.[46] There he shows how, in the absence of a strategic rationale to explain casualties, British soldiers and the Ministry of Defence have commemorated fallen soldiers not only as professionals but also, decisively, as fathers, wives, sons, and daughters. Compatible processes are observable in the United States, though often articulated in alternate ways at the local level. The domestic sphere has penetrated into the very domain of combat. Even as the two institutions retreat from each other at one level, relations between individual service-personnel and their kin have intensified precisely because the family has become a critical site of social solidarity and capital; the domestic and military spheres now interpenetrate in quite new ways.

As conditions on the FOBs improved, technological advances such as phone calls, texting, and instant messaging have allowed soldiers to step back into their roles of husbands and fathers (and wives and mothers). Soldiers in the past two centuries have looked forward to hearing their names called during mail call; families were always, of course, important to them, as Stouffer and Little show. Yet the FOB represents a profound revision of this relationship. Instant connectivity from the FOB to the family has become a necessity for both soldiers and their kin. By remaining an integral part of the family problem-solving process, soldiers continue to feel needed and valued by the family; they are able to reaffirm their domestic relations.

Although the connectivity with the family significantly increases the level of conscious awareness of familial obligations and concerns, it also most likely places additional stress on the soldier. As mentioned earlier, for instance, the soldiers' major concern for their own safety focused on the impact of their death on their family. Taken to the extreme, however, the perceived responsibilities to the family could result in a soldier's hesitation to take risk required for mission accomplishment or unit welfare. Although this study found no evidence of this occurring, it remains a possibility. Additionally, despite assurances from multitasking young soldiers that they can compartmentalize the various demands on their attention, spending time ruminating on family concerns and events can disrupt a mission focus. It is difficult to imagine that a soldier who learns from his wife during a phone call at 8 p.m. that his five-year-old is having significant behavioural problems in school, will be able to devote complete attention to a mission occurring thirty minutes later. Special Forces teams understand this and thus go into isolation prior to a deployment in order to 'get their heads into the game'. Yet, whatever its operational costs, there will be no return to the distant (but important) influence which the family once exerted on the deployed soldier in the twentieth century. In the twenty-first century, the domestic sphere has penetrated into the very heart of the armed forces even on operations to produce a novel hybrid familial–professional institution where soldiers maintain their bonds with each other simultaneously and in parallel with equally important relations with their spouses and children.

Finally, if the interconnectedness with family satisfies (or depletes) the soldier's need (or supply) of emotional support, it is possible that it comes at the expense of the social relations with comrades. Just as the FOB has removed many of the past stressors of combat, the introduction of the family via technology into the combat environment may be removing the need for—and influence of—social cohesion. The result may be a more professional, impersonal—and more task cohesion–oriented—military. Alternatively, the rise of professionalized forms of cohesion, described in this volume and in King's work on the infantry platoon, may have actually precipitated the increased importance of the family to deployed soldiers. Precisely because soldiers relate to each other as professional teammates, the family has become an increasingly important site of emotional catharsis and support; it compensates for a decline of interpersonal solidarity. At the same time, the infiltration of the family into the combat unit may have also precipitated the depersonalization of primary group relations. Since soldiers seek emotional succour in the domestic sphere, they are less emotionally dependent upon each other. Their relations can become more businesslike. It is difficult to isolate the direction of causality definitively. However, it seems certain that the current renegotiation of the domestic–military nexus has very significant effects on soldiers and their relations with each other.

Conclusion

In an odd twist, the same issues associated with fighting from a FOB were played out 7000 miles away at Nellis Air Force Base, Nevada—albeit with a slight variation. Instead of a battle rhythm of fighting the war in person and then returning to a FOB to connect virtually with families, Nellis pilots were remotely controlling Predators—Hellfire missile-armed, unmanned aerial vehicles (UAVs)—in the war zone and then commuting home to their families in person. In other words, these pilots were fighting a long-distance war via technology, but remaining at home. Despite the benefits of being home to fight the war (which paralleled the many amenities of the FOBs—safety, rest, food, and comfort), many Predator pilots reported diminished morale and more problems in their personal lives. One pilot commented, 'I get to sleep in my own bed', but then pointed out that he feels pulled in two directions—between spending more time fighting the war and being an integral part of his family. Interestingly, he added that it was easier when he was physically deployed to war because he was isolated from the emotional demands of family life.[47]

The commander of the Predator unit commented on the psychological impact of this type of warfare: 'Our people are proud they contribute to the war from home. But being at home brings some additional stresses.'[48] A journalist chronicling the Nellis unit contrasted the stresses of Predator pilots to those of deployed soldiers. She concluded that 'soldiers in the field have to cope with danger, but at least they live in one world, whereas their counterparts at Nellis commute daily from war to civilian life'.[49]

In reality, many twenty-first-century soldiers operating out of FOBs experience the same issues from commuting to war. They appreciate creature comforts like the good and abundant food, refreshing air conditioning, hot showers, and safety afforded by the FOB. But like Predator pilots, many soldiers not only experience the competition between the greedy institutions of the family and the military for a limited amount of attention, time, and emotional capital but constitute the key actors in a new domestic–military nexus. Fighting from the FOB changes the fighting capacity of combat soldiers, but it also appears to change the calculus of combat cohesion. It appears that despite the voluminous literature on the bonds between soldiers, new developments in the way of waging war make combat cohesion a topic that continues to be in need of further research and study.

Notes

1. The views expressed in this article are those of the authors and do not necessarily reflect the official policy or position of the US Army War College, the US Army, or the US Department of Defense.
2. Roger Little, 'Buddy Relations and Combat Performance', in Morris Janowitz (ed.) *The New Military: Changing Patterns of Organization* (New York: Russell Sage Foundation, 1964), 218.
3. Brigadier General Karl Horst in David Zucchino, 'The Conflict in Iraq; Comforts of Home Amid Perils of Iraq', *Los Angeles Times*, 27 March 2005, A1.
4. Conditions at individual FOBs can vary greatly. On one extreme were the huge FOBs located near Baghdad or Bagram that boasted air-conditioned sleeping and work trailers, cavernous dining facilities, a spacious post exchange (department store), cappuccino bars, well-stocked gyms, and internet cafes. On the other end of the scale, some FOBs were located in old Iraqi or Afghan bases or abandoned factories with soldiers still living in tents, food ferried from other FOBs, and more austere conditions. Overall, however, most soldiers lived on FOBs somewhere in the middle range—air-conditioned quarters, a small PX, a MWR facility with internet and phone lines, and a contracted dining facility.
5. Charles Moskos, Charles *The American Enlisted Man. The Rank and File in Today's Military* (New York: Russell Sage Foundation, 1970), 141.
6. Peter S. Kindsvatter, *American Soldiers: Ground Combat in the World Wars, Korea, and Vietnam* (Lawrence: University Press of Kansas, 2003), 49.
7. An extreme example was the 14 September 2012 attack on Camp Bastion that killed two US Marines and destroyed or severely damaged eight aircraft.
8. This chapter relies heavily on observations collected from US enlisted soldiers deployed to OPERATION IRAQI FREEDOM in May 2005—specifically over fifty structured interviews conducted with male and female soldiers in the ranks of staff sergeant and below in locations throughout Iraq and Kuwait. Interviews included combat arms as well as combat service support soldiers and involved active and reserve soldiers. The only prerequisite for inclusion in the interview sample was that the soldiers had to regularly carry out missions off the FOB (i.e., the study does not include soldiers who spend their combat tour exclusively on the FOB—and are irreverently called 'fobbits').
9. Gregory L. Belenky, *Effects of Continous Operations (CONOPS) on Soldier and Unit Performance* (Walter Reed Army Institute of Research, 1987), 2–6.
10. Samuel Stouffer, Arthur Lumsdaine, Marion Harper Lumsdaine, Robin Williams, M. Brewster Smith, Irving Janis, Shirley Star and Leonard Cottrell, *The American Soldier*, vol. 2: *Combat and Its Aftermath* (Princeton, NY: Princeton University Press, 1949), 78.
11. James R. Ebert, *A Life in a Year: The American Infantryman in Vietnam 1965–1972* (Novato, CA: Presidio, 1993), 154.
12. John Dollard, *Fear in Battle* (New Haven, CT: Institute of Human Relations, Yale University, 1943), 52.

13. Stouffer et al., *The American Soldier*, 79.
14. Deepa Babington, 'No Fries With That: Fast Food Axed at Afghan Bases', *Reuters*, 5 April 2010 <http://www.reuters.com/article/2010/04/05/us-afghanistan-fastfood-idUSTRE6342J720100405>. Accessed 2 February 2015.
15. Kindsvatter, *American Soldiers*, 95.
16. Roscoe C. Blunt Jr, *Inside the Battle of the Bulge: A Private Comes of Age* (Westport, CT: Praeger, 1994), 121, as noted in Kindsvatter, *American Soldiers*, 95.
17. Dollard, *Fear in Battle*, 42.
18. Sergeant Jeanne Crochet quoted in Antonio Castaneda, 'Between Bombs and Incoming Shells, Tanning and 'Star Wars' Offer Soldiers Respite from War', *Associated Press News*, 10 June 2005.
19. Juliet Macur, 'At War in Bomb Field, at Peace in Field of Dreams', *New York Times*, 19 August 2005, A1.
20. V-mail, online exhibit, National Postal Museum <http://postalmuseum.si.edu/exhibits/past/the-art-of-cards-and-letters/mail-call/v-mail.html>. Accessed 10 February 2015.
21. For a brief history of the Military Auxiliary Radio System program, see <http://www.ckycs.com/kyarmymars/whatismars.htm>. Accessed 10 February 2015.
22. See Larry W. Applewhite and David R. Segal, 'Telephone Use by Peacekeeping Troops in the Sinai', *Armed Forces & Society* 17(1) (1990): 124.
23. For example, see Ward S. Just, *To What End: Report from Vietnam* (Boston, MA: Houghton Mifflin, 1968), 23.
24. Rick Marshall, 'Battlefield Communications: Technology's Challenge,' *Defense & Foreign Affairs* (October 1986): 26.
25. See Morten G. Ender, 'G.I. Phone Home: The Use of Telecommunications by the Soldiers of Operation Just Cause', *Armed Forces & Society* 21(3) (1995): 440.
26. See D. Bruce Bell, Walter R. Schumm, Benjamin Knott and Morten G. Ender, 'The Desert Fax: A Research Note on Calling Home from Somalia', *Armed Forces & Society* 25(3) (1999): 504.
27. For the Gulf War, see Morten G. Ender and David R. Segal, 'V(E)-Mail to the Foxhole', *Journal of Political and Military Sociology* 24(4) (1995): 94.
28. See Dan Morse and Jesse Drucker, 'Private Messages: Soldier Helps 101st Stay Wired to Home', *Wall Street Journal*, 25 March 2003, A1.
29. Unless, of course, the unit takes casualties or they are engaged in operations reported in the media. For those occasions, soldiers are quick to communicate that they are all right.
30. Judith Donath, 'Sociable Media', in *Encyclopedia of Human-Computer Interaction*, vol. 2, ed. William Sims Bainbridge (Great Barrington, MA: Berkshire Publishing Group, 2004), 629.
31. Michael D. Doubler, *Closing With the Enemy: How GIs Fought the War in Europe, 1944–1945* (Lawrence: University Press of Kansas, 1994), 257 [emphasis added].
32. Headquarters, Department of the Army, DA Pamphlet 608–47, *A Guide to Establishing Family Support Groups* (Washington, DC, 1993), 2.
33. Jonathan Shay, *Achilles in Vietnam: Combat Trauma and the Undoing of Character* (New York: Touchstone, 1994), 86.

34. Mady Wechsler Segal, 'The Military and the Family as Greedy Institutions', *Armed Forces & Society* 13(1) (1986): 9, based on the work by Lewis A. Coser, *Greedy Institutions: Patterns of Undivided Commitment* (New York: Free Press, 1974).
35. Shattai Noy, 'Combat Stress Reactions', in Reuven Gal and A. David Mangelsdorff (eds.) *Handbook of Military Psychology* (New York: John Wiley & Sons, 1991), 511.
36. This data is derived from Ben Shalit, *Psychology of Conflict and Combat* (New York: Praeger, 1988), 11.
37. Stouffer et al., *The American Soldier*, 84.
38. Stouffer et al., *The American Soldier*, 87.
39. Little, 'Buddy Relations and Combat Performance', 218.
40. Little, 'Buddy Relations and Combat Performance', 219.
41. Current marriage are rates from Headquarters, Department of the Army <http://www.armyg1.army.mil/hr/docs/demographics/FY11_ARMY_PROFILE.pdf>. Second World War rates are from Samuel A. Stouffer, et al., *The American Soldier: Adjustment During Army Life*, vol. 1 (Princeton, NJ: Princeton University Press, 1949), 114.
42. US Bureau of the Census, *Living Arrangements of Children: 2009*, June 2011 <http://www.census.gov/prod/2011pubs/p70-126.pdf>, 4.
43. 'The New American Father', *Pew Research Social & Demographic Trends*, Pew Research Center, Washington, DC., (14 June 2013), <http://www.pewsocialtrends.org/2013/06/14/the-new-american-father/>. Accessed 2 February 2015.
44. Robert Putnam, *Bowling Alone* (London: Simon and Schuster, 2000), 23–8.
45. Putnam, *Bowling Alone*, 184.
46. Anthony King, 'The Afghan War and "Postmodern" Memory: Commemoration and the Dead of Helmand', *British Journal of Sociology* 61(1) (2010): 1–25.
47. Major Shannon Rogers as reported by Sally B. Donnelly, 'Long-Distance Warriors', *Time*, 12 December 2005, <http://www.time.com/time/magazine/article/0,9171,1137653,00.html>. Accessed 2 February 2015.
48. Lieutenant Colonel John Harris as reported in Donnelly, 'Long-Distance Warriors'.
49. Donnelly, 'Long-Distance Warriors'.

14

Beyond the Battlefield

Christopher Dandeker and Simon Wessely

Introduction

A long-standing theme in the 'war and society' research tradition is that military performance is affected by factors both on and off the field of operations. In Chapter 13, Lenny Wong and Stephen Gerras explored the changing relations between the military and the domestic sphere. Yet the interaction between civilian and political agencies and purely military activity has long been noted. Examples from the late nineteenth and early twentieth centuries include the impact of changes in industrial technology on firepower and the dispersion of the control of soldiers, and the effects of civilian legal and social norms on what kinds of discipline (for example, corporal punishments) can be justified during military operations when soldiers cannot be considered just as objects of the military machine but also as citizen subjects. The literature on combat motivation tends to generalize from the army experience, an observation that does not apply to the important work of Kellett.[1] Another, more recent, example is the extent to which just war principles affect not only how states legitimize engaging in an armed conflict but also how they and their publics hold soldiers to account for their conduct in military operations; thus some writers, like Helen McCartney, suggest that, partly because of media coverage, soldiers are being held to higher standards than before.[2] By the same token, soldiers themselves as well as their families might have higher expectations of the standards of care and support than they receive before, during, and after military operations. While it is a well-established fact that one of the factors that motivates soldiers to fight in battle is loyalty to the cohesive, small combat group, Charles Moskos contended that cohesion 'will maintain the soldier in his combat role only when he has an underlying commitment to the worth of the larger social system for which he is fighting'.[3] He referred to this as a more or less conscious commitment to a 'latent ideology', which, from the point of view of our argument, can encompass a range of considerations: from a soldier's belief in the cause for which a war is being fought, the military institution of which he or she is a part, to the family and friends to whom the soldier has commitments and wishes to return after the mission has been completed. Another element is the soldier's own desire to survive and how to temper that desire with the obligations he or she feels to others in the small combat group and wider loyalties.

This chapter examines sociopolitical—and above all the legal—factors beyond the battlefield and their effects on military performance in twenty-first-century military operations and how soldiers balance their commitments to their own survival and wider obligations. It focuses on the effects of social and legal norms on the combatant status of soldiers and the implications for their military performance. An overarching theme in the discussion is the idea of *'the duty of care'*, which has both moral and legal connotations: the moral obligation of governments (and the public) to provide medical and other support for their soldiers in operations, and legal requirements, breaches of which are justiciable, so that action may arise from this consideration by soldiers and or their families. This distinction between the moral and legal aspects of the duty of care lies at the heart of debates about the meaning of the 'military covenant'. Measures designed to reduce the risk of casualties, such as personal body armour, might also inhibit movement under fire. Casualty evacuation and the minimization of the risk of a wounded Service member becoming a death can slow up an operation. Realistic training for an operation at home or in theatre but within a base has to be constrained by human rights considerations, which now, as anticipated by some observers five years ago, are entering into the field of operations itself. Do duty of care considerations underpin or undermine the 'risk culture' that is an essential component of military performance? What can we learn about these matters from a decade of war?

The Contract of Unlimited Liability: Psychological, Legal Social, and Cultural Aspects

An important theme in contemporary civil–military relations and military ethics is the contract of 'unlimited liability' that the soldier has with the state. In addition to obedience to the legal orders stemming from his or her client—the state—the soldier has to accept that military service involves subordinating the self to collective purposes on a 24/7 basis, and that this may involve the risk of serious harm or death as he or she applies and receives lethal force in operations.[4] In short, soldiers may be required to kill and risk being killed in carrying out their duties.

This contract between the soldier and the state is not static but a dynamic one. It has a psychological aspect: soldiers seek to manage their liability or the risks of death they face in combat in different historical situations. This is a striking theme in Ben Shephard's excellent *A War of Nerves*.[5] A fundamental characteristic of military life is what may be termed the soldier's dilemma. The soldier, engaged in battle, not only has a sense of duty and sacrifice but also an instinct for personal survival. The soldier's dilemma is how to balance the demands of the organization with his or her personal needs. How the soldier's dilemma is resolved, and whether the soldier wears out or his or her 'nerve' breaks, will depend upon a number of factors. These include: the proximity to and the experience of combat, including the fighting environment (on land, at or beneath the sea, or in the air); the length of time spent in stressful combat operations; the social dynamics of military groups—at both the micro-level of small groups as well as at higher levels of military organization; the provision of social support from military and non-military organizations, including the families of military personnel, and, not least, the nature of military leadership.

Another aspect of the dynamics of the soldier's dilemma is that it is shaped by power struggles between those at either end of the chain of command. As Shephard shows, military history is replete with examples of the tensions that will, necessarily, arise between frontline troops and staff organizations. Referring to the experience of RAF bombing crews engaged in missions over Germany during the Second World War, he argues that '(a)t times of great strain a battle of wills would develop between the aircrews and the staff attempting to keep the attack going'.[6]

The dynamics of the soldier's dilemma and breaking point are also influenced by the wider society. For example, the British armed services could rely on ideals of patriotism and duty during the First World War. However, this was less the case in 1939, when patriotism had become more *conditional*. Citizens recruited to the armed services were better fed, in better physical condition than their 1914 counterparts; they were also better educated and politically aware, and both able and willing to question whether their psychological contract with the Services was perceived to be fair. As Paul Fussell argues, they also were very much aware of the First World War experience. Appeals to uncritical patriotism had less resonance directly because of that experience, and indeed it tended to demotivate rather than motivate Fussell.[7] This trend towards 'conditional patriotism' led to a century-long discussion about whether citizens were becoming less tolerant of the physical and emotional demands of warfare. More recently, this has led to debates on whether civilian populations and the armed services are increasingly casualty-shy and risk-averse—and indeed, as in the US, whether the latter is more cautious than the former—a point considered further below.[8]

These debates have also focused on the social and military significance of masculinity (note, all the material Shephard uses is on male personnel) and the consequences of what he refers to as the 'feminisation of public life'.[9] By this he means the tendency (perhaps most evident in Anglo-American culture) for an increased willingness to display emotions in public; to admit that existential problems trouble one and would benefit from the interventions of therapists. From a military point of view, Shephard mused on whether this cultural trend would make today's younger generation less able or willing to take on the stresses and strains of the soldiers' dilemma.[10] Some senior military personnel are themselves concerned about the impact of risk aversion on military performance; but this may well have independent roots in the fear of currently serving personnel that their decisions, which necessarily risk life, will attract litigation from the personnel whose lives are affected by them.

Shephard argues that the profession of military psychiatry created a discrete field of expertise, which was defined by a cluster of syndromes during the twentieth century. Starting with debates over the nature and causes of 'shell-shock', it culminated with the formulation of Post-Traumatic Stress Disorder (PTSD) in the wake of the Vietnam War. PTSD was to be found in both military and civilian contexts (such as the aftermath of catastrophic accidents) and it also became evident that an adequate psychiatry had to explore the wider sociological context of medical health and illness. However, this argument is not without its problems. Wessely suggests that what happened with PTSD was exactly the opposite: while the Second World War and immediate Second World War literature, especially in the US, was both sociologically and analytically driven, the arrival of PTSD, with its emphasis on the purely personal and individual, meant that social and group influences almost entirely disappeared from the trauma narratives. Something like PTSD was indeed a restatement of the causes of breakdown—removing the issues of character, personality, and predisposition thought so important to the First World War and Second World War generations, but also moving away from the concepts of the critical role of group cohesion, membership, and motivation that arose out of the Second World War experiences and the nature of fighting groups, replacing it instead with a remarkable emphasis on the meaning of the trauma to the individual, and the individual alone.[11]

PTSD was also linked with changing social attitudes to health and personal responsibility and, in the military context, shifts in the power relationships between soldiers (including veterans) and the chain of command. Today, in the US, UK, and elsewhere, soldiers and veterans are increasingly able and willing to contest with the armed forces on whether the latter have been negligent in meeting their responsibility to provide a duty of care by controlling the risks to which personnel are subject: soldiers are redrawing the psychological contract; they are asking more in return for their providing commitment and sacrifice. In their dealings with soldiers, since the middle of the twentieth century, the chain of command has had to recognize that their judgements need to take into account assessments of military psychiatry, although this was tempered by considerable scepticism from that quarter. Indeed, it was only at the end of the century, and especially post-Vietnam, that few commanders would dare question (at least publically) the need to take direct note of mental health issues and engage with mental health services. Even if private scepticism remains, it would be for many, we suggest, a career-ending move to publicly question this. In any event, by the end of the twentieth century the autonomy of military judgement itself, on a whole range of medical and other social issues, was being eroded by the scrutiny of civilian courts (in the UK case, by legal authorities outside the UK such as the European Court of Justice and the European Court of Human Rights). Such are the consequences of the trend towards conditional patriotism.

The above developments should be understood in the broader historical context, especially the changing character of war and public perceptions of armed conflict and what the public considers to be appropriate conduct by soldiers and governments. Helen McCartney argues that '(i)n western liberal societies, the legitimacy of the use of force is increasingly being questioned'.[12] She suggests that this long-term trend has been accentuated by the wars of 9/11, which despite efforts by President George W. Bush, and for a brief period President Barack Obama, could not (especially after the fall of Saddam Hussein and the beginning of the reconstruction and counter-insurgency period) be sold to the public as existential conflicts or 'wars of necessity'. As Patrick Porter has shown, the popular distinction between wars of choice and those of necessity is misleading: even so-called existential conflicts involve choices. However, the material point here is that the public is cautious and sceptical about the wars of 9/11 and whether they were worth fighting and incurring the huge costs; and their scepticism is rooted in the sense that they were wars that were more discretionary than governments argued both at the time and since.[13]

Furthermore, public doubts about the rationale for war—*ius ad bellum*—coincides with potentially unrealistic demands for the military to apply the just war *ius in bello* principles of proportionality, discrimination, and minimal collateral damage to civilian non-combatants. These additional pressures on military personnel are reinforced by technology—intense media scrutiny of most but not all military operations (but even 'black' and Special Forces operations cannot avoid some media questioning) and changing attitudes to risk and risk taking in matters far broader than war and international security. McCartney suggests that the public have reflected on the impact of the technology of war, and increasingly expect bloodless campaigns and ones in which the precision of munitions as well as military strategy means that toleration of error is less than it was. An interesting point in this connection is made by Mark Urban concerning the contrast between the realities of the accuracy of modern weapons and frequency of incidents of 'friendly fire' and public perceptions that these are regular occurrences.[14]

In all of this we need to be cautious about using the term 'public', as some of the pressures from this quarter come from its more or less organized sectors of NGOs, pressure groups, and media (now especially social media) elements. As with questioning of the legitimacy of force and demands to implement just war principles more stringently in the conduct of war, the public's attitude towards risk is rooted not only in events of the past decade but reflects a longer term trend. Notwithstanding the advances in science and technology that will continue to transform our lives for what most would regard as the better in terms of longevity and standards of living, there is a dark side to this sense of 'progress'. There is a growing popular culture of distrust in the truth claims of scientific experts and whether they do more harm than good. The literature on risk and trust is directly relevant to the changing position of the professions, including the military profession.[15] This culture of distrust is encouraged in part by the proliferation of sources of professional and pseudo-professional advice, especially via the internet,[16] on the nature and management of risk, which can promote complex and contradictory feelings of being able to control one's own destiny as well as feelings of bewilderment and powerlessness in managing one's own risks and life course. These feelings can, in turn, lead to a variety of responses: some may experience a paralysis of risk aversion, others may want to exercise their freedom but also wish to have advantages of support from the litigation and the 'blame and compensation culture' when their choices 'go wrong'; and these individuals can include those who positively seek the 'buzz' of risk in controlled environments. A decade ago, the Henley Management Centre 2002 highlighted this paradoxical situation very well in one of its predictions for 2020: that 'limited risk is a major leisure activity but the real thing seriously scares people'.[17] It seems likely that the domestication of the armed forces to which Leonard Wong and Stephen Gerras allude might be seen as part of this process. Precisely because soldiers are increasingly understood in familial terms and their relations to their spouses and children have become an increasingly salient element of military (and of course, civilian) culture, casualties are weighted with far greater significance than in the past.

So far as military affairs are concerned, the key question is to what extent does the new risk-averse/individualized culture affect cohesion on the frontline and combat performance? Are risk-averse tendencies presenting a challenge to the traditional military values of courage, 'can do', and risk taking? Will future recruits be reluctant to join the most risky components of the armed services (such as infantry and armour roles) and/or be keen to enjoin their military employers to deploy them in such a way as to minimize the risks they take, even at the cost of collateral damage to others?[18] The evolution of what can be termed risk culture in wider society has not gone unnoticed by soldiers and politicians. Some soldiers (and the MOD) think that a climate of risk avoidance and blame and compensation culture undermine a key element of military organization and the beliefs that underpin it: the need to take calculated risks in the prosecution of operations—such calculated risk is very much not recklessness and has to take into account the possibility of casualties. Casualty free operations, especially in environments such as Iraq and Afghanistan are, however, unrealistic scenarios, although government perceptions that such scenarios figure strongly in the public mind could lead them to choose armed conflicts and how to fight in them in ways that do indeed minimize the risk of home casualties. And they have to take into account public resistance to any perceived cavalier attitude to non-combatant and to combatant casualties. These considerations, together with austerity suggest that Libya and Mali point to the future than do Iraq and Afghanistan—what Christopher Dandeker has called a shift to 'labor-lite' operations.[19]

This discussion reflects the dynamics of the contract of unlimited liability that has underpinned the chapter up to now: that the contract is a relationship of reciprocity, with the terms of exchange between the parties subject to alteration in different historical contexts. Thus, reciprocity is not only based on pay but also, as reflected in the Military Covenant, now enshrined in law, on an expectation that in return for sacrifices and the foregoing of certain rights such as the right to strike, the soldier will receive support and services 'reward and recognition'. Indeed, in the UK, the government's intention is that the soldier will not suffer unfair disadvantage relative to civilians, and in some instances be given special consideration, for example bereavement or injury (physical and psychological).[20]

An additional element of the contract, as the recently developed concept of the 'military covenant' makes clear, is that this relationship is not dyadic but Trinitarian, extending to the wider public who, in return for the sacrifices soldiers make on their behalf, is expected by the military and government to offer support and understanding directly to the armed forces. In addition, they are expected to deliver more indirect support to the armed forces through paying taxes for defence, supporting government measures to underpin the health and welfare of Service and ex-Service personnel, as well as contribute to the third or Charitable sector through organizations such as The Royal British Legion, the Soldiers Sailors RBL, Soldiers', Sailors' and Airmen's Families Association, which when merged with Forces Help Society was rebranded as SSAFA in 2013, Help for Heroes, Combat Stress, and other smaller actors.

In turn, the armed forces, in order to receive support from society as well as government, must live up to their own covenant obligations: to understand

and apply the rules stemming from just war principles and also those aspects of civilian law that apply to them as citizens even in a demanding environment. Breaches of this dimension of the Covenant bring the military into disrepute and, as with any other tainted brand, can lead to withdrawal of public support. Thus, a further aspect of the Military Covenant pertains to the relationship the soldier has with the officers in authority over him or her: the soldier owes obedience and loyalty to the officers who in return are supposed to care for those in their charge. Breaches of this relationship, for example bullying of personnel (or more extreme instances of unauthorized uses of force against POWs or civilians), can damage the reputation of the armed forces and can undermine the support and trust that the military expects and needs from the public to buttress its legitimacy as a social institution. This legitimacy and reputation rests not only on the support of domestic public opinion but wider international publics because what the armed forces do is increasingly available to the scrutiny of 'global surveillance'.[21]

The Military Covenant is a problematic component of public and military policy. Originating in army doctrine and having migrated to the public and now enshrined in law, it provides an important, but unstable, means of military and government attempts to frame civil–military relations and the respective roles of the government, the public, and the armed forces. As Sarah Ingham has shown, the British Army sought to use the Covenant not only to remind soldiers of their obligations—their contract of unlimited liability—but to request what they regarded as appropriate levels of support and undertaking from public and government—not least equipment and resources. Indeed, it was an attempt to defend itself and its professional space from potentially intrusive and limiting legal reforms and subsequently to justify what it was doing in Basra and Helmand.[22] It was the Covenant that converted hitherto more or less tacit understandings of the contract of unlimited liability into an explicit, formalized statement. However, once out of the military doctrinal context, the Covenant can be and has been used by a variety of actors—government and non-government—including Service charities. Although the Army sought to protect itself by legalizing the relationship between the soldier and society and the soldier and his or her commanders into a contract it opens a once implicit understanding into a litigable issue. In other words,the Military Covenant actually institutionalizes the much legalized culture which it ostensibly opposes; it provides an example of army 'governmentality'—where the very institution coming under external control adopts the mindset of its controllers.[23] We can now consider recent issues in the Trinitarian interaction that have been influenced by the Military Covenant. As will be shown, these interactions constitute spaces or sites of contestation over the terms of the military contract. In that context we can ask what has changed over the past decade of war.

Military Spaces and Contestation over the Military Contract of Unlimited Liability

As mentioned above, the different aspects of the military contract interact in interesting ways. In broad terms, there are three sites or spaces in which these interactions occur:[24] home garrisons, bases and training installations within the UK and allied territories (such as Germany and Cyprus); bases overseas from which operations are launched, which can be subdivided into those that are more or less adjacent to combat operations; and out of base on patrol. It is here that, until recently, one could find an approximation to 'pure military space' in which soldiers and the chain of command operated, from a legal point of view, in conditions of combat immunity, especially immunity from charges based on alleged breaches of human rights or claims of negligence.[25] As indicated earlier, depending on the circumstances, combat immunity could be considered both as a form of protection for military personnel from charges arising from what they might do on a battlefield but also a protection of the MoD from charges brought by Service personnel (and/or their families if a soldier was a fatality)—as, for example in the argument that the MOD has a duty of care to provide the best possible conditions for survival and safety in operations.

Different groups—military and civilian—can have diverging interests and strategies with regard to what goes on in these spaces. For example, both soldiers and their families, as well as civilians affected by military action, can place social, moral, and legal pressure on the military—especially the Ministry of Defence (MoD)—to seek redress of perceived wrongs ranging from bullying and abuse in training centres, inadequate equipment and force protection leading to unnecessary death or injury, and of course alleged abuses committed by soldiers themselves against detainees.[26] These issues can involve media campaigns and parliamentary initiated quasi-judicial enquiries as in that conducted by Nicholas Blake QC, in response to a request from the House of Commons in 2006, into the controversy over the standards of training and care in the Deepcut training establishment.[27] There have also been expressions of interest from within the legal process itself such as the role of coroners, especially in connection with the question of whether there have been deficiencies in equipment provided to the armed forces and whether the MoD is liable for any injuries or working to Service personnel that can be attributed to such provision.[28] Meanwhile, yet other civilian groups, as well as government sensitive to public concerns, hold soldiers to account for breach of just war principles—*ius in bello*—in connection with a range of issues including: abuse of detainees in Iraq, breaches of rules of engagement, taking of war trophies, abuse of wounded and killed combatants. Government, meanwhile, has sought to maximize the conditions under which the armed forces can prosecute military objectives within the constraints of just war principles, but not restricted by further intrusions into military space by considerations flowing from either human rights legislation or claims for negligence against them or the chain of command. Some groups can therefore seek to defend soldiers from what are seen as inappropriate or overly stringent applications of the law to their actions on the battlefield, while others are seeking to hold soldiers to strict account on the same grounds. Meanwhile, the MoD, while clearly having to hold its personnel to account for their actions in battle, has been at the same time concerned that there should be limits to this process—that in the outside bases and in the field of battle considerations of human rights should not apply and nor should the government be liable to negligence claims for any purported equipment failures or other decisions made in the prosecution of a military operation.

The activities of different interest groups focused on the safety of military personnel can lead to paradoxical effects.[29] For example, many contemporary tactics are dictated by the amount of weight that soldiers carry. Much of that weight comprises body armour and helmet (as well as water and an extra weapon). Body armour results not just from technological innovation but is driven by perceptions of the need to limit home side casualties. These perceptions are complex, involving government judgements of what the public is prepared to accept, as well as what soldiers themselves feel is reasonable protection. However that may be, body armour has effects on collective performance in combat. It is too hot to run about in body armour in Helmand and, because of the threat of improvised explosive devices (IEDs), many contacts involve prodigious levels of firing unaccompanied by movement.[30] So far as the UK is concerned and other countries except the US, if one takes a casualty, that tends to mean the end of the operation if it is a patrol or small operation, and if a major operation then a major slowdown in its progress until the casualty is evacuated.

There are other, specific, cases of these paradoxical effects. The Commanding Officer of 2 PARA in Sangin in 2008 remarked (in a talk at the Royal United Services Institution [RUSI]) that it would take a lot of consideration for him to tell his men not to wear body armour even though he knew it reduced their effectiveness. This was because if one of them was killed he would potentially be liable—reflecting the relevance here of the Military Covenant. The nature of combat in Helmand has also produced its signature wound—the multiple amputee. This is a symbol of the medical advances that have radically changed the ratio of very seriously wounded and killed in action over the past decade.[31] Another relevant case is that of the four British soldiers drowning in a canal in June 2010. They were in a Mastiff, secured by seatbelts; both the vehicle and the belts are defence against IEDs. However, in this case, the Mastiff slipped off the small road as it moved out of the way of an Afghan National Army vehicle. It tipped up and went into the canal; the three soldiers in the back drowned.[32] They could not get out. This example shows how legal requirements—stemming from the postmodern individualized culture—can have very material, paradoxical effects on the ground.

Turning from armour and safety to the psychological health of soldiers and duty of care issues, it is ten years since the class action of Service personnel against the MoD, known as the PTSD case.[33] As the authors pointed out the 'class action was brought by more than 2000 British military personnel (the claimants) who had served in a number of major operations prior to 1996... including Northern Ireland, the Falklands, the Gulf War and Bosnia'.[34] As one of the authors argued in a review of the lessons of the case three years after the court's decision, the 'claimants argued that the MoD was negligent in failing to take measures to prevent, detect or treat the development of psychiatric illness in general and PTSD in particular'.

On the key issues the MoD won the case. Thus the ruling by Justice Owen meant that the main or primary function of the military is to fight and to win in battle, not to prevent psychological disorder. Furthermore, the idea that the military could reasonably prevent mental health disorders arising out of combat was similar to the idea that they could prevent all physical injury—something devoutly to be wished, but impossible to achieve. Instead, Mr Justice Owen concluded that the MoD must establish reasonable systems of care for mental health problems (and concluded that they had done so within the limits set by the knowledge available at the time). As for screening for psychological vulnerability, this was currently impractical, and indeed the numbers of false positives (those falsely identified as being likely to breakdown but in practice do not) meant it should probably not even be tried, especially because of the potential to stigmatize those who are prevented from deploying for no good reason.[35] However, the appeal of screening to the media, politicians, and public alike mean that the issue will resurface from time to time. In addition, the military is right to be paternalistic and to cultivate a culture of toughness among Service personnel. Indeed, for combat effectiveness there is a continuing requirement to subordinate the needs of the individual to the needs of the military group and wider organization. Mr Justice Owen also extended the High Court decision in civil cases that an employer is entitled to assume that an employee is reasonably healthy unless and until it is told otherwise—those 'in trouble' themselves have a duty to say so.

The MoD did not win on every matter. For example, there was a poor transfer of Service records to the NHS. Although the MoD and Services operated to the best of their ability given the state of knowledge at the time, there was a clear inference that the quality of the organizational memory of the military on mental health issues needed to be addressed. Likewise, the charge that both the chain of command and the defence medical services had failed to learn the 'lessons of Vietnam' was rejected, on the grounds that there was continuing debate as to what these lessons were, and likewise how relevant these were outside the United States.[36]

At the time, the Owen ruling was recognized to be a conservative one; the authors wondered then if these issues were revisited a decade on, the outcome might well be different. This view was shaped by the influence of human rights legislation, especially in the health sector, which it was felt would be likely to increase over time. In this context, a paradox arises: the MoD has a greater duty of care to make provision for the needs of those who are injured psychologically, even if the act of injury is not litigable (due to battlefield immunity, the primacy of military objectives, and group needs). Precisely because such cases might be non-litigable then it might and probably will be argued that the MoD has a higher duty of care to make good injuries sustained. In this regard there is an important distinction between two aspects of legitimacy: (a) conforming to what is legally required; and (b) conforming with or meeting expectations of wider social groups about what they think would be just and equitable for the military to do for their personnel in terms of sacrifices they give to the military and the country. Thus this is an area where the primacy of the collective and the imperatives of war have to be finessed with attention to needs of the individual—indeed doing this might forestall or slow the process of legal challenges to this core feature of military culture in future.

Yet the challenges to the primacy of 'fit to fight' and combat immunity continue. Over the past few years it has been established that the Human Rights Act applies to bases and hospitals/garrisons but not on the battlefield, that is, once soldiers had left their bases. However, the idea that once soldiers were in what we referred to earlier as 'pure military spaces' they and the chain of command are insulated from negligence-based claims is no longer acceptable—at least socially and now legally. The courts stated that the MoD could indeed be sued for negligence (for example in regard to deficiencies in equipment, procedures and the like), which the MoD challenged. Nevertheless, the courts also ruled that despite pressures, especially from Service personnel's families, the Human Rights Act did not apply to these spaces. Thus, while the MoD challenged the ruling on negligence, the families did the same with regard to the human rights question. The MoD argued that if human rights did not apply off base in theatre then negligence claims should not either. It adopted what might be called an 'all or nothing' approach to seeking a ruling on these two issues.[37] Yet this approach was ultimately unsuccessful at the UK Supreme Court.

The three key principles established by the Supreme Court ruling in June 2013 were: that the MoD could be sued for negligence; the scope of human rights was extended in military space, particularly operational areas; and there was a narrowing of the interpretation of the idea of combat immunity. With regard to the case of 'friendly fire 'involving Challenger tanks in Iraq, the MoD could be sued by the families of fatalities because the events were 'far removed from active operations against the enemy' and thus the idea of combat immunity did not apply to decisions made in that context. So far as provision of inadequate equipment in the form of snatch Land Rovers, also in Iraq, was concerned, Service personnel and or their families could sue the MoD under Article 2 of the European Convention of Human Rights concerned with the Right to Life.[38]

Unsurprisingly, there are concerns in military and MoD circles that the ruling will lead to a flood of litigation and have deleterious effect on training and equipment. Furthermore, some think that the philosophy of command and decision-making based on risk taking and the 'can-do' approach will be undermined, damaging operational effectiveness by introducing excessive caution and doubt in situations requiring decisiveness rather than fear of having one's actions reviewed by those who are able to draw on the advantages of second sight in a committee room or court of law. In this connection, it is important to note the Supreme Court's minority ruling: the 'realities of war have to be recognised and that the work of the armed forces in the national interest should not be impeded by the threat of litigation if things go wrong'.

This formulation echoed the concerns expressed the previous year (June 2012) by the Army's Chief of General Staff, when he remarked that

> The very close scrutiny to which the government and the forces have been subjected by the media, by public enquiries, and by coroner's inquests for example – stimulated in no small part by the domestic political atmosphere in which the Iraq and Afghan campaigns have been conducted, as well as the high human and financial cost, has reset the appetite for and handling of risk in military operations.

He went on, 'A variety of awkward legal, ethical, human rights and equipment issues have been exposed. There will be an expectation in some circles in society that the sort of zero-risk culture that is understandably sought in many other walks of life ought to be achievable on the battlefield.'[39] These remarks should not be accepted uncritically. While risk taking and a 'can-do' attitude are often viewed as an indispensable cultural foundation of operational effectiveness in the context of the unique demands of the battlefield, there are important qualifications to be made. Scrutiny and accountability with regard to ensuring adequate provision of equipment, and reasonable decision-making that encompasses reasonable risks but excludes recklessness, is an inevitable part of today's social climate and legal norms.[40] Another way of looking at this point is to see it as an opportunity for the refining of military judgement and a fuller development of professionalism and the removal of any attitudes that do not underpin this principle. It marks the culmination of a long-term trend in which military personnel have rights as well as obligations on the battlefield: in the soldier's dilemma, the balance between self-preservation and the interest of the organization have shifted to the former and his or her family members.

It is important to bear these qualifications in mind when the Army seeks to counter the perceived negative effects of risk aversion, not least because senior officers are likely to return to this theme frequently in the coming years. As an example, in December 2013, the Chief of the Defence Staff (CDS) made some pertinent comments on risk aversion and how it places political constraints on military operations, pleading that the 'courageous instinct' should not be lost within the armed forces, or amongst the public and politicians. These comments were understandable given the reluctance to intervene militarily in Syria in the summer of 2013 and the sense that the UK would not find it easy to persuade the public to behave similarly to and operate with France in military operations in Africa. Post-Iraq and the withdrawal from Afghanistan the problem of persuading the public that the armed forces are an essential and useful tool of national power is palpable: 'My stark conclusion is that, when you compare the international security context to the UK domestic scene, then one of my great military challenges as CDS is to help re-validate the utility of the military instrument of national power in the minds of government and the wider public.'[41] It is evident that the problem of persuading publics to support future operations will not be easy to solve, especially in the conditions including: an open-ended mission in terms of time and mandate; a difficulty in identifying any existential threat to the UK; the lack of a UN resolution of authorization; and kinetic operations in countries from which some sections of the UK population trace their roots.[42]

Conclusion

It is evident that soldiers and governments are being held to account for their actions by publics who know about and use the principles of just war. It seems the stringency of that accountability has increased as the opportunities to scrutinize operations have extended and the spaces or jurisdictions of the military have been penetrated by liberal, legal, and social norms from wider civilian society. Of how soldiers themselves feel about these changes, we do not know enough; although we can assume that they want and expect as good a quality of health care before, during, and after operations as possible, what their current detailed views on the soldier's dilemma are remains a good research question.

Meanwhile, sections of the public interested in soldiers' welfare, not least Service families, will continue to seek redress through the courts or other channels for perceived wrongs that have led to the death or injury of military personnel; and soldiers themselves will do so too, as was the case in the PTSD class action of 2003. After the 2013 Supreme Court rulings, the MoD will continue to have to provide 'duty of care' even in non-litigable issues because of rising social expectations of what that duty comprises. And it will not just have to do 'the right thing' but be *seen* by the public to be doing so. The military's 'can-do' culture will have to compromise with the values of safety and risk avoidance. Combat immunity in pure military spaces has gone, at least socially (and is being interpreted legally more narrowly, a process that is likely to continue), and will not return. No military should place any value on recklessness in the planning or execution of military operations. The question is whether managing risk, saving soldiers' lives and protecting them from injury, as well as requiring them to conform strictly to *ius in bello* just war principles will add another quotient to the friction of war and lessen the military and political success of future campaigns.

Notes

1. Anthony Kellett, *Combat Motivation: The Behaviour of Soldiers in Battle* (Boston, MA: Kluwer Nijhof, 1982).
2. Helen McCartney, 'The Military Covenant and the Civil–military Contract in Britain', *International Affairs* 86(2) (2010): 411–28; Anthony Forster, 'The Military Covenant and British Civil–Military Relations: Letting the Genie out of the Bottle', *Armed Forces & Society* 38(3) (2012): 273–90; David Fisher, *Morality and War, Can War be Just in the Twenty-first Century?* (Oxford: Oxford University Press, 2011).
3. Charles C. Moskos, Jr *The American Enlisted Man: The Rank and File in Today's Military* (New York: Russell Sage Foundation, 1970).
4. General Sir John Hackett, *The Profession of Arms, Lees Knowles Lectures Trinity College Cambridge* (London: Times Publishing Company, 1962), 63; Richard Swain, Professionalism: Service Unsullied by Partisanship, National Defense University 2010. <http://www.ndu.edu/INSEL/docUploaded/obligations%20of%20military%20professionalism.pdf>. Accessed 10 February 2013.
5. Ben Shephard, *Soldiers and Psychiatrists in the Twentieth Century: A War of Nerves* (Pimlico: Random House, 2002).
6. Shephard, *Soldiers and Psychiatrists*, 286.

7. Paul Fussell, *The Great War and Modern Memory* (Oxford: Oxford University Press, 1975).
8. Christopher Gelpi, Peter Feaver, Jason Reifler, 'Success Matters: Casualty Sensitivity and the War in Iraq, *International Security'*, 30(3) (2005), 7–46.
9. Shephard, *Soldiers and Psychiatrists*, 397.
10. Some years before, in the late Cold War period, Antony Beevor argued that, across the Army in 1987, there was an awareness of change in not only quality of recruits but also changes in their attitudes: 'They are not necessarily insubordinate, one sergeant explained, they just don't understand what authority or rank mean. Like many instructors, he felt they had little grasp of reality and that their expectations of the Army were based on film fantasies.' Antony Beevor (1990), quoted in John Baynes, 'Recruiting the Professional Army', in Hew Strachan (ed.) *The British Army Manpower and Society into the Twenty First Century* (London: Frank Cass 2000), 57.
11. Simon Wessely, 'Twentieth-Century Theories on Combat Motivation and Breakdown', *Journal of Contemporary History* 41(2) (2006): 269–86; Tristan McGeorge, Jamie Hacker-Hughes and Simon Wessely, 'The MoD PTSD Decision: A Psychiatric Perspective', *Occupational Health Review* 122 (July/August 2006): 21–8, 21.
12. McCartney, 'The Military Covenant and the Civil–Military Contract in Britain', 3.
13. Patrick Porter, 'A Matter of Choice: Strategy and Discretion in the Shadow of World War II', *Journal of Strategic Studies* 35(3) (2012): 27.
14. Mark Urban, The Annual War Studies Lecture: 'War and the British Media: A View from Behind the Hesco', 28 January 2014 <http://www.youtube.com/watch?v=XxIrW85NRqo>. Accessed 5 February 2015.
15. James Burk, 'Expertise, Jurisdiction and Legitimacy of the Military Profession', in D.M Snider, G.L Watkins and L. Matthews (eds.) *The Future of the Army Profession* (New York: McGraw-Hill, 2002), 19–38.
16. Professionals often have to deal with that first encounter when the client quizzes them on what they think of the analysis and recommendations provided by the client based on a trawl of more or less reputable internet sources.
17. Briefing on Future Social Trends provided to Adjutant General seminar by the Henley Centre Ltd, UK, January 2002.
18. Simon Wessely, 'Risk, Psychiatry and the Military', *British Journal of Psychiatry* 186 (2005): 459–66. This issue was highlighted in the use of air power in Kosovo in such a way as to ensure as near to zero losses of pilots as possible, and the apparent reluctance of the US to countenance the use or threat to use ground forces. See the related discussion in Paul W. Kahn, 'War and Sacrifice in Kosovo', *Philosophy and Public Policy Quarterly* 19, Spring/Summer2/3 (1999): 1–6. Evidence from recent history confirms that US and other national publics are prepared to suffer casualties so long as they are proportionate to the cause being fought (and so if there is doubt about that cause a government will be in trouble), are not futile, and are being prosecuted in a campaign that appears to be well led and likely to be successful. For an early statement of this line of argument, see James Burk, 'Public Support for Peacekeeping in Lebanon and Somalia: Assessing the Casualties Hypothesis', *Political Science Quarterly* 114(1) (1999): 53–80; Philip Everts and Pierangelo Isernia, *Public Opinion and the Use of Force* (London: ECPR/Routledge,

2001). On the theme of 'success matters'—indeed matters most—and the question of the 'rightness' or 'wrongness' of the wars fought by governments in public attitudes towards the cost of military operations, see Gelpi et al. 'Success Matters: Casualty Sensitivity'. For a government, a particularly difficult headache would be to conduct a 'wrong' war the success of which is doubted—some would argue that this sums up Bush's legacy to Obama in Afghanistan, which is why the latter has sought to ensure that that page of armed conflict is turned over firmly in 2014. Also see Christopher Dandeker, 'The End of Western Military Intervention? What "success" means in Afghanistan, Iraq and Libya', in James Burk (ed.), *How 9/11 Changed our Ways of War* (Stanford, CA: Stanford University Press, 2013), 116–48.

19. Christopher Coker, *War in an Age of Risk* (Cambridge: Polity Press, 2009); Martin Shaw, *The New Western Way of War: Risk-Transfer War and its Crisis in Iraq* (Cambridge: Polity, 2005); Christopher Dandeker, 'The End of Western Military Intervention?'.
20. UK Government 2013, The Armed Forces Covenant <https://www.gov.uk/the-armed-forces-covenant>. Accessed 5 February 2015.
21. Shaw, *The New Western Way of War*.
22. Sarah Ingham, *The Military Covenant: Its Impact on Civil-Military Relations in Britain* (Basingstoke: Ashgate, 2014).
23. McCartney, 'The Military Covenant and the Civil–Military Contract in Britain'; Forster, 'The Military Covenant and British Civil–Military Relations'; Sarah Ingham and Christopher Dandeker, 'The Covenant We Must Protect From the Lawyers', *Parliamentary Brief*, 24 June 2010.
24. Burk, 'Expertise, Jurisdiction and Legitimacy of the Military Profession'; Forster, 'The Military Covenant and British Civil–Military Relations'.
25. Peter Rowe, *The Impact of Human Rights Law on Armed Forces* (Cambridge: Cambridge University Press, 2006).
26. Jonathan Aitken, *The Aitken Report: 2008, An Investigation into Cases of Deliberate Abuse and Unlawful Killings in Iraq in 2003 and 2004* (British Army, 2008).
27. Nicholas Blake, *A Review of the Circumstances surrounding the deaths of four soldiers at Princess Royal Barracks, Deepcut between 1995 and 2002* (London: HMSO, 2006) <http://www.official-documents.gov.uk/document/hc0506/hc07/0795/0795.pdf>. Accessed 5 February 2015.
28. Deborah Haynes, 'Army treatment of bullied soldier "unforgivably bad", coroner says', *The Times*, 3 March 2014 <http://www.thetimes.co.uk/tto/news/uk/defence/article4021657.ece>. Accessed 5 February 2015.
29. The next two paragraphs stem from a discussion with Anthony King.
30. Email conversation with Anthony King.
31. On the evolution of military medicine, war, and casualty rates, see Tanisha M. Fazal, 'Dead Wrong? Battle Deaths, Military Medicine, and Exaggerated Reports of War's Demise', *International Security* 39(1) (2014): 95–125.
32. Thomas Harding, 'Four British soldiers drown in Afghanistan canal', 24 June 2010, *The Telegraph* <http://www.telegraph.co.uk/news/newstopics/onthefrontline/7851207/Four-British-soldiers-drown-in-Afghanistan-canal.html>. Accessed 5 February 2015.

33. Multiple Claimants, 2003, *Multiple PTSD Claimants v MoD*—2003 EWHC 1134 (QB); Tristan McGeorge, Jamie Hacker-Hughes and Simon Wessely, 'The MoD PTSD Decision: a psychiatric perspective', *Occupational Health Review* 122 (July/August 2006): 21–8, 21.
34. McGeorge, Hacker-Hughes and Wessely, 'The MoD PTSD Decision', 21.
35. Roberto Rona et al., 'Screening for Physical and Psychological Illness in the British Armed Forces: III The Value of a Questionnaire to Assist a Medical Officer to Decide Who Needs Help', *Journal of Medical Screening* 11 (2004): 158–63. Roberto Rona et al., 'Mental Health Screening in Armed Forces Before the Iraq War and Prevention of Subsequent Psychological Morbidity: Follow Up Study', *British Medical Journal* 333 (2006): 991–4.
36. Simon Wessely and Edgar Jones, 'Psychiatry and the Lessons of Vietnam: What Were They and are They Still Relevant?', *War and Society* 22 (2004): 89–103.
37. 'MoD in Latest Appeal Against Iraq Negligent Equipment Claims', LeighDay website, 15 February 2013 <http://www.leighday.co.uk/News/2013/February-2013/MoD-in-Latest-Appeal-Against-Iraq-Negligent-Equipm>. Accessed 2 February 2015.
38. Richard Norton-Taylor, 'Supreme Court MoD ruling "will have huge impact on military operations"', *The Guardian*, 13 June 2013 <http://www.theguardian.com/law/2013/jun/19/supreme-court-mod-ruling-impact>. Accessed 5 February 2015.
39. Cited in Norton-Taylor, 'Supreme Court MoD ruling "will have huge impact on military operations"'.
40. This point arose in discussions about risk and military culture in a session of the Inter-University Seminar on Armed Forces and Society Biennial Conference in October 2013. We are indebted to a conversation with Vincent Connelly at that meeting.
41. Jones, Sam 'UK losing "courageous instinct", warns general', *The Financial Times* 18 December 2013 <http://www.ft.com/cms/s/0/ea62dafe-6827-11e3-8ada-00144feabdc0.html#axzz2suWvSuZ8>. Accessed 12 March 2014.
42. House of Commons Defence Committee, *Towards the next Defence and Security Review, Part 1, 7th Report of Session 2013–14*, Vol. 1, p. 15, paras 20, 21 <http://www.publications.parliament.uk/pa/cm201314/cmselect/cmdfence/197/197.pdf>. Accessed 8 February 2014.

15

The Future of Cohesion

Anthony King

Introduction

In a monograph on urban warfare, Stephen Graham has sought to expose the dangers of recent developments in military technology:

> One should read these emerging US high-tech urban warfare and counterinsurgency programmes as symptoms of desire – technophilic dreams and feshistic urges for mastery and control, adjusted to the new imperatives of urban counterinsurgency warfare and declining US political and economic power. These programmes and imaginings also reflect deep-seated, long-standing tendencies within US military culture to seek all-conquering superweapons that can annihilate all enemies – if possible from a distance. Superior technology is thus the key to all locks, the silver bullet.[1]

Although Graham is certainly correct to suggest that, since at least the middle of the twentieth century, the United States has sought to replace soldiers on the battlefield with firepower and has developed remarkable surveillance systems, he significantly exaggerates the extent to which US forces believe in a purely technological solution to war, still less an omnipotent one.

Yet, while perhaps rhetorical, his suggestion that war might become precise and clean usefully represents a much wider current of contemporary thought. For instance, both Martin Shaw and Christopher Coker have discussed the rise of 'risk-transfer warfare' in which advanced Western and, above all, American forces employ technology in order to minimize their own casualties while simultaneously maximizing combat power.[2] In turn, Shaw and Coker represent an extension of a range of scholarship which appeared in the 1990s which declared that war had become a 'spectator-sport' or 'virtual';[3] indeed, in the case of Jean Baudrillard, he opined that the Gulf War had not taken place at all.[4] War was now a bloodless televisual illusion generated by precision weapons and the media.

Graham fears that in the future, by means of their technophilic fantasies, the armed forces will be able to dissect and amputate cities, monitoring them ubiquitously and striking their enemies at will: 'it is but a short step from imagining and designing systems to automatically "acquire targets" within a city, to developing robotized weapon systems designed to kill or destroy those targets under the ever-diminishing human supervision'.[5] To this end, 'even more chillingly, the notion of swarms of minuscule aerial vehicles or even armed, robotic insects is now being widely explored in blue-sky military thinking and research'.[6] It is somewhat ironic that while Graham fears the emergence of an omniscient military able to dominate the cityscape, most other commentators draw quite different conclusions from their observations of recent fighting in cities. They believe that urban warfare is a highly likely prospect in the future and that far from being a domain for the frictionless exertion of military power, it will be a desperate struggle precisely because Western technological advantages will be substantially negated in this environment. Indeed, while Alice Hills is fully cognizant of the potential of military technology, she claims that urban warfare, conducted by small groups of infantry soldiers, will remain 'premodern' relying not primarily on technology but on determination, morale, and brutality.[7] For her, street fighting is always savage. The struggles in Grozny, Fallujah, Sangin, Homs, Khobane, and east Ukrainian towns seem to confirm the point.

It would seem very unwise to suggest that war can be appeased by technology, as Stephen Graham claims; war will not become surgical. While the application of new forms of digital technology to military problems has materially altered the armed forces and the way they organize themselves to conduct war, warfare will continue to involve small groups of combatants in a brutal contest for terrain and the people who live there—and more so if the next conflict is indeed an urban one. The collective skill and solidarity of these small bands of warriors will be a crucial factor in deciding the outcome of battle. It is precisely because war will inevitably remain a collective, social endeavour that this volume has focused on the question of small-group cohesion in the twenty-first century. The small group provides a perspicuous insight into some of the micro-realities of contemporary conflict.

Of course, at the same time, the analysis of cohesion indirectly fulfills another subordinate aim. The volume attempts to speak to the social sciences more generally. The small military unit remains a potentially rich but under-exploited resource for the social sciences and sociology in particular; in combat, many of the universal processes of social interaction and solidarity are demonstrated with rare lucidity. The exploration of group dynamics in combat elaborated upon here may ideally be useful in understanding processes of group formation and inter-group dynamics in the twenty-first century generally. While the fundamental processes of collective action may indeed be universal, their manifestation is always historically particular. This exploration of the combat performance of professional soldiers in the twenty-first century—and the distinctive relationships which have emerged between them—may illustrate the distinctive process of group formation today much more widely.

The Findings

Towards Military Practice

Many of the chapters in this volume should be read as a response to the classical studies of cohesion and, above all, to Janowitz and Shils' rightly celebrated 1948 article on the Wehrmacht. The importance of this literature is not in question; it represents some of the finest sociological work that has ever been done on the armed forces. However, there was an obvious and major shortcoming with it. Although Janowitz and Shils ostensibly explored the combat performance of the Wehrmacht, they ultimately made a significant deviation; rather than exploring the actual fighting methods of the German primary group, specific battles, or operations in which German units were involved, they prioritized what individual soldiers reported about their personal relations with other soldiers, not how platoons actually fought together in combat. The combat performance and the cohesion of the primary group was explained in terms of the professed dispositions of individual soldiers towards each other and the motivations which arose from those relations. Effectively, Janowitz and Shils demilitarized the military.

In Chapter 2, Tarak Barkawi has argued that, in fact, their error was much more profound and widespread. Military scholars have repeatedly invoked the primary group or political motivation, as master factors in explaining military performance in ways which are essentialist and Eurocentric. Indispensable though some of this analysis is, Barkawi claims that the particular configuration of military units and the dynamics of the battlefield itself disappear from view, even though they constituted the immediately precipitating conditions for what soldiers do. Writing from within the emergent paradigm, he shows how interactional dynamics in military units are very general, if not universal.

This volume follows Barkawi's injunctions. It follows the 'practical turn' evident across the social and political sciences from the 1970s. It accepts the primary group and political motivation theses but seeks to situate cohesion and combat performance in their institutional and micro-social contexts; this collection examines the collective performance of professional soldiers in the twenty-first century in barracks, training, and, above all, on operations and in combat. Accordingly, a quite different archive of evidence—and methods—is demanded. Individual and collective post-action interviews remain an important resource, as they were in the 1940s, but discussions might more usefully focus on what soldiers do and have done together as soldiers and, above all, on training, rather than on what they thought or felt about their comrades; these discussions should not ignore informal socializing, but they should focus on formal military activity, since this is what distinguishes the soldier from the civilian. At the same time, fieldwork and participant observation as well as archival descriptions of military practice and specific cases of combat become critical.

Naturally, this focus necessarily involves various problems of representativeness and reliability. It requires an accurate account of combat, but precisely because the frontline is a confusing and frightening place, accounts of actions there have to be treated with great caution. Rob Johnson sought to show the methodological difficulties of analysing combat, especially when situated in a historical context; recurrent patterns are certainly observable in combat, but, in addition to the problem of reliability, war is also experienced, interpreted, and represented in alternative ways in different eras and by different groups. Crucially, Johnson begins to suggest that the way in which citizen soldiers experienced and recorded combat in the twentieth century may be different from today's professionals. It is not always easy for scholars to differentiate the two and to develop a coherent account of the fighting itself. It is perhaps precisely because it is difficult to develop a coherent understanding of what actually happens on the battlefield that monocausal explanations have often been attractive to social scientists as they study cohesion; simplistic appeals to the primary group or political motivation arise where detailed knowledge of what has actually taken place in a firefight are lacking.

Professional Cohesion

Notwithstanding the methodological problems of reconstructing practice (especially in combat), if the analytical focus has shifted to performance—especially of the small unit—then this volume has produced substantial evidence about the critical factors in its production in a professional army today. Traditional forms of motivation such as patriotism may not be irrelevant to some forces, but the overwhelming emphasis of the chapters presented here is that training and the inculcation of drills have been the primary means of ensuring cohesion in combat. For instance, Eyal Ben-Ari illustrated the centrality of impersonal drills and indeed an impersonalized form of cohesion to the performance of the Israeli Army in urban combat. In these operations, especially during the Second Intifada, 'organic' combat units were reassembled and merged due to the exigencies of specific missions and troop availability.[8] In place of social familiarity, Israel Defense Forces (IDF) soldiers relied on 'swift trust' to generate cohesion. They were able to cooperate with each other by reference to common tactics and procedures, adducing whether their new partners were competent and trustworthy in executing these tactics by means of accelerated processes of 'mutual testing'. Significantly, and against the classical theory of military cohesion, units based on 'swift trust' seemed to be as—even perhaps more—effective than organic units.

In her chapter on the French Foreign Legion, Mikaela Sundberg provides a very different but commensurable example of how impersonal drills are central to the unification of military units today. Precisely because of its multinational system of recruitment, the Legion cannot presume a common cultural background among its troops. Indeed, the generation of discrete primary groups, or mafias, among particular national groupings is regarded as actively dangerous for the cohesion of the Legion. Accordingly, the Legion prioritizes drills, hierarchy, and submission to orders to an extreme which is quite unique in Western forces today. The result may be that the organization is less adaptable than some of its professional peers. Yet it would be difficult to question the combat performance of the Legion. For instance, in January 2013, *2ème regiment étranger de parachutists* (2 REP) successfully completed their mission, which included a parachute drop, in remote Mali against Islamicist insurgents despite the extreme environment in which they were operating. The Legion is a distinctive professional force but the primacy of formal procedure—of following orders however absurd they might seem—usefully cautions against analytical over-investment in the personal relations between soldiers. In the Legion, soldiers are explicitly bound not by their private friendships but by their formal corporate obligations and procedures.

The British Army may be less austere than the Legion, but the description of a platoon's use of drills to extract itself from an ambush on Waterloo Day in Chapter 9 illustrates the centrality of trained responses to the professional military just as clearly. Garth Pratten's analysis of the contemporary Australian digger equally affirms the point. Professional Australian soldiers of the twenty-first century often appeal to an imagined, even invented, military tradition; they see themselves as diggers whose citizen predecessors sacrificed themselves in the First and Second World Wars. However, in reality, they understand their mission and the dangers it involves in terms of professionalism. They identify professional drills and training as critical to their combat performance and respect comrades who show high levels of competence. The point is confirmed by many other soldiers today. As one British officer observed: 'You train two three four times. Once you have trained six or seven times it becomes a drill.' For these soldiers, drills and training are primary to cohesion and combat performance not friendship.

This volume recommends that scholars pay close attention to established drills, since they are so central to military performance. This injunction might be read to imply that personal bonds between soldiers in a trained, professional force are irrelevant. This would be a serious mistake. Drills are critical to military performance because, especially in a highly trained professional force, they are the choreographies which soldiers instinctively follow in combat. Yet, however deeply ingrained these drills are in combat teams, soldiers must be motivated to execute them. In the face of enemy resistance, even the most drilled troops could choose to retreat rather than to execute their assault drills. Individual and collective motivation is utterly essential to combat performance, therefore. Here, interpersonal bonds become important. The central point of the chapter I co-wrote with Patrick Bury was to explore not only the execution of drills but the intense emotional attachments which operate in combat and which play a critical role in determining what soldiers actually do; we sought to plot the micro-context in which particular emotional responses motivated specific actions. We describe an incident when one of Patrick Bury's corporals agreed to follow an order even though it was extremely dangerous and could theoretically have been refused by him. The corporal consented willingly to the order because of his emotional dependence upon his commander to whom he wanted to prove his allegiance and dedication. Since the combat zone remains a terrifying and disorientating domain even for the most specialized professional, visceral emotional responses remain critical to social action there. However, while fully recognizing the central role which emotions and emotional bonds play in combat, our chapter also sought to demonstrate that these emotions cannot be understood in generic individual terms. The love which Bury's corporal professed—and then demonstrated—was not a pristine personal attachment but the product of a specific and identifiable institutional context; above all, the corporal's desire for professional recognition and credibility from his peers and superior. His 'love' was comprehensible only within the highly professionalized culture of his regiment; it was an extreme manifestation of his own professional identity and his professional relations with the members of his platoon.

As the example of Bury's corporal demonstrates, one of the most powerful emotions at work in a professional force is shame. In my chapter on different disciplinary patterns in the citizen and all-volunteer army, I explored the critical role which shame plays in the professional force where formal discipline, especially for combat refusal, cowardice, and desertion, has been substantially eroded. The chapter was intended to show how professionalism is motivating in and of itself. Professionalism becomes a collective reference point for soldiers today, and they are motivated to perform by the threat of professional shame and the pursuit of honour. Today's professional forces do not court-martial those soldiers defined as combat refusers or those who might have been defined as cowards in the past; rather, they shame them. An internalized system of honour and shame enforced by professional soldiers on each other—and on themselves—has replaced the external judicial process of the court martial. Formal discipline has become less necessary precisely because soldiers are imbued with a sense of professional honour. The primary motivating factor for soldiers in combat is that they will be respected by their peers for upholding professional standards. At the same time, a powerful sanction of ridicule and ostracism also operates on those who fall short of professional expectations. Thus, while soldiers are no longer shot for cowardice in the professional army, they can be annihilated professionally by a failure to perform on the frontline. They face professional extinction rather than actual execution. It has to be stressed in the strongest terms that professional soldiers are extraordinarily sensitive to their reputations with their peers and any suggestion that they have acted inappropriately engender the most extreme emotional responses from them; they are deeply susceptible to the most visceral sense of shame.

The Core Group

A central claim of contemporary military scholarship, followed in this volume, is that cohesion has become impersonal; soldiers rely on drills which they can perform with individuals with whom they have had very little interaction (against classical understandings of military cohesion). This might suggest that regimental identities are now archaic and the armed forces could become absolutely fungible organizations able to organize personnel in any configuration for any particular mission. Ben-Ari's concept of swift trust and the appearance of ad hoc units in the IDF might be read in this way. However, this would be an error. Although professional soldiers do not need intense social bonds to perform—and they certainly do not necessarily need to be close friends—it is false to suggest that all-volunteer armies have become transposable, fluid military organizations in which any individual can work with any other individual to form instantly effective teams.

In the Second Intifada, IDF units and subunits did not merge randomly; established subunits with stabilized teams were attached to other subunits which were similarly internally stable. Ad hoc units were amalgams of already existing teams not agglomerations of random individuals, none of whom knew each other at all. Similarly, in Afghanistan, operations have not randomly thrown together individual soldiers into ad hoc units. Rather, operations have been built around a core unit or subunit, typically a company or a platoon, to which specialist personnel or subteams are attached. The core unit provides the framework for the operation. It is in this context that augmentees can integrate most easily. It is important to emphasize that the solidarity within the framework company or battlegroup is professional; it is constituted primarily by competence not by pristine bonds of friendship. However, in the core group, long-standing mutual trust has been built up as a result of the members' recurrently proven professional competence; their collective drills and procedures have been worked up to a high level. This core group is hugely beneficial and, indeed, indispensable for attached subgroups because it provides a clear and identifiable pattern of practice which the attached can adopt and to which they can adjust. The existence of a core group eliminates the need for time-consuming negotiations, necessary in the genuinely ad hoc formation, about what constitutes established practice. In a professional military, united teams of soldiers who have trained together for long periods and who constitute nodes of collective expertise generate significant efficiencies.

Crucially, the creation and maintenance of a core group does not simply mean that the established units have refined their drills. Because professional soldiers in established units have worked with each other over a long period and know that they will work with each other in the future, the force of shame and honour is magnified; emotional commitment is amplified. The professional reputation of the soldier will stay with them throughout their career to be invoked by their fellows. Professional soldiers can surely operate with soldiers whom they do not know on the basis of established procedures, but personnel stability and the existence of units are vital for the professional military, not only refining collective performance but accentuating the mechanism of shame and honour which motivates soldiers to perform well. In this context, regimental identity becomes a means of encouraging good performance. For instance, Patrick Bury's Royal Irish platoon was attached to the 2nd Battalion, The Parachute Regiment, during his tour of Helmand and he records how he was very conscious of ensuring the reputation of the Royal Irish with the airborne by operating professionally. It is a recurrent theme. Regimental identity can consequently engender very high levels of performance and indeed facilitate the kind of ad hoc, instant cohesion which Ben-Ari describes. The teams which form ad hoc units want to prove the worthiness of the organization from which they come.

The Female Soldier

The professionalization of the armed forces has had a profound—if sometimes overlooked—impact on the way in which soldiers interact with each other and conduct themselves on the battlefield. Impersonal drills and training have become central to any explanation of performance. One of the most striking exemplifications of this transformation is the integration of women into the combat arms in Iraq and Afghanistan. Although the UK and US formally maintained a ban on females serving in combat units throughout Iraq and Afghanistan, many American and British women served in combat in these campaigns, sometimes with distinction. Recognizing this new reality, on 24 January 2013, Leon Panetta, the US Secretary of Defence, announced that all combat roles would be open to women from 2016. Recent announcements suggest that the United Kingdom, the last NATO country which enforces gender restrictions on ground combat roles, will soon lift its ban on women serving in the combat arms.[9] At this point, gender restrictions will have been lifted in all NATO forces; women will be able to serve in all branches of the armed forces. This is a remarkable, almost historically unique, development.

The accession of minority groups and now women into the armed forces can be understood as a manifestation of professionalism. While the advance of egalitarian labour laws have, of course, played an unignorable role, female accession has been facilitated in the last ten years by the professional culture of the armed force. It is worth contrasting today's forces with those of the twentieth century. In the mass-citizen armies of the twentieth century, it would have been difficult for women to serve as combat soldiers because armies relied so heavily on masculinity to motivate and unite their soldiers, as the work of Janowitz and Shils, Stouffer et al., Little, and Eisenhart shows.[10] With the emergence of a more impersonalized, competence-based, professionalized cohesion evident in Iraq and Afghanistan, women have been able to be incorporated into the infantry and to serve successfully in combat. There, when their accession has been successful, they have been judged not by their gender but by their performance, just like their male peers. Major Eleanor Taylor of 1 Royal Canadian Regiment, who was the first women in NATO to command an infantry company in combat, represents an important case of how professionalism has facilitated female accession. Taylor was one of the first women to be assigned to the infantry, and her career is instructive. Having initially failed her final infantry training course on the penultimate day, she re-took the course to finish in the top third of her cohort; a fact which has been very important to her professional credibility. She has remained at that standard throughout her career. In 2010, she was involved in a series of major combat operations around the south-west of Kandahar city in which her performance was excellent. Taylor's experiences accord with those of many other women in the armed forces who have been similarly accepted by male soldiers on the basis of their competence.

Of course, it is important not to exaggerate or idealize this process of integration. Despite the success of female soldiers like Major Taylor, the opposition to women in combat roles remains in the armed forces; in some cases, these have involved serious cases of harassment, abuse, and assault.

Women are routinely denigrated as 'sluts' (open to sexual advances) or 'bitches' (not available), in either of which case they are accused of undermining cohesion. However, while these categories remain institutionalized within the armed forces, a third definition has emerged in the last decade; that of the honorary man. Individual women who have performed their role successfully have been accorded full membership of the professional group; they attain male status. Naturally, there are significant problems with this category. Only individual women can be ascribed this status, not women in general. In addition, it is very difficult for women to maintain this status because any perceived indiscretion or mistake leads very easily to the re-inscription of the bitch–slut binary. In many cases, since male soldiers determined how a woman was to be categorized, females could find themselves defined as sluts or bitches irrespective of their behaviour.

Nevertheless, although the accession of women to the combat arms and the emergence of the category of the honorary man with it represents a minor revision rather than a major reformation of military culture, it is important evidence of the advance of professionalism in the military. In the relatively recent past of the 1990s, it was more or less inconceivable, even in the Canadian Forces, which had lifted all gender bans at that time, that women could serve in combat. In Iraq and Afghanistan, a significant number of women have been accepted into combat units on the basis of their competence and have served in combat on the frontline. The impersonal competence-based cohesion of the professional force may be less attractive to many commentators than the citizen army of the twentieth century unified by civic virtues and personal bonds, but it seems to have allowed for the successful and egalitarian integration of diverse social groups. Critically, the accession of women into the armed forces and the normality of a once unconscionable policy usefully illustrates the distinct divide which separates military forces in this century from those of the twentieth century.

Deviant Professionalism

The professionalization of the armed forces has generated a distinctive kind of solidarity among troops in which formal drills and procedures play a crucial role and are invested with immense social and emotional significance. As a result, the performance of professional troops in the last ten years has often been superior to that of their conscript predecessors. Yet it is important to avoid ameliorism here. Professional forces require core groups which often coalesce around unit identity. However, overdeveloped regimental identities can become problematic; hyper-professionalism can generate deviant forms of practice and solidarity. Garth Pratten explores the way in which the professionalized identity of Australian soldiers has sometimes immanently undermined discipline. Precisely because they aspire to the accentuated professionalism of the Special Forces, regular Australian troops have sometimes also liked to adopt some of the more informal aspects of Special Forces culture. Pratten records the way in which Australian soldiers in Oruzgan were reprimanded for wearing flip-flops and hitting golf balls from their FOBs. The emphasis on pure performance and drills can sometimes encourage a disrespect for apparently superfluous elements of military formality. The result is that a false sense of professionalism and its priorities can ironically actually undermine cohesion and performance.

Chiara Ruffa's chapter addresses another problem, if indirectly. She documents the way in which the professional ethos of the Alpini has been inverted so that, while they demonstrate high levels of corporate solidarity, the regiment are far more effective on peace-support operations than on conventional military ones. Professionalism for the Alpini is constituted through individual mountaineering and skiing skills rather than through combat performance. Moreover, precisely because they are a professional organization, enjoying significant political prestige and economic autonomy from the state and civil society, they have been able to preserve their privileged position for decades. The Alpini may be unusual and their war-fighting skills are viewed with some scepticism by their allied partners, yet they illustrate a wider issue which has sometimes been noted of professional forces in the last decade. Precisely because these forces have been deployed on missions of choice, rather than national survival, professional troops in Iraq and Afghanistan have sometimes preferred the safety which passivity offers over energetic engagement with the population and, therefore, hostile forces. Many national task forces in Afghanistan have been notably inactive. It would be difficult to describe the British effort in Helmand as reluctant. Yet, while the British have been typically robust, they have also been deeply concerned with casualties, to the point that a single casualty has sometimes led to the cancellation of a mission. In this way, today's professional armies have become similar to European armies of the eighteenth century whose commanders were equally reluctant to expend them in battle since they were all but irreplaceable. Inspired by a heightened sense of patriotism to avenge the 9/11 attacks, only US troops seem, in remarkable contrast to the 1990s, to have been (temporarily) inured to the problem of professional self-protection. Professionalism may bring organizational benefits and does seem to have improved performance, especially in combat, but the potential pathologies of professional organizational culture should not be ignored.

Future Research

The 9/11 wars are now coming to an end and Western forces are returning to a state they call 'contingency'. It is possible that the intense and large-scale engagement which has characterized the last decade and more will be avoided in the near future, though recent events in central and western Africa, Ukraine, Syria, and Iraq suggest that some form of military intervention remains likely more or less permanently for Western troops. This new situation raises many important and fascinating questions for scholars.

As the chapters of Christopher Dandeker and Simon Wessely and Leonard Wong and Stephen Gerras emphasize, the last ten years have involved a quite radical transformation in social–military relations. More specifically, the relations between the family and the armed forces are undergoing a profound, complex, even paradoxical change. Once incorporated almost entirely into service life, families are becoming more autonomous from the military, primarily because wives (who once gave up their jobs) now increasingly work too; in Europe, this domestic–military divide is accentuated by the declining provision of family welfare in the form of housing and educational support for military families. The armed forces are becoming less of a total institution for many service-personnel and their families, especially in Europe, Canada, and Australia, and more a public-sector workplace like any other. However, simultaneously, as Wong and Gerras's analysis of the FOB demonstrates, the relations between individual service-personnel and their wives and children are being intensified. These relations are infiltrating the military profession and even operations; domestic relations are becoming increasingly important factors in the working lives of soldiers. A new domestic–military nexus is observable, reflecting a wider transformation of gender relations. These important changes have only begun to be analysed, but they promise to be hugely significant for the armed forces in the future.

There are many other areas of interest. If French operations in Mali are any guide, the question of combat cohesion—and therefore the central issue which has animated this volume—is likely to remain relevant and important in the coming era. However, it may be wise for scholars to expand the field of inquiry from combat troops to other areas of the armed forces. There seem to be a number of topics which might be fertile areas of investigation. Cohesion has typically referred to the solidarity of the small primary group, the squad, platoon, and, perhaps, the company; it refers to the performance of military units in which soldiers know all the members of their group and work closely with them. The armed forces rely on these unified groups to prosecute their campaigns. However, the power of modern armed forces lies not only in these small groups but crucially in the ability to coordinate the actions of numerous platoons, companies, and battalions and direct them to organizational goals. Here the headquarters is critical. In *The Transformation of Europe's Armed Forces*, I began to explore the emergence of Rapid Reaction Corps in Europe; these headquarters remain important to this day. However, at this moment, Western forces, especially the United States, the UK, and France, having learnt from Iraq and Afghanistan, are reinvesting in the divisional level. The division emerged as a military formation in the Seven Years War but was consolidated in its modern combined arms form, consisting of three brigades of about 20,000 soldiers, in the First World War. The contemporary division is a very different entity, but its headquarters still commands up to three brigades widely dispersed on a complex battlespace. The divisional headquarters attempts to generate organizational cohesion at the higher level by coordinating and directing its subordinate units. Given the investment which Western armies are making in the divisional level, it seems an appropriate and, perhaps, fertile avenue of study. The investigation of headquarters is highly recommended as a research programme into the future.

There are many other important changes which are equally worthy of investigation but which have tended to be ignored in the past. During the Second World War, General Patton famously declared: 'Gentlemen, the officer who doesn't know his communications and supply as well as his tactics is totally useless'. In staff colleges today, this has been reduced to the common aphorism: 'Amateurs talk tactics, professionals talk logistics.' Since the indispensability of logistics to all forms of war-fighting is so thoroughly recognized by the armed forces, it is surprising how relatively little work has been done on logistics, especially in the social sciences. Martin van Creveld's work, *Supplying War*,[11] which traces the different ways in which the armed forces have supported themselves historically, is an obvious and important exception. Yet generally, scholars, including military historians, have tended to focus on combat and the combat arms. The result is that there is little systematic analysis of how armies supplied themselves in Iraq and Afghanistan and, indeed, how logistics has undergone a radical transformation since the end of the Cold War with just-in-time and outsourcing techniques being widely and creatively used. Indeed, the logistical effort in Afghanistan, where supplies were transported by sea to Karachi, through Pakistan to the Khyber Pass and on into Afghanistan, was one of the more remarkable military achievements of the recent past. Its operational significance (for both the Allies who relied on it and the Taliban who seem to have taxed but not interdicted the lorries carrying their enemy's supplies in a paradoxical process of mutual self-perpetuation) has been severely under-examined. Studies of contemporary logistics, which focuses on the actual methods used by the armed forces to procure, store, move, and track supplies, would be hugely revealing about military organizations and their capabilities.

It was widely reported that while one in one hundred personnel were contractors during the Gulf War of 1991, one in ten US personnel were contractors at the time of the Iraq Invasion in 2003. Scholars rightly took this fact as evidence of rapid and extensive military privatization, especially of the US armed forces. Since that time, extensive work has been carried out on the emergence of private military and security companies and their distinctive cultures. Yet the increasing use of outsourcing after the end of the Cold War is part of a wider transformation of the armed forces; a transformation which has involved the emergence of what might be termed a post-Fordist military.[12] Like commercial and industrial companies, the post-Fordist military consists of a core of highly trained professional personnel typically dedicated to central military functions and, above all, to combat; they have been the primary focus of this volume. This core is now supported by a peripheral, contracted, civilian workforce, often employed part-time to provide support services. Yet there is a semi-peripheral workforce which both provides crucial support services but which has also in some cases performed important combat functions in Iraq and Afghanistan: the Reserves. As Western forces reorganize themselves after Afghanistan, they are restructuring the Reserves and the relationship between the Reserves and the regular forces. They are reconfiguring the institutional relations between the professional core and the semi-peripheral Reserves, in many cases seeking (somewhat paradoxically) to establish the Reserves as part of the regular force. The process raises huge organizational questions but also speaks directly to the question of corporate identity, solidarity, and combat performance. The reformation of the Reserves constitutes a new area for the study of cohesion. There are numerous other potential areas of investigation.

However, whatever line of inquiry scholars adopt, they should remember that, ultimately, however large the armed forces apparently are, all of their prodigious powers are reducible to relatively small groups of people making decisions, planning, or acting together in specific locations and at particular times in headquarters, bases, or on the battlefield. As Erving Goffman rightly noted, the everyday face-to-face interactions of people is where the action finally is.[13] Even the largest organization with the most sophisticated communications systems ultimately relies on numerous small teams and sub-groupings. The armed forces comprise a vast and intricate hierarchical nexus in which diverse formations, units, and groups are constantly interacting, cooperating, and competing with each other to generate extraordinary collective phenomena. It is to this rich, almost infinite reality, the lifeworld of the armed forces, which social scientists should attend. Military scholars should look at what soldiers actually do and try to explain their actions by reference to the situational, institutional, and historic context in which it occurs. This volume has sought in a small way to contribute to that goal and to encourage other scholars to consider the armed forces seriously. There is no suggestion that the future of the armed forces will be any less interesting or troubling than the last fifteen years.

Notes

1. Stephen Graham, *Cities under Siege: The New Military Urbanism* (London: Verso, 2010), 162.
2. Martin Shaw, *The New Western Way of War* (Cambridge: Polity, 2005); Christopher Coker, *War in an Age of Risk* (Cambridge: Polity, 2009).
3. James Der Derian, *Virtuous War* (Oxford: Westview, 2001); Colin McInnes, *Spectator Sport Warfare* (London: Lynne Rienner, 2002); Paul Virilio and Sylvie Lotzinger *Pure War* (Paris: Semiotexte, 1997).
4. Jean Baudrillard, *The Gulf War Did Not Take Place* (Indiana: Indiana University Press, 1995).
5. Graham, *Cities under Siege*, 167.
6. Graham, *Cities under Siege*, 173.
7. Alice Hills, *Future Wars in Cities: Re-thinking a Liberal Dilemma* (London: Frank Cass, 2004), 26.
8. Eyal Ben-Ari, Zeev Lerer, Uzi Ben-Shalom, and Ariel Vainer, *Rethinking Contemporary Warfare: A Sociological View of the Al-Aqsa Intifada* (New York: SUNY Press, 2010), 74.
9. *The Guardian*, 'British army should consider giving women close-combat roles, says general', 6 April 2014 <http://www.theguardian.com/uk-news/2014/apr/06/ british-army-women-combat-roles-general>. Accessed 10 February 2015; *The Daily Mail*, 'Let women fight on the front line: Defence Secretary tells Army to end macho image', 9 May 2014 <http://www.dailymail.co.uk/news/article-2623429/Time-ditch-macho-image-allow-women-fight-frontline-Defence-Secretary-Philip-Hammond-says.html>. Accessed 10 February 2015.
10. Roger Little, 'Buddy Relations and Combat Performance', in M. Janowitz (ed.) *The New Military* (New York: Russell Sage Foundation, 1964); Samuel Stouffer, Arthur Lumsdaine, Marion Harper Lumsdaine, Robin Williams, M. Brewster Smith, Irving Janis, Shirley Star, and Leonard Cottrell, *The American Soldier*, vol. 2: *Combat and Its Aftermath* (Princeton, NJ: Princeton University Press, 1949); Donna Winslow, *The Canadian Airborne Regiment: A Socio-Cultural Inquiry* (Ottawa, Canada: Minister of Public Works and Government Services, 1997); Helena Carreiras, *Gender and the Military* (London: Routledge, 2006); Nora Kinzer Stewart, *Mates and Muchacos: Unit Cohesion in the Falklands/Malvinas War* (New York: Brasseys Inc, 1991). OF-4, female, US Army, email communication, 12 February 2014.
11. Martin van Creveld, *Supplying War* (Cambridge: Cambridge University Press, 1977).
12. Anthony King, 'The Post-Fordist Military', *Journal of Political and Military Sociology* 34(1) (2006): 339–54.
13. Erving Goffman, *Where the Action Is* (London: Allen Lane, 1969).

Bibliography

Ackerman, Robert K. 'Training, Not Technology, Is the Key to Urban Warfare', *Signal Magazine* May (2001) <http://www.us.net/signal/Archive/may01/training-may.html>. Accessed 12 May 2012.

Addison, Paul and Angus Calder (eds.). *Time to Kill: The Soldier's Experience of War in the West 1939–1945* (London: Pimlico, 1997).

Ahrne, Göran. *Social Organizations. Interaction Inside Outside and Between Organizations* (London: Sage, 1994).

Ahrne, Göran and Nils Brunsson. 'Organization Outside Organizations: The Significance of Partial Organization', *Organization* 18 (2011): 83–104.

Aitken, Jonathan. *The Aitken Report: 2008, An Investigation into Cases of Deliberate Abuse and Unlawful Killings in Iraq in 2003 and 2004* (British Army, 2008).

Allport, Gordon W. *The Nature of Prejudice* (Cambridge: Addison-Wesley, 1954).

Ambrose, Stephen. *Citizen Soldiers* (New York: Touchstone, 1997).

Anderson, Benedict. *Imagined Communities: Reflections on the Origin and Spread of Nationalism* (London: Verso, 1991).

Audoin-Rouzeau Stephané and Annette Becker. *1914–1918: Understanding the Great War* (London: Profile Books, 2002).

Andreski, Stanislav. 'Causes of the Low Morale of the Italian Armed Forces in the Two World Wars', *Journal of Strategic Studies* 5(2) (1982): 276–7.

Applewhite, Larry and David Segal. 'Telephone Use by Peacekeeping Troops in the Sinai', *Armed Forces & Society* 17(1) (1990): 117–26.

Aran, Gideon. 'Parachuting', *American Journal of Sociology* 80(1) (1974): 123–52.

Arkin, William and Lynne Dobrofsky. 'Military Socialization and Masculinity', *Journal of Social Issues* 34(1) (1978): 151–66.

Arnold, David. 'Bureaucratic Recruitment and Subordination in Colonial India: The Madras Constabulary, 1859–1947', in Ranajit Guha (ed.) *Subaltern Studies IV: Writings on South Asian History and Society* (Delhi: Oxford University Press, 1985).

Ascoli, David. *A Day of Battle: Mars-la-Tour, August 16, 1870* (London: Harrap, 1987).

Ashley, David. 'Army's Spirit', *Australian Army Journal* 10(3) (2013): 203–12.

Ashworth, Tony. *Trench Warfare 1914–1918, The Live & Let Live System* (London: Macmillan Press, 1965).

Azari, Jaz, Christopher Dandeker, and Neil Greenberg. 'Cultural Stress: How Interactions With and Among Foreign Populations Affect Military Personnel', *Armed Forces & Society* 36(4) (2010): 585–603.

Babington, Anthony. *For the Sake of Example: Capital Courts Martial 1914–1918, the Truth* (London: Martin Secker and Warburg, 1983).

Bach, André. *Fusillé pour l'Example 1914–15* (Paris: Talladier, 2013).

Bachelet, Jean-René. 'La formation des militaires à l'éthique dans le métier des armes', *Droits Fondamentaux* 6 (janvier–décembre 2006): 1–10.

Bachelet, Jean-René. *Pour une éthique du métier des armes. Vaincre la violence.* (Paris: Vuibert-Espace éthique, 2006).

Balck, Wilhelm. *Taktik,* 6 vols., 4th edn. (Berlin: R. Eisenschmidt, 1908–12).

Balck, William. *Tactics,* 2 vols., 4th rev. edn., trans. Walter Krueger (Ft. Leavenworth, KS: US Cavalry Association, 1911–12).

Balkind, Jon. 'A Critique of Military Sociology: Lessons from Vietnam', *The Journal of Strategic Studies* 1(2) (1978): 235–59.

Barnett, Michael and Emanuel Adler. 'Studying Security Communities in Theory, Comparison and History', in Emanuel Adler and Michael Barnett (eds.) *Security Communities* (Cambridge: Cambridge University Press, 1999).

Barrera, Paola. 'Crisis in the Military: Rethinking Conscription and the Military Code', in Robert Y. Leonardi, Raffaella Nanetti, and Piergiorgio Corbetta (eds.) *Italian Politics: A Review* (vol. 2) (London: Pinter, 1988).

Bartov, Omer. 'The Conduct of War: Soldiers and the Barbarization of Warfare', *Journal of Modern History* 64, Supplement (December 1992): 32–45.

Bartov, Omer. *Hitler's Army* (Oxford: Oxford University Press, 1992).

Bartov, Omer. *The Eastern Front, 1941–45: German Troops and the Barbarisation of Warfare* (Oxford: Oxford University Press, 1985).

Battilossi, Stefano. 'Armed Forces', in Giovanni Moliterno (ed.) *Encyclopedia of Italian Political Culture* (New York: Routledge, 2000), 41–3.

Battistelli, Fabrizio. 'Peacekeeping and the Postmodern Soldier', *Armed Forces and Society* 23(3) (1997): 467–84.

Baum, Dan. 'Battle Lessons: What the Generals Don't Know', *The New Yorker,* 17 January 2005, 1–7. <http://www.newyorker.com/fact/content/?7050117fa_fact>. Accessed 12 May 2012.

Bauman, Zygmunt. *Modernity and the Holocaust* (Cambridge: Polity Press, 1989).

Baynes, John. *Morale: A Study of Men and Courage* (London: Avery, 1967).

Baynes, John. 'Recruiting the Professional Army', in Hew Strachan (ed.) *The British Army Manpower and Society into the Twenty First Century* (London: Frank Cass, 2000).

Barrera, Paola. 'Crisis in the Military: Rethinking Conscription and the Military Code', in Robert Y. Leonardi, Raffaella Nanetti, and Piergiorgio Corbetta (eds.) *Italian Politics: A Review* (vol. 2) (London: Pinter, 1988).

Baudrillard, Jean. *The Gulf War Did Not Take Place* (Indiana: Indiana University Press, 1995).

Beal, Daniel, Robin Cohen, Michael Burke and Christy McLendon. 'Cohesion and Performance in Groups: A Meta-Analytic Clarification of Construct Relations', *Journal of Applied Psychology* 88(6) (2003): 998–1004.

Beard, George M. *American Nervousness: Its Causes and Consequences* (New York: Putnam's, 1881).

Beaumont, Joan. 'Remembering the Heroes of Australia's Wars: From Heroic to Post-heroic Memory', in Sibylle Scheipers (ed.), *Heroism and the Changing Character of War* (Basingstoke: Palgrave Macmillan, 2014).

Bedeschi, Giulio. *Il Segreto degli Alpini* (Mursia: Milano, 2004).

Beevor, Antony. *Berlin: The Downfall 1945* (London: Penguin, 2003).

Beevor, Antony. *D-Day* (London: Penguin, 2010).

Beevor, Antony. *Inside the British Army* (London: Chatto and Windus, 1990).

Belenky, Gregory. *Effects of Continuous Operations (CONOPS) on Soldier and Unit Performance* (Washington, DC: Walter Reed Army Institute of Research, 1987).

Bell, Bruce, Walter Schumm, Benjamin Knott, and Morten Ender. 'The Desert Fax: A Research Note on Calling Home from Somali', *Armed Forces & Society* 25(3) (1999): 509–21.

Bell, Catherine. *Ritual Theory, Ritual Practice* (New York: Oxford University Press, 1992).

Bellavia, David. *House to House* (London: Simon and Schuster, 2007).

Ben-Ari, Eyal. *Mastering Soldiers: Conflict, Emotions, and the Enemy in an Israeli Military Unit.* (New York: Berghahn Books, 1998).

Ben-Ari, Eyal. 'What is Worthy of Study? The Sociology of the Armed Forces in Current-Day Conflicts', paper presented at a conference on the 'Production of Knowledge on Peace and Conflict', Exeter University, June 2012.

Ben-Ari, Eyal and Meytal Eran-Jona. 'The Israeli Military in Palestinian Civilian Contexts: Hybrid Conflicts, Multiple Logics and Organizational Transitions', unpublished manuscript.

Ben-Ari, Eyal, Zev Lehrer, Uzi Ben-Shalom and Ariel Vainer. *Rethinking the Sociology of Combat: Israel's Combat Units in the Al-Aqsa Intifada* (Albany: State University of New York Press, 2010).

Ben-Shalom, Uzi, Zeev Lehrer, and Eyal Ben-Ari. 'Cohesion During Military Operations: A Field Study on Combat Units in the Al-Aqsa Intifada', *Armed Forces and Society* 32(1) (2005): 63–79.

Bernstein, Douglas A., Alison Clarke-Stewart, Edward J. Roy, and Christopher D. Wickens. *Psychology* (Boston, MA: Houghton Mifflin, 1997).

Biddle, Stephen. *Military Power: Explaining Victory and Defeat in Modern Battle* (Princeton, NJ: Princeton University Press, 2004).

Black, Jeremy. 'Determinisms and Other Issues', *Journal of Military History* 68 (2004): 122–7.

Black, Jeremy. *Rethinking Military History* (London: Routledge, 2004).

Blake, Nicholas. *Review of the Circumstances surrounding the deaths of four soldiers at Princess Royal Barracks, Deepcut between 1995 and 2002* (London: HMSO, 2006) <http://www.official-documents.gov.uk/document/hc0506/hc07/0795/0795.pdf>. Accessed 5 February 2015.

Blaxland, John C. *The Australian Army from Whitlam to Howard* (Melbourne: Cambridge University Press, 2013).

Blunt, Roscoe. *Inside the Battle of the Bulge: A Private Comes of Age* (Westport, CT: Praeger, 1994).

Boff, Jonathan. 'The Morale Maze: the German Army in late 1918', *Journal of Strategic Studies* 37 (6–7): 855–78.

Boff, Jonathan. *Winning and Losing on the Western Front: The British Third Army and the Defeat of Germany in 1918* (Cambridge: Cambridge University Press, 2012).

Bogart, Leo (ed.). *Social Research and the Desegregation of the U.S. Army* (Chicago: Markham, 1969).

Boot Max. *Invisible Armies: An Epic History of Guerrilla Warfare from Ancient Times to the Present* (New York: Liveright Publishing, 2013).

Bourke, Joanna. *An Intimate History of Killing: Face-to-Face Killing in 20th Century Warfare* (London: Granta, 1999).

Bourne, Peter. *Men, Stress and Vietnam* (Boston, MA: Little Brown, 1970).

Bramley, Vince. *Excursion to Hell* (London: Pan, 1991).

Briefing on Future Social Trends provided to Adjutant General seminar by the Henley Centre Ltd, UK, January 2002.

Brimley, Shawn. 'Crafting Strategy in an Age of Transition', *Parameters* 38(4) (2009): 27–42.

Britt, Thomas W., Carl Andrew Castro, and Amy B. Adler (eds.). *Military Life: The Psychology of Serving in Peace and Combat*, vol. 1: *Military Performance* (Westport, CT: Praeger, 2006).

Britt, Thomas W. and James M. Dickinson. 'Morale during Military Operations: A Positive Psychology Approach', in Britt, Castro, and Adler, *Military Life.*

Brown, J., 'Fifty Shades of Grey: Officer Culture in the Australian Army', *Australian Army Journal* 10(3) 2013: 244–54.

Brown, Malcolm. *Tommy Goes to War* (London and Melbourne, J. Dent and Sons, 1978).

Browne, A.T.A. 'A Study of the Anatomy of Fear and Courage in War', *The Army Quarterly and Defence Journal* 106(3) (1976): 3–4.

Browning, Christopher. 'Ordinary Men or Ordinary Germans', in Robert R. Shandley (ed.) *Unwilling Germans? The Goldhagen Debate* (Minneapolis: University of Minnesota Press, 1998).

Browning, Christopher R. *Ordinary Men. Reserve Police Battalion 101 and the Final Solution in Poland* (New York: Harper Perennial, 1993).

Bucholz, Arden. *Hans Delbrück & The German Military Establishment: War Images in Conflict* (Iowa City: University of Iowa, 1985).

Buckley, John. *British Armour in the Normandy Campaign 1944* (London: Frank Cass, 2004).

Buckley, John. *Monty's Men: The British Army and the Liberation of Europe* (London: Yale University Press, 2013).

Bullard, Maj. R.L. 'A Moral Preparation of the Soldier for Service and Battle', *Journal of the Military Services Institute* XXXI(120) (1902): 779–92.

Burgress, Mark. 'Marine's Three-Block War in Iraq', Center for Defense Information (2003) <http://www.cdi.org/friendlyversion/printversion.cfm?documentID=1834>. Accessed 5 June 2012.

Burk, James. 'Public Support for Peacekeeping in Lebanon and Somalia: Assessing the Casualties Hypothesis', *Political Science Quarterly* 114(1) (1999): 53–80.

Burk, James. 'Expertise, Jurisdiction and Legitimacy of the Military Profession', in D.M. Snider, G.L. Watkins, and L. Matthews (eds.) *The Future of the Army Profession* (New York: McGraw-Hill, 2002): 19–38.

Bury, Patrick. *Callsign Hades* (London: Simon and Schuster, 2010).

Caforio, Giuseppe (ed.). *Handbook of Military Sociology* (New York: Kluwer, 2003).

Calvert, Michael. *Prisoners of Hope* (London: Leo Cooper, 1996).

Cameron, Craig. *American Samurai: Myth, Imagination and the Conduct of Battle in the First Marine Division, 1941–1951* (Cambridge: Cambridge University Press, 1994).

Cantwell, John. 'A Price Too High', *The Monthly*, October 2012.

Cantwell, John with Greg Bearup. *Exit Wounds: One Australian's War on Terror* (Carlton: Melbourne University Press, 2012).

Cappuri, Jean-Paul. 'Tribunal correctionnel de Bastia Le sergent-instructeur du 2e REP condamné à deux ans ferme', *Corse Matin*, 12 May 2010 <http://www.corsematin.com/article/corse/tribunal-correctionnel-de-bastia-le-sergent-instructeur-du-2e-rep-condamne-a-deux-ans-.3362.html>. Accessed 9 May 2013.

Cappuri, Jean-Paul. 'Une vie brisée pour une autorité bafouée', *Corse Matin*, 17 March 2011 <http://www.corsematin.com/article/corse/une-vie-brisee-pour-une-autorite-bafouee.307694.html>. Accessed 9 May 2013.

Caputo, Philip. *A Rumour of War* (London: Pimlico, 1999).

Carreiras, Helena. *Gender and the Military* (London: Routledge, 2006).

Ceci, Laura. *La Resistenza Dei Militari* (Roma: Biblink, 2006).

Chakrabarty, Dipesh. *Provincializing Europe: Postcolonial Thought and Historical Difference* (Princeton, NJ: Princeton University Press, 2000).

Chakravarty, Gautam. *The Indian Mutiny in the British Imagination* (Cambridge: Cambridge University Press, 2005).

Chaudhuri, Rhudra and Theo Farrell. 'Campaign disconnect: operational progress and strategic obstacles in Afghanistan, 2009–2011', *International Affairs* 87(2) (2011): 271–96.

Chodoff, Elliot P. 'Ideology and Primary Groups', *Armed Forces & Society* 9(4) (1983): 569–93.

Christensen, Claus Bundgaard, Niels Bo Poulsen, and Peter Scharff Smith. *Under Hagekors og Dannebrog: Danskere i Waffen SS 1940–1945*. (Copenhagen: Aschehoug, 1998).

von Clausewitz, Carl. *On War*, trans. M. Howard and P. Paret (Princeton, NJ: Princeton University Press, 1989).

Clayton, Anthony. *Warfare in Woods and Forests* (Bloomington: Indiana University Press, 2012).

Clodfelter, Mark. 'Aiming to Break Will: America's World War II Bombing of German Morale and its Ramifications', *Journal of Strategic Studies* 33(3) (2010): 401–35.

Cockerham, William. 'Attitudes Towards Combat Among US Army Paratroopers', *Journal of Political and Military Sociology* 6 (Spring) (1978): 1–15.

Cohen, Hannah. *Let Stephen Speak* (London: Sylvan Press, 1943).

Cohen, Yves. *Le siècle des chefs. Une histoire transnationale du commandement et de l'autorité (1890–1940)* (Paris: Editions Amsterdam, 2013).

Cohn, Bernard S. *Colonialism and Its Forms of Knowledge: The British in India* (Princeton, NJ: Princeton University Press, 1996).

Coker, Christopher. *War in an Age of Risk* (Cambridge: Polity, 2009).

Coker, Christopher. *The Warrior Ethos: Military Culture and The War on Terror* (Abingdon: Routledge, 2007).

Coleman, Stephen. *Military Ethics. An Introduction with Case Studies* (Oxford: Oxford University Press, 2012).

Colin, Jean. *L'Éducation Militaire de Napoleon* (Paris: Librairie Militaire R. Capelet et Cie, 1900).

Colin, Jean. *L'Infanterie au XVIIIe Siecle: la tactique* (Paris, Berger-Levrault, 1907).

Colin, Jean. *The Transformation of War*, trans. L Pope-Hennessy (London: Hugh Rees, 1912).

Colineau, Stéphane and William Assayag. 'Affaire Mahé : un colonel déstabilisé', *20 Minutes*, 27 April 2006.

Collins, Randall. *Interaction Ritual Chains* (Princeton, NJ: Princeton University Press, 1994).

Collins, Randall. *Violence: a Micro-Sociological Theory* (Princeton, NJ: Princeton University Press, 2008).

Connery, David, David Cran, and David Evered. *Conducting Counterinsurgency: Reconstruction Task Force 4 in Afghanistan* (Newport: Big Sky Publishing for the Australian Army History Unit, 2012).

Connolly, Peter. *Counterinsurgency in Uruzgan 2009* (Canberra: Land Warfare Studies Centre, 2011).

Copp Terry and Bill McAndrew. *Battle Exhaustion: Soldiers and Psychiatrists in the Canadian Army, 1939–1945* (London: McGill-Queen's University Press, 1990).

Copp, Terry. *Fields of Fire: The Canadians in Normandy* (London: University of Toronto Press, 2003).

Corns, Cathryn and John Hughes-Wilson. *Blindfold and Alone* (London: Pheonix, 2005).

Coser, Lewis. *Greedy Institutions: Patterns of Undivided Commitment* (New York: Free Press, 1974).

Coss, Edward. *All for the King's Shilling: the British Soldier under Wellington, 1808–1814* (Norman: University of Oklahoma Press, 2010).

Creveld, Martin van. *Fighting Power: German and US Army Performance, 1939–1945* (Westport, CT: Greenwood Press, 1982).

Creveld, Martin van. *Supplying War* (Cambridge: Cambridge University Press, 1977).

Crew, Francis Albert Eley. *The Army Medical Services, Campaigns*, vol. 4: *North-West Europe* (London: Her Majesty's Stationery Office, 1962).

Crichton-Browne, Sir James. 'La Vieillesse,', *RS* 49 (1892): 168–78.

What 'success' means in Afghanistan, Iraq and Libya', in James Burk (ed.) *How 9/11 Changed our Ways of War* (Stanford, CA: Stanford University Press, 2013): 116–48.

David, J. 'Close Combat: Army's Fundamental Skill', *Australian Army Journal*, 2010 7(1): 153–63.

Davies, Christie. 'Itali sunt imbelles', *Journal of Strategic Studies* 5(2) (1982): 266–9.

De Amicis, Edmondo. *Cuore* (Torino: Einaudi, 1974).

De Caroli, Giacinto and Maria Luisa de Caroli, *Viaggio al Fronte*. (Cuneo: BB Europa Edizioni, 2011).

Department of Defence Census 2011 Public Report (Canberra, Australia: Department of Defence, 2012).

der Derian, James. *Virtuous War* (London: Basic Books, 2001).

Dinter, Elmar. *Hero or Coward: Pressures Facing the Soldier in Battle* (London: Frank Cass, 1985).

Dirks, Nicholas B. *Castes of Mind: Colonialism and the Making of Modern India*. (Princeton, NJ: Princeton University Press, 2001).

Dollard, John. *Fear in Battle* (New Haven, CT: Institute of Human Relations, Yale University, 1943).

Doubler, Michael. *Closing With the Enemy: How GIs Fought the War in Europe, 1944–1945* (Lawrence: University Press of Kansas, 1994).

Donaldson, Mark. *The Crossroad* (Sydney: Macmillan, 2013).

Dower, John W. *War without Mercy: Race and Power in the Pacific War* (New York: Pantheon Books, 1986).

Duffy, Christopher. *The Military Experience in the Age of Reason* (London: Routledge and Kegan Paul, 1987).

Du Picq, Ardant. *Battle Studies: Ancient and Modern Battle*, trans. John N. Greely and Robert C. Cotton (Harrisburg, PA: Military Service Publishing, 1947).

Du Picq, Ardant. *Battle Studies*, trans. John. N. Greely and Robert C. Cotton (Charleston, SC: Bibliobazaar, 2006).

Dupuy, Trevor N. 'Theory of Combat', in Franklin D. Margiotta (ed.) *Brassey's Encyclopaedia of Military History and Biography* (London: Brassey's, 1994).

Durch, William J. and Madeline L. England. *The Purpose of Peace Operations* (New York University: Center on International Cooperation, 2009).

Durning, Kathleen P. 'Women at the Naval Academy: An Attitude Survey', *Armed Forces & Society* 4(4) (1978): 569–88.

Dyer, Gwynne. *War* (London: The Bodley Head, 1996).

Ebert, James. *A Life in a Year: The American Infantryman in Vietnam 1965–1972*, (Novato, CA: Presidio, 1993).

Echenberg, Myron. *Colonial Conscripts: The Tirailleurs Sénégalais in French West Africa, 1857–1960* (Portsmouth: Heinemann, 1991).

Echevarria II, Antulio J. *After Clausewitz: German Military Thinkers before the Great War* (Lawrence, KS: University Press of Kansas, 2000).

Ehrenreich, Barbara. *Blood Rites: Origins and History of the Passions of War* (London: Virago, 1997).

Ehrenreich, Barbara. 'War Without Humans: Modern Blood Rites Revisited' <http://ehrenreich.blogs.com/barbaras_blog/2011/07/war-without-humans-modern-bloodrites-revisited.html>.

Eisenhart, Ronald. 'You Can't Hack it Little Girl: A Discussion of the Covert Psychological Agenda of Modern Combat Training', *Journal of Social Issues* 31(4) (1975): 13–24.

Ellis, John. *Brute Force: Allied Strategy and Tactics in the Second World War* (London: André Deutsch, 1990).

Ellis, John. *The Sharp End: The Fighting Man in World War II* (London: Compendium, 1993).

Eltinge, Capt. Leroy. 'Psychology of War', *Journal of the United States Cavalry Association* XXI(84) (May 1911): 1015–52.

Ender, Morten. 'G.I. Phone Home: The Use of Telecommunications by the Soldiers of Operation Just Cause', *Armed Forces & Society* 21(3) (1995): 435–53.

Ender, Morten and David Segal. 'V(E)-Mail to the Foxhole', *Journal of Political and Military Sociology* 24 (Summer 1995): 83–104.

English, John. *Lament for an Army* (Toronto: Canadian Institute of International Affairs, 1998).

Evans, Humphrey. *Thimayya of India: A Soldier's Life* (New York: Harcourt, Brace, and Co., 1960).

Evans, Michael. *City Without Joy*, Occasional Papers No. 2 (Canberra: Australian Defence College, 2007).

Evans, Michael and Alan Ryan (eds.), *The Human Face of Warfare: Killing, Fear and Chaos in Battle* (St Leonards: Allen and Unwin, 2000).
Everts, Philip and Pierangelo Isernia. *Public Opinion and the Use of Force* (London: ECPR/ Routledge, 2001).
Falcon, R.W. *Handbook on Sikhs for Regimental Officers* (Allahabad: Pioneer Press, 1896).
Farwell, Byron. *Queen Victoria's Little Wars* (New York: W.W. Norton, 1985).
Faulk, Henry. *Group Captives: The Re-Education of German Prisoners of War in Britain 1945–1948* (London: Chatto & Windus, 1977).
Fay, Peter Ward. *The Forgotten Army: India's Armed Struggle for Independence 1942–1945* (Ann Arbor: The University of Michigan Press, 1993).
Fehrenbach, Thomas. *This Kind of War* (New York: Macmillan, 1963).
Fennell, Jonathan. *Combat and Morale in the North African Campaign: The Eighth Army and the Path to El Alamein* (Cambridge: Cambridge University Press, 2011).
Fennell, Jonathan. 'In Search of the 'X' Factor: Morale and the Study of Strategy', *Journal of Strategic Studies* 37(6–7) (2014): 799–828.
Fennell, Keith. *Warrior Brothers* (North Sydney: Bantam, 2009).
Ferguson, Niall. 'Prisoner Taking and Prisoner Killing in the Age of Total War: Towards a Political Economy of Military Defeat', *War in History* 11(2) (2004): 149–92.
Festinger, Leon, Stanley Schachter, and Kurt Back. *Social Pressures in Informal Groups: A Study of Human Factors in Housing* (New York: Harper, 1950).
Fewster, Kevin. 'Jacka, Albert (1893–1932)', *Australian Dictionary of Biography*, National Centre of Biography, Australian National University, <http://adb.anu.edu.au/biography/jacka-albert-6808/text11779>. Accessed 5 December 2013.
Fisher, David. Morality and War, Can War be Just in the Twenty-first Century? (Oxford: Oxford University Press, 2011).
Fleming, Colin M. 'New or Old Wars? Debating a Clausewitzian Future', *Journal of Strategic Studies* 32(2) (2009): 213–41.
Fogu, Claudio. '*Italiani Brava Gente*: The Legacy of Fascist Historical Culture on Italian Politics of Memory', in Richard Ned Lebow, Wulf Kansteiner, and Claudio Fogu (eds.) *The Politics of Memory in Postwar Europe* (Durham, NC: Duke University Press, 2006), 147–76.
Foner, Jack D. *Blacks and the Military in American History* (New York: Praeger, 1974).
Fontane, Theodor. *Der Krieg gegen Frankreich*, 4 vols. (Berlin: R. v. Decker, 1877–78), I, 380.
Forrest, Alan. 'Conscription as Ideology: Revolutionary France and the Nation in Arms', in Lars Mjøset and Stephen van Holde (eds.) *The Comparative Study of Conscription in the Armed Forces. Comparative Social Research*, vol. 20 (Bradford: Emerald Group Publishing, 2002), 95–115.
Forster, Anthony. 'The Military Covenant and British Civil–Military Relations: Letting the Genie out of the Bottle', *Armed Forces & Society* 38(2) (2012): 273–90.
Fox, Richard G. *Lions of the Punjab: Culture in the Making* (Berkeley: University of California Press, 1985).
Frangipane, Ettore. *A come Alpini* (Trento: Curcu e Genovese, 2011).
Franklin, John Hope. *From Slavery to Freedom: A History of American Negroes* (New York: Knopf, 1948).
Frederick, Jim. *Black Hearts* (London: Macmillan, 2010).

French, David. '"Tommy is no Soldier": The Morale of the Second British Army in Normandy, June–August 1944', *Journal of Strategic Studies* 19(4) (1994): 154–78.

Fritz, Stephen G. *Frontsoldaten: The German Soldier in World War II* (Kentucky: University of Kentucky Press, 1995).

Fuller, John G. *Troop Morale and Popular Culture in the British and Dominion Armies 1914–18* (Oxford: Clarendon Press, 1990).

Gal, Reuvan and Franklin D. Jones. 'A Psychological Model of Combat Stress', in Franklin D. Jones, L.R. Sparacino, Victoria L. Wilcox, James M. Rothberg, and James W. Stokes (eds.) *War Psychiatry* (Washington, DC: TMM Publications, 1995), 133–48.

Gal, Reuven and David A. Mangelsdorff (eds.). *Handbook of Military Psychology* (New York: John Wiley & Sons, 1991).

Gal, Reuven and Frederick J. Manning. 'Morale and its Components: A Cross National Comparison', *Journal of Applied Social Psychology* 17(4) (1987): 369–91.

Gammage, Bill. *The Broken Years: Australian Soldiers in the Great War* (Ringwood: Penguin, 1985).

Gaunt, D. 'Vikings VM'ingS in the "Gan"', *Globe and Laurel* (March/April 2007): 107.

Gellner, Ernest. *Nations and Nationalism* (London: Wiley-Blackwell, 2006).

Gelpi, Christopher, Peter Feaver, and Jason Reifler. 'Success Matters: Casualty Sensitivity and the War in Iraq', *International Security* 30(3) (2005): 7–46.

Gettleman, Jeffrey. 'Soldier Accused as Coward Says He Is Guilty Only of Panic Attack', *New York Times*, 6 November 2003 <http://www.nytimes.com/2003/11/06/us/soldier-accused-as-coward-says-he-is-guilty-only-of-panic-attack.html>. Accessed 22 January 2015.

Giorgi, Rosa. *I Santi* (Milano: Electa, 2002).

Giradet, Edward. *Afghanistan: The Soviet War* (London: Croom Helm, 1985).

Gisick, M. 'Citizen-Soldier, Warrior Caste: American Soldiers in the "War on Terror"', PhD proposal (Australian National University, 2013).

Glass, Charles. *Deserter: The Last Untold Story of the Second World War* (London: HarperCollins, 2013).

Goffman, Erving. *Where the Action is* (London: Allen Lane, 1969).

Goldhagen, Daniel Jonah. *Hitler's Willing Executioners: Ordinary Germans and the Holocaust* (New York: Alfred R. Knopf, 1996).

Goltz, Gen. d. Inf. Freiherr. v.d. 'Moralisches Heimweh', *Deutsche Revue* (January 1903): 1–5.

Goltz, C. Freiherr v.d. 'Luxus im Heere', *Deutsche Revue* (December 1903): 257–63.

Gooch, John. 'Italian Military Competence', *Journal of Strategic Studies* 5(2) (1982): 257–65.

Gooch, John. 'Italy during the First World War', in Allan R. Millett and Williamson Murray (eds.) *Military Effectiveness*, vol. 1: The First World War (Cambridge: Cambridge University Press, 2010).

Glenn, Russell W. and Gina Kingston. *Urban Battle Command in the 21st Century* (Santa Monica, CA: Rand Corporation, 2005).

Graham, Stephen. *Cities under Siege: The New Military Urbanism* (London: Verso, 2010).

Granatstein, Jack. *Canada's Army: Waging War and Keeping the Peace* (Toronto: University of Toronto Press, 2004).

Graves, Robert. *Goodbye to All That* (London: Jonathan Cape, 1929).

Gray, Colin S. *Modern Strategy* (Oxford: Oxford University Press, 1999).

Grey, Jeffrey. *A Military History of Australia* (third edition) (Melbourne: Cambridge University Press, 2008).

Griffin, Lieut. Joseph T. 'Military Instruction for Schoolboys', *Journal of the Military Services Institute* 35 (1905): 61–7.

Griffith, James. 'Further Considerations Concerning the Cohesion-Performance Relation in Military Settings', *Armed Forces and Society* 34(1) (2007): 138–47.

Griffith, Paddy. *Forward into Battle: Fighting Tactics from Waterloo to the Near Future* (Novato, CA: Presidio, 1991).

Griffith, Paddy. 'Fighting Spirit: Leadership and Morale in the "Empty Battlefield" of the Future', in Michael Evans and Alan Ryan (eds.) *The Human Face of Warfare: Killing, Fear and Chaos in Battle* (London: Allen and Unwin, 2000), 109–23.

Grossman, David. *On Killing: The Psychological Costs of Learning to Kill in War and Society* (Boston, MA: Back Bay Books, 1996).

Guibert, Comte Jacques de. *Essai Général de Tactique* (Paris: Economica, 2004).

Guion, Robert M. 'Industrial Morale (A Symposium): 1. The Problem of Terminology', *Personnel Psychology* 11(1) (1958): 60–2.

Gusterson, Hugh. 'Can the Insurgent Speak?', in Tarak Barkawi and Keith Stanski (eds.) *Orientalism at War* (London: Hurst & Co., 2012).

Hackett, General Sir John. The Profession of Arms, Lees Knowles Lectures Trinity College Cambridge (London: Times Publishing Company, 1962).

Hallo, Jean. *Monsieur Légionnaire* (Paris: Lavazuelle, 2007).

Hammett, J. 'We Were Soldiers Once... The Decline of the Royal Australian Infantry Corps?', *Australian Army Journal* 5(1) (2008): 39–50.

Handel, Michael I. *Masters of War: Classical Strategic Thought* (London: Frank Cass, 2001).

Hanson, Fergus. *Australia and New Zealand in the World: Public Opinion and Foreign Policy* (Sydney: Lowy Institute, 2012).

Hanson, Victor David. *The Soul of Battle: from Ancient Times to the Present Day* (New York: Free Press, 1999).

Harding, Thomas. 'Four British soldiers drown in Afghanistan canal', *The Daily Telegraph*, 24 June 2010 <http://www.telegraph.co.uk/news/newstopics/onthefrontline/7851207/Four-British-soldiers-drown-in-Afghanistan-canal.html>. Accessed 12 February 2014.

Harnden, Toby. *Dead Men Risen: The Reality of Britain's War in Afghanistan* (London: Quercus, 2011).

Harel, Amos and Avi Isacharoff. *The Seventh War: How We Won and Why We Lost the War with the Palestinians* (Tel Aviv: Yedioth Aharonot, 2004) (Hebrew).

Harris, Kira, Gringart Eyal, and Drake Deirdre. 'Military Retirement: Reflections from Former Members of Special Operations Forces', *Australian Army Journal* 10(3) (2013): 97–112.

Hart, Stephen Ashley. Colossal Cracks: Montgomery's 21st Army Group in Northwest Europe, 1944–45 (Mechanicsburg, PA: Stackpole, 2007).

Hasse, Else. 'Nervosität und Geistesleben', *Der Türmer* 13 (1911): 577–87.

Hasluck, Paul. *The Government and the People 1942–1945* (Canberra: Australian War Memorial, 1970).

Hauser, William L. *America's Army in Crisis: A Study in Civil-Military Relations* (Baltimore, MD: Johns Hopkins University Press, 1973).

Haynes, Deborah. 'Army treatment of bullied soldier "unforgivably bad", coroner says', *The Times*, 3 March 2014 <http://www.thetimes.co.uk/tto/news/uk/defence/article4021657.ece>

Hedges, Chris. *War is a Force That Gives us Meaning* (New York: Public Affairs, 2002).

Heilig, Gerhard. 'Fabrikarbeit und Nervenleiden', *Medizinische Reform* 16(31) (1908): 369–97.

Hellpach, Willy. *Nervosität und Kultur* (Berlin: J. Räde, 1902).

Helmer, John. *Bringing the War Home: The American Soldier in Vietnam and After* (New York: The Free Press, 1974).

Helmus, Todd C. and Russell W. Glenn. *Steeling the Mind: Combat Stress Reactions and Their Implications for Urban Warfare* (Santa Monica, CA: Rand Corporation, 2004).

Henderson, Darryl. *Cohesion: the Human Element* (Washington, DC: National Defence University Press, 1985).

Henderson, William Darryl. *Why the Vietcong Fought: A Study of Motivation and Control in a Modern Army in Combat* (Westport, CT: Greenwood Press, 1979).

Henriksen, Rune. 'Warriors in Combat', *Journal of Strategic Studies* 30(2) (2007): 187–223.

Herodotus. *The Histories*, trans. Alfred Godley (Cambridge: Harvard University Press, 1920).

Hills, Alice. *Future Wars in Cities: Re-Thinking a Liberal Dilemma* (London: Frank Cass, 2004).

Hirschman, Albert. *Exit Voice and Loyalty. Responses to Decline in Firms, Organizations and States* (Cambridge, MA: Harvard University Press, 1970).

Hirst, Paul. *Space and Power: Politics, War and Architecture* (London: Polity, 2005).

Hobsbawm, Eric. *Nations and Nationalism Since 1780: Programme, Myth, Reality* (Cambridge: Cambridge University Press, 2012).

Hockey, John. 'Switching On: Sensory Work in the Infantry', *Work Employment and Society* 23(3) (2009): 477–93.

Hoenig, Fritz. *Untersuchungen über die Taktik der Zukunft: Entwickelt aus der neueren Kriegsgeschichte*, 4th edn. (Berlin: R. Felix, 1894), 141–2; trans. as Fritz Hoenig, *Inquiries into the Tactics of the Future. Developed from Modern Military History*, 4th edn., trans. Carl Reichmann (Kansas City, MO: Hudson-Kimberly, 1898).

Holmes, Richard. *Acts of War: The Behaviour of Men in Battle* (London: HarperCollins, 2004).

Holmes, Richard. *Firing Line* (London: Penguin, 1985).

Holmes, Richard and John Keegan. *Soldiers: A History of Men in Battle* (London: Konecky and Konecky, 1997).

Holt, Tonie and Valmai Holt. *In Search of the Better Ole: The Life, The Works and the Collectibles of Bruce Bairnsfather* (Portsmouth: Milestone, 1985).

Howard, Michael. *Franco-Prussian War* (London: Methuen, 1981).

Howard, Michael. 'Men against Fire: The Doctrine of the Offensive in 1914', in *Makers of Modern Strategy from Machiavelli to the Nuclear Age*, Peter Paret, ed., (Princeton, NJ: Princeton University Press, 1986), 519.

Howard, Michael. 'When are Wars Decisive?', *Survival* 41(1) (1999): 128.

House of Commons Defence Committee, *Towards the next Defence and Security Review, Part 1, 7th Report of Session 2013–14*, vol. 1, p. 15, paras 20, 21 <http://www.publications.parliament.uk/pa/cm201314/cmselect/cmdfence/197/197.pdf>. Accessed 8 February 2014.

Hughes, Daniel J. *Moltke on the Art of War: Selected Writings* (Novato, CA: Presidio, 1993).

Hull, Isabel V. *Absolute Destruction. Military Culture and the Practices of War in Imperial Germany* (Ithaca, NY: Cornell University Press, 2005).

Huntington, Samuel P. *The Soldier and The State: The Theory and Politics of Civil-Military Relations* (Cambridge, MA: Harvard University Press, 1957).

Ignazi, Piero, Giampiero Giacomello and Fabrizio Coticchia. *Italian Military Operations Abroad- Just Don't Call It War* (London: Palgrave MacMillan, 2012).

Ingham, Sarah. *The Military Covenant: Its Impact on Civil-Military Relations in Britain* (Basingstoke: Ashgate, 2014). Ingham, Sarah and Dandeker, Christopher. 'The Covenant We Must Protect From the Lawyers', *Parliamentary Brief*, 24 June 2010.

'Inquiry Officer's Report into the facts and circumstances surrounding a shooting incident in Afghanistan that resulted in the deaths of LCPL. S. Milosevic, SPR J. T. Martin and PTE R. H. F. Poate and injuries to other Australian soldiers on or about 29 AUG 12', <http://www.defence.gov.au/coi>. Accessed 26 September 2013.

Irondelle, Bastien. 'L'armée française et l'éthique dans les conflits de post-Guerre Froide', *Critique internationale* 41(4) (2008): 119–36.

Italian Taurinense Brigade. *Alpini In Afghanistan. Diario dell'Ultima Missione della Taurinense nel Paese degli Aquiloni* (Torino: SusaLibri, 2013).

Janowitz, Morris. *The Professional Soldier* (New York: Free Press, 1960).

Janowitz, Morris and Edward A. Shils. 'Cohesion and Disintegration in the Wehrmacht in World War II', in Morris Janowitz (ed.) *Military Conflict* (London: Sage, 1975).

Janowitz, Morris and Edward A. Shils. 'Cohesion and Disintegration in the Wehrmacht in World War II', *Public Opinion Quarterly* (Summer 1948): 280–315.

Janowitz, Morris and Stephen D. Wesbrook (eds.). *The Political Education of Soldiers* (Beverly Hills: Sage, 1983).

Jaulmes, Adrien. 'Afghanistan: Les légionnaires du 2e REP pacifient la vallée de Tagab', *Le Figaro*, 26 April 2010 <http://www.asafrance.fr/archives/223-le-2e-rep-pacifie-la-vallee-de-tagab->. Accessed 6 February 2014.

Jennings, Christopher and Adrian Weale. *Green-Eyed Boys* (London: HarperCollins, 1996).

Johns, John H. et al. *Cohesion in the US Military* (Washington, DC: National Defense University Press, 1984).

Johnson, Rob. *The Afghan Way of War* (London: Hurst, 2011).

Johnson, Rob. *Outnumbered, Outgunned and Undeterred* (London: Thames and Hudson, 2011).

Johnston, Mark. *At the Frontline: Experiences of Australian Soldiers in World War II* (Melbourne: Cambridge University Press, 1996).

Johnston, Mark. *Fighting the Enemy: Australian Soldiers and Their Adversaries in World War II* (Melbourne: Cambridge University Press, 2000).

Jones, Sam. 'UK losing "courageous instinct", warns general', *The Financial Times*, 18 December 2013 <http://www.ft.com/cms/s/0/ea62dafe-6827-11e3-8ada-00144feabdc0.html#axzz2suWvSuZ8>. Accessed 12 March 2014.

Jünger, Ernst. *Storm of Steel* (Harmondsworth: Penguin, 2003).

Junger, Sebastian. *War* (London: Fourth Estate, 2011).

Kahn, Paul W. 'War and Sacrifice in Kosovo', *Philosophy and Public Policy Quarterly* 19, Spring/Summer 2/3 (1999): 1–6.

Kaye, John William. *A History of the Sepoy War in India 1857–1858*, vol. 1 (London: W.H. Allen & Co., 1864).

Keegan, John. *The Face of Battle* (London: Penguin, 1976).

Keegan, John. 'Towards a Theory of Combat Motivation', in Paul Addison and Angus Calder (eds.) *Time to Kill: The Soldier's Experience of War in the West 1939–1945* (London, Pimlico 1997).

Kellett, Anthony. Combat Motivation: The Behaviour of Soldiers in Battle (Boston, MA: Kluwer Nijhof, 1982).

Kelman, Herbert C. and V. Lee Hamilton. *Crimes of Obedience: Towards a Social Psychology of Authority and Responsibility* (New Haven, CT and London: Yale University Press, 1989).

Kelleher, Catherine. 'Mass Armies in the 1970s: the Debate in Western Europe', *Armed Forces & Society* 5(1) (1978): 3–30.

Kern, Stephen. *The Culture of Time and Space 1880–1918* (Cambridge, MA: Harvard University Press, 1983).

Kahn, Paul W. 'War and Sacrifice in Kosovo', *Philosophy and Public Policy Quarterly* 19, Spring/Summer2/3 (1999): 1–6.

Kier, Elizabeth. 'Homosexuality in the US Military: Open Integration and Combat Effectiveness', *International Security* 23(2) (1998): 5–39.

Kiernan, V.G. *Colonial Empires and Armies 1815–1960* (Stroud: Sutton, 1998).

Kiley, Sam. *Desperate Glory* (London: Bloomsbury, 2009).

Killingray, David and David Omissi (eds.). *Guardians of Empire* (New York: St. Martins, 1999).

Kimerling, Rachel, Amy E. Street, Joanne Pavao, Mark W. Smith, Ruth C. Cronkite, Tyson H. Holmes, and Susan M. Frayne, 'Military-Related Sexual Trauma among Veterans Health Administration Patients Returning from Afghanistan and Iraq', *American Journal of Public Health* 100 (August 2010): 1409–12.

Kimmel, Melvin J. and Francis E. O'Mara. 'The Measurement of Morale', US Army Research Institute for the Behavioural and Social Sciences (1981).

Kimmel, Melvin J., Francis E. O'Mara, and Nehama Babin. 'The Development of a Unit Morale Measure for Army Battalions', Technical Report 617, US Army Research Institute for the Behavioural and Social Sciences (March 1984).

Kindsvatter, Peter. *American Soldiers: Ground Combat in the World Wars, Korea, and Vietnam* (Lawrence: University Press of Kansas, 2003).

King, Anthony. *The Ethos of the Royal Marines: The Precise Application of Will*, unpublished report commissioned by Commander, Commando Training Centre, 2004.

King, Anthony. *The Ethos of 16 Air Assault Brigade: The Struggle for Self-Definition*, unpublished report commissioned by Commander 16 Air Assault Brigade, 2006.

King, Anthony. 'The Post-Fordist Military', *Journal of Political and Military Sociology* 34(1) (2006): 339–54.

King, Anthony. 'The Word of Command: Communication and Cohesion in the Military', *Armed Forces and Society* 32(1) (2006): 493–512.

King, Anthony. 'The Afghan War and "Postmodern" Memory: Commemoration and the Dead of Helmand', *British Journal of Sociology* 61(1) (2010): 1–25.

King, Anthony. 'Understanding the Helmand Campaign: British Military Operations in Afghanistan', *International Affairs* 86(2) (2010): 311–32.

King, Anthony. *The Combat Soldier: Infantry Tactics and Cohesion in the Twentieth and Twenty-First Centuries* (Oxford: Oxford University Press, 2013).

King, Anthony. *The Transformation of Europe's Armed Forces: from the Rhine to Afghanistan* (Cambridge: Cambridge University Press, 2014).

Kinzer Stewart, Nora. *Mates and Muchachos: Unit Cohesion in the Falklands/Malvinas War* (New York: Brasseys Inc, 1991).

Kirke, Charles. 'Group Cohesion, Culture, and Practice', *Armed Forces & Society* 35(4) (2009): 745–53.

Knox, MacGregor. 'The Italian Armed Forces: 1940–3', in Allan R. Millett and Williamson Murray (eds.) *Military Effectiveness* (Boston, MA: Unwin Hyman, 1988), 136–79.

Korb, Lawrence J., Sean E. Duggan, and Laura Conley. 'Implementing the Repeal of "Don't Ask, Don't Tell" in the U.S. Armed Forces' (Washington, DC: Center for American Progress, March 2010).

Krakauer, Jon. *Where Men Win Glory. The Odyssey of Pat Tillman* (New York: Doubleday, 2009).

Kramer, Eric H., Bart van Bezooijen, and Roos Delhaij. 'Sense-making During Operations and Incidents', in Joseph Soeters, Paul C. van Femema, and Robert Beers (eds.) *Managing Military Organizations: Theory and Practice* (London: Routledge, 2010).

Krulak, Charles C. 'Forward: *Ne Cras*: Not Like Yesterday', in Richard H. Schultz Jr. and Robert L. Pfaltsgraff Jr. (eds.) *The Role of Naval Forces in 21st –Century Operations* (Washington, DC: Brassey's, 2000), xi–xvii.

Kuss, Han. *Schiksal einer Generation* (Selent: Pour la Mérite, 2008).

Laffin, John. *Digger: The Story of the Australian Soldier* (London: Cassell, 1959).

Lake, Marilyn and Henry Reynolds (eds.). *What's Wrong with Anzac: The militarization of Australian history* (Sydney: University of New South Wales Press, 2010).

Land Warfare Doctrine 1: The Fundamentals of Land Warfare (Australian Army, Puckapunyal, 1998).

Langlois, Colonel Hippolyte. *Lessons from Two Recent Wars*, Translated for the General Staff (London: His Majesty's Stationery Office, 1909), 89, 94.

Larroumet, Marie. *Mythe et images de la Légion étrangère* (Paris: L'Harmattan, 2004).

Latimer, Jon. *Burma: The Forgotten War* (London: John Murray, 2004).

Lears, T.J. Jackson. *No Place of Grace: Antimodernism and the Transformation of American Culture 1880–1920* (New York: Pantheon, 1981).

Lee, Jonathan L. 'The Comprehensive Review Working Group and Don't Ask, Don't Tell Repeal at the Department of Defense', in James E. Parco and David A. Levy (eds.). *Evolution of Government Policy towards Homosexuality in the US Military* (New York: Routledge, 2014), 167–91.

Leonhard, Robert R. *The Principles of Warfare for the Information Age* (Novato, CA: Presidio, 2000).

Levi Martin, John. *Social Structures* (Princeton, NJ: Princeton University Press, 2009).

Levy, Yagil. *Israel's Death Hierarchy: Casualty Aversion in a Militarized Democracy* (New York: New York University Press, 2012).

Liddell Hart, Basil. *The Other Side of the Hill* (London: Cassells, 1951).

Liddell Hart, Basil. *Memoirs, I* (London: Cassell and Co., 1967).

Liddell Hart, Basil. *Strategy: The Indirect Approach* (London: Faber and Faber, 1967).

Linderman, Gerald F. *The World Within War. America's Combat Experience in World War II* (Cambridge, MA: Harvard University Press, 1997).

Lindqvist, Sven. *Exterminate All the Brutes* (London: Granta, 1997).

Lindqvist, Sven. *A History of Bombing* (London: Granta, 2001).

Linn, Brian M. and Russell F. Weigley. 'The "American Way of War" Revisited', *Journal of Military History* 66(2) (2002): 501–33.

Little, Roger. 'Buddy Relations and Combat Performance', in Morris Janowitz (ed.), *The New Military: Changing Patterns of Organization* (New York: Russell Sage Foundation, 1964), 195–224.

Lopasic, Alexander. 'Italian Military Performance in the Second World War: Some Considerations', *Journal of Strategic Studies* 5(2) (1982): 270–5.

Lucey, Terry. *Quantitative Techniques* (London: Continuum, 1996).

Lupi, Tonino. *Ricordi di Guerra di un Alpino* (Genova: Scuola Tipografica SS. Vergine di Pompei, 1957).

Lutz, Tom. *American Nervousness, 1903: An Anecdotal History* (Ithaca, NY: Cornell University Press, 1991).

Lynn, John A. *Battle: A History of Combat and Culture from Ancient Greece to Modern America* (Boulder, CO: Westview Press, 2003).

Lynn, John A. *The Bayonets of the Republic: Motivation and Tactics in the Army of Revolutionary France* (Oxford: Oxford University Press, 1996).

Machiavelli, Niccolò. *The Art of War*, trans. Neal Wood (New York: Bobbs Merrill, 1965).

McCartney, Helen. 'The Military Covenant and the Civil–Military Contract in Britain', *International Affairs* 86(2) (2010): 411–28.

MacCoun, Robert. 'What is Known about Unit Cohesion and Military Performance', in National Defence Research Institute, *Sexual Orientation and US Military Personnel Policy: options and assessment* (Washington, DC: RAND, 1993).

MacCoun, Robert, Elizabeth Kier, and Aaron Belkin. 'Does Social Cohesion Determine Motivation in Combat?', *Armed Forces and Society* 32(4) (2006): 646–54.

MacCoun, Robert and William Hix. 'Cohesion and Performance', in *Sexual orientation and U.S. military policy: An update of RAND's 1993 study* (Santa Monica: RAND, 2010).

McChrystal, Stanley. *My Share of the Task* (London: Penguin).

McCoy, Bryan. *The Passion of Command* (Quantico, VA: Marine Corps Association, 2007).

Macdonald, Lynn. *1914* (London: Macmillan, 1987).

McGorman, Ewan. *Life in the French Foreign Legion: How to Join and What to Expect When You Get There* (Hellgate Press, 2000).

McInnes, Colin. *Spectator Sport Warfare* (London: Lynne Rienner, 2002).

McKernan, M. and M. Browne (eds.). *Australia: Two Centuries of War and Peace* (Canberra: Australian War Memorial, 1988).

Macklin, Robert. *Jacka VC* (Crows Nest: Allen & Unwin, 2006).

MacLeod, David I. 'Socializing American Youth to Be Citizen-Soldiers', in Manfred F. Boemeke, Roger Chickering, and Stig Förster (eds.) *Anticipating Total War* (Cambridge: Cambridge University Press, 1999), 137–66.

McMahan, Jeff. *Killing in War* (Oxford: Oxford University Press, 2009).

MacMunn, Lt. Gen. Sir George. *The Armies of India* (London: Adam and Charles Black, 1911).

MacMunn, Lt. Gen. Sir George. *The Martial Races of India* (London: Sampson, Low, Marston & Co., n.d. [1933]).

McNeill, William H. *Keeping Together in Time: Dance and Drill in Human History* (Cambridge: Harvard University Press, 1995).

McPherson, James M. *For Cause and Comrades. Why Men Fought In The Civil War* (Oxford: Oxford University Press, 1997).

Manning, Frederick. 'Morale, Cohesion and Esprit de Corps', in Gal Reuven and Mangelsdorff A. David (eds.) *Handbook of Military Psychology* (New York: John Wiley & Sons, 1991).

Marlantes, Karl. *Matterhorn* (London: Corvus, 2010).

Marshall, Samuel Lyman Attwood. *Men against Fire: The Problem of Battle Command in Future War* (New York: The Free Press, 1947).

Marshall, Samuel Lyman Attwood. *Bird* (New York: Warner Books, 1968).

Marshall, Samuel Lyman Attwood. *Notes on Urban Warfare* (Aberdeen Proving Ground, Maryland: US Army Materiel Systems Analysis Agency, 1973).

Marshall, Samuel Lyman Attwood. *Men against Fire: The Problem of Battle Command in Future War* (Norman: University of Oklahoma, 2000).

Marston, Daniel P. *Phoenix from the Ashes: The Indian Army in the Burma Campaign* (London: Praeger, 2003).

Martin, Michel. 'Conscription and the Decline of the Mass Army in France, 1960–75', *Armed Forces & Society* 1977 3(3): 355–406.

Massobrio, Giovanni and Giorgio Rochat. *Breve Storia Dell'esercito Italiano Dal 1861 Al 1943* (Torino: Einaudi, 1978).

Masters, Chris. *Uncommon Soldier: Brave, Compassionate and Tough, The Making of Australia's Modern Diggers* (Sydney: Allen & Unwin, 2012).

Maylor, Rob with Robert Macklin. *SAS Sniper: The World of an Elite Australian Marksman* (Sydney: Hachette Australia, 2010).

Mazower, Mark. *Hitler's Empire: Nazi Rule in Occupied Europe* (London: Penguin, 2009).

Mazower, Mark. 'Military Violence and Nationalist Socialist Values: The *Wehrmacht* in Greece 1941–1944', *Past and Present* 134 (February 1992): 129–58.

Médecins sans frontiers. *'People are left defenceless'—Testimonies from Ivory Coast: An Overview 2003–2006* (Operational Center of Brussels, 2006).

Menning, Bruce W. *Bayonets before Bullets: The Imperial Russian Army, 1861–1914* (Bloomington: Indiana University, 1992).

Merchet, Jean-Dominique. 'L'armée en mots troubles', *Libération*, 5 January 2006 <http://www.liberation.fr/grand-angle/010134446-l-armee-en-mots-troubles>

Merchet, Jean-Dominique. 'Affaire Mahé : la défiance régnait chez les officiers…', *Marianne Blog*, 27 November 2012 <http://www.marianne.net/blogsecretdefense/Affaire-Mahe-la-defiance-regnait-chez-les-officiers_a852.html>. Accessed 25 December 2012.

Mershon, Sherie and Steven Schlossman. *Foxholes & Color Lines: Desegregating the U.S. Armed Forces* (Baltimore, MD: The Johns Hopkins University Press, 1998).

Meyers, Samuel L. and Albert D. Biderman (eds.). *Mass Behavior in Battle and Captivity: The Communist Soldier in the Korean War* (Chicago: The University of Chicago Press, 1968).

Meyerson, Debra, Karl E. Weick, and Roderick M. Kramer. 'Swift Trust in Temporary Groups', in Roderick M. Kramer and Tom R. Tyler (eds.) *Trust in Organizations: Frontiers of Theory and Research* (Thousand Oaks, CA: Sage, 1996), 166–95.

Middleton, Karen. *An Unwinnable War: Australia in Afghanistan* (Melbourne: Melbourne University Press, 2011).

Milgram, Stanley. *Obedience to Authority: An Experimental View* (New York: Harper and Row, 1974).

Miller, William Ian. *The Mystery of Courage* (Cambridge, MA: Harvard University Press, 2000).

Miller, Laura and John Allen Williams. 'Do Military Policies on Gender and Sexuality Undermine Combat Effectiveness?', in Peter D. Feaver and Richard H. Kohn (eds.) *Soldiers and Civilians: The Civil-Military Gap and American Security* (Cambridge, MA: MIT Press, 2001), 361–402.

Millet, Allan R. and Williamson Murray (eds.). *Military Effectiveness*, vol. 1: *The First World War* (Cambridge: Cambridge University Press, 1988).

Mjøset, Lars and Stephen Van Holde. 'Killing for the State, Dying for the Nation: An Introductory Essay on the Life Cycle of Conscription into Europe's Armed Forces', in Lars Mjøset and Stephen van Holde (eds.) *The Comparative Study of Conscription in the Armed Forces. Comparative Social Research*, vol. 20 (Bradford: Emerald Group Publishing, 2002), 3–94.

'MoD in Latest Appeal Against Iraq Negligent Equipment Claims', LeighDay website, 15 February 2013 <http://www.leighday.co.uk/News/2013/February-2013/MoD-in-Latest-Appeal-Against-Iraq-Negligent-Equipm>. Accessed 2 February 2015.

Moore, Bruce. *A Lexicon of Cadet Language* (Canberra: Australian National Dictionary Centre, 1993).

Moore, Darren. *The Soldier* (London: Icon Books, 2009).

Moore, Harold and James Galloway. *We Were Soldiers Once…And Young* (London: Corgi, 2002).

Moran, Lord. *The Anatomy of Courage* (Garden City Park, NY: Avery Publishing Group, Inc., 1987).

Moreman, T.R. *The Jungle, The Japanese and the British Commonwealth Armies at War, 1941–45* (Abingdon: Frank Cass, 2005).

Moskos, Charles. 'The American Combat Soldier in Vietnam', *Journal of Social Issues* 31(4) (1975): 25–37.

Moskos, Charles. *The American Enlisted Man. The Rank and File in Today's Military* (New York: Russell Sage Foundation, 1970).

Moskos, Charles. 'Why Men Fight: American Combat Soldiers in Vietnam', *Trans-Action* 7(1) (1969): 13–23.

Moskos, Charles. *Soldiers and Sociology* (United States Army Research Institute for Behaviour and Social Sciences, 1988).

Moskos, Charles C., John Allen Williams, and David R. Segal (eds.). *The Postmodern Military: Armed Forces after the Cold War* (New York: Oxford University Press, 2000).

Motowidlo, Stephan J., B.E. Dowell, Borman Walter, P.D. Johnson, and M.D. Dunnette. *Motivation, Satisfaction, and Morale in Army Careers: A Review of Theory and Measurement* (Arlington, VA: US Army Research Institute for Behavioral and Social Sciences, 1976).

Muir, Rory. *Tactics and the Experience of Battle in the Age of Napoleon* (New Haven, CT: Yale University, 1998).

Mullen, Brian and Carolyn Copper. 'The Relation between Group Cohesiveness and Performance: An Integration', *Psychological Bulletin* 115 (1994).

Mullins, Nathan. *Keep Your Head Down: One Commando's Brutally Honest Account of Fighting in Afghanistan* (Sydney: Allen & Unwin, 2011).

Multiple Claimants, 2003, Multiple PTSD Claimants v. MoD—2003 EWHC 1134 (QB).

Murray, Maj. Stewart L. *The Future Peace of the Anglo-Saxons* (London: Hugh Rees, 1905).

Neitzel, Sonke and Harald Welzer. *Soldaten: On Fighting, Killing and Dying: The Secret Second World War Tapes of German POWs* (London: Simon and Schuster, 2013).

Newsome, Bruce. 'The Myth of Intrinsic Combat Motivation', *Journal of Strategic Studies* 26(4) (2003): 24–46.

Normann, Roderick de. 'Infantry Regiment 950—Germany's Indian Legion', *Journal of the Society for Army Historical Research* 75(303) (1997): 172–90.

Norton-Taylor, Richard 'Supreme Court MoD ruling "will have huge impact on military operations"', The *Guardian*, 19 June 2013 <http://www.theguardian.com/law/2013/jun/19/supreme-court-mod-ruling-impact>. Accessed 2 February 2015.

Nosworthy, Brent. *Battle Tactics of Napoleon and His Adversaries* (London: Constable, 1995).

Nuti, Leopoldo. *L'esercito italiano nel secondo dopo guerra, 1945–1950: la sua ricostruzione e l'assistenza militare alleata* (Roma: Ufficio Storico, Stato Maggiore dell'Esercito, 1989).

Oliver, Alex. *Australia in the World: Public Opinion and Foreign Policy* (Sydney: Lowy Institute, 2013).

Palazzo, A. 'No Casualties Please, We're Soldiers', *Australian Army Journal* 5(3) (2008): 65–78.

Palmer, J.A.B. *The Mutiny Outbreak at Meerut in 1857* (Cambridge: Cambridge University Press, 1966).

Peano, Nazzareno. 'Salimmo in montagna per fare qualcosa', *Rivista dell'Istituto Storico della Resistenza e della Societa' Contemporanea* 81 (2012): 207–21.

Pedersen, P. '"Recock and Refigure": Trooper Mark Donaldson VC speaks', *Wartime* 46 (2009): 9–15.

Pedersen, P. 'The Falling Leaves of Tizak', *Wartime* 57 (2012): 8–15.

Petersen, Ib Damgaard (ed.). *Faldne i Danmarks Frihedskamp 1940–45* (Copenhagen: Gyldendal, 1970).

Pettit, Lieut.-Col. James S. 'How Far Does Democracy Affect the Organization and Discipline of Our Armies, and How Can its Influence Be Most Effectually Utilized?' (Seaman Prize Essay), *Journal of the Military Services Institute* 38 (1905): 1–38.

Picken, Bruce. *Viet Nam Diggers' Language* (Loftus: Australian Military History Publications, 2006).

Porter, Patrick. *Military Orientalism* (London and New York: Hurst & Co., 2009).

Posen, Barry R. 'Urban Operations: Tactical Realities and Strategic Ambiguities', in Michael C. Desch (ed.) *Soldiers in Cities: Military Operations in Urban Terrain* (Carlisle, PA: Strategic Studies Institute, 2001), 149–66.

Pouvreau, Ana. *Le système Légion: Un modèle d'intégration des jeunes étrangers* (Sceaux: L'esprit du livre editions, 2008).

Pratten, Garth. M. *Australian Battalion Commanders in the Second World War* (Melbourne: Cambridge University Press, 2009).

Pratten, Garth. M. and Glyn Harper (eds.). *Still the Same: Reflections on Active Service from Bardia to Baidoa* (Georges Heights: Army Doctrine Centre, 1996).

Priest, Robert F., Alan G. Vitters, and Howard T. Prince. 'Coeducation at West Point', *Armed Forces & Society* 4(3) (1978): 589–606.

Pross, Harry (ed.). *Die Zerstörung der Deutschen Politik: Dokumente 1871–1933*, trans. Richard S. Levy (Frankfurt am Main: Fischer Bücherei, 1959), 29–31.

Proud, M. 'The COIN Environment: The Integration and Execution of Lethal and Non-Lethal Effects at the Tactical Level – Platoon Team Lessons from MRTF-2 in Afghanistan', *Australian Army Journal* 8(1) (2011): 23–38.

Putkowski, Julian and Julian Sykes. *Shot at Dawn* (Barnsley: Pen and Sword, 1988).

Quigley, Hugh. *Passchendaele and the Somme* (London: Methuen, 1928).

Quimby, Robert. *The Background of Napoleonic Warfare: The Theory of Military Tactics in 18th Century France* (New York: Columbia University Press, 1957).

Raineri, Giovanni. *Congedati ma sempre Alpini anche nel nuovo millennio, Cronaca dei terzi venticinque anni della Sezione di Mondovi' dell'Associazione Nazionale Alpini* (Cuneo: SSL Tipografia, 2003).

von Ranke, Leopold. Geschichte der romanischen und germanischen Völker von 1494 bis 1514 (1824), preface.

Rebuilding Afghanistan: The story of Australia's 2nd Reconstruction Task Force (Salisbury: Australian Defence Force, 2008).

Reconstruction Task Force 3, October 2007–April 2008 (Townsville: 3rd Combat Engineer Trust Fund, 2008).

Renna, Mario. *Ring Road. Sei mesi con gli Alpini in Afghanistan* (Milano: Mursia, 2011).

Revelli, Nuto. *Mai Tardi. Diario di un Alpino in Russia* (Cuneo: Panfilo Editore, 1943).

Rona, Roberto, Margaret Jones, Lisa Hull, Oded Horn, Dominic Murphy, Matthew Hotopf, and Simon Wessely. 'Screening for Physical and Psychological Illness in the British Armed Forces: III The Value of a Questionnaire to Assist a Medical Officer to Decide Who Needs Help', *Journal of Medical Screening* 11 (2004): 158–63.

Rona, Roberto, Margaret Jones, Lisa Hull, T. Brown, Oded Horn, Dominic Murphy, Matthew Hotopf, and Simon Wessely. 'Mental Health Screening in Armed Forces Before the Iraq War and Prevention of Subsequent Psychological Morbidity: Follow Up Study', *British Medical Journal* 333 (2006): 991–4.

Roosevelt, Theodore. 'Twisted Eugenics', *Outlook* 106 (1904): 30–4.

Rosen, Stephen Peter. *Societies and Military Power: India and Its Armies* (Ithaca, NY: Cornell University Press, 1996).

Ross, Jane. *The Myth of the Digger: The Australian Soldier in Two World Wars* (Sydney: Hale & Iremonger, 1985).

Rostker, Bernard D and Harris, Scott (eds.). *Sexual Orientation and U.S. Military Personnel Policy: Options and Assessment* (Washington, DC: RAND, 1993).

Rowe, Peter. *The Impact of Human Rights Law on Armed Forces* (Cambridge: Cambridge University Press, 2006).

Rowe, Peter. 'Military Misconduct during International Armed Operations: "Bad Apples" or Systemic Failure?', *Journal of Conflict and Security Law* 13(2) (2008): 165–89.

Royal, Benoît. *L'éthique du soldat français: la conviction d'humanité.* (Paris: Economica-Collection 'Stratégies et doctrines', 2010).

Rubin, Elizabeth. 'Battle Company is Out There', *New York Times Magazine*, 24 February 2008 <http://www.nytimes.com/2008/02/24/magazine/24afghanistan-t.html?pagewanted=all&_r=0>. Accessed 20 February 2013.

Ruffa, Chiara. *Imagining War and Keeping Peace? Military cultures and peace operation effectiveness*. Doctoral dissertation (Florence: European University Institute, 2010).

Ryan, Joseph W. *Samuel A. Stouffer and The GI Survey* (Knoxville, TN: University of Tennessee Press, 2013).

Said, Edward. W. *Orientalism* (New York: Vintage Books, 1979).

Saint-Fuscien, Emmanuel. À *Vos Ordres? La relations d'autorité dans l'armée française de la Grand Guerre* (Paris: Edition de l'Ecole des hautes etudes en sciences sociales, 2011).

Salazar, Jaime. *Legion of the Lost* (New York: Berkeley Caliber, 2006).

Savage, Paul L. and Richard A. Gabriel. 'Cohesion and Disintegration in the American Army: An Alternative Perspective', *Armed Forces & Society* 2(3) (1976): 340–76.

Saxe, Maurice du. *Reveries on the Art of War*, trans. Thomas Phillips (New York: Dover Publications, 2007).

Scheff, Thomas. 'Shame and Conformity: The Deference-Emotion System', *American Sociological Review* 53(3) (1988): 395–406.

Schlieffen, Alfred Graf von. 'Krieg der Gegenwart', *Deutsche Revue* (January 1909): 13–23.

Schultz, J.H. 'Einige Bemerkungen uber Feindschaftgefuhle im Kriege', *Neurologisches Centralblatt* 34(11) (1 June 1915): 376.

von Scherff, Wilhelm. *Studien der neuen Infanterie-Taktik* (Berlin: A. Bath, 1872–74).

Segal, David and Meyer Kestnbaum. 'Professional Closure in the Military Market: A Critique of Pure Cohesion', in D. Snider (ed.) *The Future of the Army Profession* (New York: Primus, 2002).

Segal, David and Joseph Lengermann. 'Combat Effectiveness', in Sam Sarkesian (ed.) *Combat Effectiveness: Cohesion, Stress, and the Volunteer Military* (Beverly Hills: Sage, 1980), 154–84.

Segal, Mady Wechsler. 'The Military and the Family as Greedy Institutions', *Armed Forces & Society* 13(1) (1986): 9–38.

'Sergeant R'. 'Life on the Ground in the 'Ghan', *Wartime* 51 (2010): 12–17.

Severloh, Hein. *WN 62: erinnerungen an Omaha Beach Normandie, 6. June 1944* (Garbsen: Hek Creativ, 2010).

Shabtai, Noy, 'Combat Stress Reactions', in Gal, Reuven and Mangelsdorff A. David (eds.) *Handbook of Military Psychology* (New York: John Wiley & Sons, 1991).

Shalit, Ben. *Psychology of Conflict and Combat* (New York: Praeger, 1988).

Shaw, Martin. *War and Genocide: Organized Killing and Modern Society* (Cambridge: Polity, 2003).

Shaw, Martin. *The New Western Way of War* (Cambridge: Polity, 2005).

Shay, Jonathan. *Achilles in Vietnam: Combat Trauma and the Undoing of Character* (New York: Touchstone, 1994).

Shen, Dana. 'The Psychology of Operational Joining: A Case-Study of a Dog Dispatch Unit', *Military Psychology* 2 (2003): 75–130 (Hebrew).

Shephard, Ben. *Soldiers and Psychiatrists in the Twentieth Century: A War of Nerves* (Pimlico: Random House, 2002).

Shephard, Ernest. *A Sergeant Major's War: From Hill 60 to the Somme* (Ramsbury: Crowood Press, 1987).

Shibutani, Tamotsu. *The Derelicts of Company K: A Sociological Study of Demoralization* (Berkeley: University of California Press, 1978).

Shuurman, Bart. 'Clausewitz and the 'New Wars' Scholars', *Parameters* (Spring 2010): 89–100.

Siebold, Guy L. 'Key Questions and Challenges to the Standard Model of Military Group Cohesion', *Armed Forces and Society* 37(3) (2011): 448–68.

Siebold, Guy L. 'The Essence of Military Group Cohesion', *Armed Forces & Society* 33(2) (2007): 286–95.

Smith, K.R. 'Understanding Morale: With Special Reference to the Morale of the Australian Infantryman in Vietnam', *Defence Force Journal* 52 (May/June 1985): 53–62.

Smith, Rupert. The Utility of Force: The Art of War in the Modern World (London: Penguin, 2006).

Smoler, Frederic. 'The Secret of Soldiers Who Wouldn't Shoot', *American Heritage* 40 (March 1989): 36–45.

Snyder, Jack. The Ideology of the Offensive: Military Decision Making and the Disasters of 1914 (Ithaca, NY: Cornell University Press, 1984).

Snyder, Timothy. *Bloodlands: Europe Between Hitler and Stalin* (London: The Bodley Head, 2010).

Somale, Francesco. 'Le Compagnie Alpine', *Rivista Militare* 4 (1878).

Sondhaus, Lawrence. *Strategic Culture and Ways of War* (London: Routledge, 2006).

Sookermany, Anders McD. 'The Embodied Soldier: Towards a New Epistemological Foundation of Soldiering in the (Post) Modernized Norwegian Armed Forces', *Armed Forces & Society* 37(3) (2011): 469–93.

Speirenburg, Pieter. 'Faces of Violence: Homicide Trends and Cultural Meanings', *Journal of Social History* 17 (1994): 701–16.

Spiller, Roger. 'Sharp Corners: Combat Operations in Urban Areas', in Michael Evans, Russell Parkin, and Alan Ryan (eds.) *Future Armies, Future Challenges* (London: Allen and Unwin, 2004), 82–95.

Stallmeister Walter and Lück Helmuth E (ed.). *Willy Hellpach: Beiträge zu Werk und Biographie* (Frankfurt am Main: P. Lang, 1991).

Steed, Brian. *Armed Conflict: The Lessons of Modern Warfare* (New York: Ballantine Books, 2003).

Stein, George H. *The Waffen SS: Hitler's Elite Guard at War* (Ithaca, NY: Cornell University Press, 1966).

Storia Fotografica della Brigata Alpina Taurinense, 60esimo 1952–1992 (Susa: Edizioni Susalibri, 2012).

Stouffer, Samuel A. *Studies in Social Psychology in The Second World War* (Princeton, NJ: Princeton University Press, 1949).

Stouffer, Samuel, Arthur Lumsdaine, Marion Harper Lumsdaine, Robin Williams, M. Brewster Smith, Irving Janis, Shirley Star, and Leonard Cottrell. *The American Soldier*, vol. 2: *Combat and Its Aftermath* (Princeton, NJ: Princeton University Press, 1949).

Strachan, Hew. 'The Soldier's Experience in Two World Wars: Some Historiographical Comparisons', in Paul Addison and Angus Calder (eds.) *Time to Kill. The Soldier's Experience of War in the West* (London: Pimlico, 1997), 369–78.

Strachan, Hew. 'The Lost Meaning of Strategy', *Survival* 47(3) (2005): 33–54.

Strachan, Hew. 'Training, Morale and Modern War', *Journal of Contemporary History* 41(2) (2006): 211–27.

Strachan, Hew. 'A Clausewitz for Every Season', *American Interest* 2(6) (July/August 2007).

Strachan, Hew. 'Strategy or Alibi? Obama, McChrystal and the Operational Level of War', *Survival* 52(5) (2010): 157–82.

Sullivan, Brian. 'The Italian Armed Forces, 1918–40', in Allan R. Millett and Williamson Murray (eds.) *Military Effectiveness* (Boston, MA: Unwin Hyman, 1988).

Sullivan, Brian. 'The Italian Soldier in Combat, June 1940–September 1943: Myths, Realities and Explanations', in P. Addison and A. Calder (eds.) *Time to Kill: The Soldier's Experience of the War in the West, 1939–45* (London: Pimlico, 1997).

Swain, Richard. 'Professionalism: Service Unsullied by Partisanship', National Defense University 2010. <http://www.ndu.edu/INSEL/docUploaded/obligations%20of%20military%20professionalism.pdf>. Accessed 10 February 2013.

Tai Yong, Tan. *The Garrison State: The Military, Government and Society in Colonial Punjab, 1849–1947* (New Delhi: Sage, 2005).

Tapper, Jake. *The Outpost* (Boston, MA: Little, Brown and Company, 2013).

Taussig, Michael. 'Culture of Terror—Space of Death: Roger Casement's Putumayo Report and the Explanation of Torture', *Comparative Studies in Society and History* 26(3) (1984): 467–97.

Tawney, Richard. *The Attack and Other Essays* (London: Allen and Unwin, 1953).

Téxier, Mélanie. 'Identités et Langues à la Légion étrangère', *Res Militaris* 1(3) (2011). Available at <http://resmilitaris.net/index.php?ID=1014409>. Accessed 2 November 2012.

Thomlinson, Damien with Michael Cowley. *Without Warning: A Soldier's Extraordinary Journey* (Sydney: Harper Collins, 2013).

Thompson, Mark. *The White War: Life and Death on the Italian Front 1915–1919* (London: Faber and Faber, 2004).

Tillman, Mary (with Narda Zacchino). *Boots on the Ground By Dusk. My Tribute to Pat Tillman* (New York: Modern Times, 2008).

Travers, Timothy. *The Killing Ground: The British Army, the Western Front and the Emergence of Modern War 1900–1918* (Barnsley: Pen and Sword, 1987).

Trench, Charles Chenevix. *The Indian Army and the King's Enemies 1900–1947* (London: Thames and Hudson, 1988).

Truscott, Lucian. *Mission Command: A Personal Story* (Novato, CA: Presidio, 1990).

Unia, Gerardo. *Il Secondo Reggimento Alpini. Dalle origini a Kabul* (Cuneo: L'Arciere, 2006).

Urban, Mark. The Annual War Studies Lecture: 'War and the British Media: A View from Behind the Hesco', 28 January 2014 <http://www.youtube.com/watch?v=XxIrW85NRqo>. Accessed 5 February 2015.

Urry, John. *Sociology Beyond Societies: Mobilities for the Twenty-first Century* (London: Routledge, 2007).

Vinocker-Hai, Vered and Yotam Amitai. 'The War from the Perspective of Lower-level Commanders in the Continued Conflict in the Gaza Strip: An Interim Assessment', *Ma'arachot* 2003 389: 30–5 (Hebrew).

Virilio, Paul and Sylvie Lotzinger. *Pure War* (Paris: Semiotexte, 1997).

Walzer, Michael. 'Two Kinds of Military Responsibility', in Michael Walzer, *Arguing About War* (New Haven, CT: Yale University Press, 2004), 315–16.

Watson, Alexander. 'Culture and Combat in the Western World, 1900–1945', *Historical Journal* 51(2) (2008): 529–46.

Watson, Alexander. *Enduring the Great War: Combat, Morale and Collapse in the German and British Armies, 1914–1918* (Cambridge: Cambridge University Press, 2008).

Watson, Alexander. 'Fighting for Another Fatherland: The Polish Minority in the German Army, 1914–18', *English Historical Review* CXXVI(522) (2011): 1137–66.

Wesbrook, Stephen D. 'The Potential for Military Disintegration', in Sam Sarkesian (ed.) *Combat Effectiveness: Cohesion, Stress and the Volunteer Military* (Beverly Hills & London: Sage, 1980), 251–61.

Wessely, Simon. 'Risk, Psychiatry and the Military', *British Journal of Psychiatry* 186 (2005): 459–66.

Wessely, Simon. 'Twentieth-century Theories on Combat Motivation and Breakdown', *Journal of Contemporary History* 41(2) (2006): 269–86.

Wessely, Simon and Edgar Jones. 'Psychiatry and the Lessons of Vietnam: What Were They and Are They Still Relevant?', *War and Society* 22 (2004): 89–103.

Weill, Henri. *Légionnaires* (Paris: Pascal Galodé, 2011).

Weinberg, Gerhard L. 'Stalingrad and Berlin: Fighting in Urban Terrain', in Michael C. Desch (ed.), *Soldiers in Cities: Military Operations in Urban Terrain* (Carlisle, PA: Strategic Studies Institute, 2001), 17–29.

Welch, David and Jo Fox (eds.). *Justifying War* (London: Routledge, 2012).

Wilcox, Vanda. 'Morale and Battlefield Performance at Caporetto, 1917', *Journal of Strategic Studies* 37(6–7) (2014): 829–854.

Wilkinson, Spenser. *The French Army before Napoleon* (Oxford: Clarendon Press, 1915).

Williams, A. T. *A Very British Killing. The Death of Baha Mousa* (London: Jonathan Cape, 2012).

Williams, Buzz. *Spare Parts: A Marine Reservist's Journey from Campus to Combat in 38 Days* (New York: Gotham Books, 2004).

Wilmot, Chester. *The Struggle for Europe* (Bungay: Richard Clay and Company, 1954).

Wilson, H.W. 'The Human Animal in Battle', *Fortnightly Review* summarized and reviewed in *Journal of the United States Cavalry Association* VIII (1895): 44–6.

Winslow, Donna. *The Canadian Airborne Regiment in Somalia: A Socio-Cultural Inquiry* (Ottawa: Canadian Government Publishing, 1997).

Winslow, Donna. 'Rites of Passage and Group Bonding in the Canadian Airborne', *Armed Forces & Society* 25(3) (1999): 429–57.

Winter, Denis. *Death's Men* (London: Allen Lane, 1978).

Wittgenstein, Ludwig. *Philosophical Investigations*, trans. G. Anscombe (Oxford: Blackwell, 1976).

Wolseley, Garnet Field Marshal. 'War and Civilization', *United Services Magazine* 14 (March 1897): 560–77.

Wong, Leonard, Thomas Kolditz, Raymond Millen, and Terrence Potter. *Why They Fight: Combat Motivation in the Iraq War* (Carlisle, PA: Strategic Studies Institute, 2003).

Wright, Evan. *Generation Kill: Devil Dogs, Iceman, Captain America and the New Face of American War* (New York: Berkeley Caliber, 2004).

Yaeger, Deborah, Naomi Himmelfarb, Alison Cammack, and Jim Mintz. 'DSM-IV Diagnosed Posttraumatic Stress Disorder in Women Veterans With and Without Military Sexual Trauma', *Journal of General Internal Medicine* 21 (2006): 65–9.

Yeaman, Stuart. *Afghan Sun: Defence, Diplomacy, Development and the Taliban* (Brisbane: Boolarong Press, 2013).

Yousafzai, Sami and Ron Moreau. 'The Taliban in Their Own Words', *Newsweek*, 5 October 2009.

Yousafzai, Sami and Ron Moreau. 'How the Taliban Lost its Swagger', *Newsweek International*, 7 March 2011.

Zemon Davis, Natalie. 'Rites of Violence: Religious Riot in Sixteenth Century France', *Past and Present* 59 (May 1973): 51–91.

Index

absence without leave, and morale 140, 141, 142, 144
ad hoc units 315
  cohesion 85–6
  swift trust 84
  urban combat 83–5
adaptability 86–8
Afghanistan 5, 61
  British Army's performance in 200–1
  combat refusal in British forces 102–3
  courts martial for abuse cases 103–4
  emergence of international military culture 188
  micro-fronts 4
  see also 7 Platoon, Royal Irish Regiment; Australian Army in Afghanistan
Akhunzada (Taliban commander) 58
Alliot-Marie, Michelle 234, 245
Alpini Regiment (Italian Army) 251, 262–3, 319
  in Afghanistan 261–2
  alpine values 258
  connection with local communities 259, 262
  deployment in First and Second World Wars 255
  deviant professional cohesion 251
  effectiveness in non-military operations 253, 257, 261–2
  elements of culture 258
  ethos of self sacrifice 259
  historical background 251–2
  internal cohesion 251, 253, 259
  joins resistance in Second World War 255–6
  legitimacy of 257
  media coverage of 256
  national recruitment 257–8
  non-military image 258
  origins of 253–4
  perceived as non-military 256–7
  political autonomy 253, 255, 258–9, 263
  poor performance 251, 253
  postwar reconstitution of 256
  representations of 259–60
  role in Italian society 258
  symbols and traditions 260–1
  territorial recruitment 254
altruism, and motivation 176–7
Ambrose, Stephen 53
Annales School 16
anti-militarism, in nineteenth century 122, 123
Aran, Gideon 8
Ariès, Philippe 16
Arkin, William 8
Ashworth, Tony 60
asymmetric warfare 58–9, 201
atrocities
  bureaucratic processes 37
  First World War 48, 54
  soldiers' motivations 36, 37–8
Audoin-Rouzeau, Stephane 54
Australian Army
  battle cunning 181
  change in archetypal military hero 169
  digger stereotype 168–9, 184–5
  digger virtues 169
  'I'm an Australian Soldier' initiative 169–70
  official ideal of the Australian soldier 169–70
  professional ethos 170
Australian Army in Afghanistan 167, 186, 313, 318–19
  altruistic motivations 176–7
  amenities and personal comfort 184, 188
  appearance and uniformity 183–4
  characteristics of soldiers 181
  cohesion 173, 175
  concerns over influence of digger stereotype 182
  digger stereotype 167, 172
  distinction between combat/non-combat personnel 178
  effectiveness of 181–2
  erosion of standards of discipline 184
  experience of combat 174–5
  failings of 182
  fatalities and casualties 171
  gulf between soldiers and society 185–7
  influences on soldiers' behaviour 167–8, 182–3
  larrikin 172–3, 181

Australian Army in Afghanistan (*cont.*)
limitations on 171
mateship 175, 178, 188
media management 187
media victim imagery 185, 186
Mentoring and Reconstruction Task Force 171
Mentoring Task Force 171
motivations of soldiers 173–4, 175–8, 188
number of personnel 170
operations in 170–1
public attitudes towards 187
public's discomfort with violence 186–7
Reconstruction Task Force 170
relations between special and conventional forces 180–1
reservists 174, 180
revenge as motivating factor 175–6
service as part of historical tradition 172–3, 187–8
special forces 170, 171, 175, 179–80, 185
Special Operations Task Group 179–80
task cohesion 174
Team Australia 178
unrepresentative of Australian society 185
warrior caste 185, 187, 188
warrior code 168, 179
Avery, Ryan 175

Bairnsfather, Bruce 49–50
Bales, Robert 104
Barkawi, Tarak 61
Bartov, Omer 10, 36, 216
battle exhaustion, and morale 141, 144
Baudrillard, Jean 309–10
Bauman, Zygmunt 37
Baynes, John 137
bayonet charge 19–20
Beard, George M 123–4
Becker, Annette 54
Ben Youssouf, Lianrifou 234
Ben-Ari, Eyal 11, 205
Ben-Shalom, Uzi 11
Bergdahl, Bowe 101–2
berserk state 277
Biddle, Stephen 134, 136, 152
Biderman, Albert 34, 35
Black, Jeremy 59
Blake, Nicholas 299
blame culture 296, 297
Bloch, Ivan 120
body armour 300
Boer War (1899–1902) 121–2
Boff, Jonathan 140
Bourdieu, Pierre 16
Bourke, Joanna 56
Boy Scouts 128
Brimley, Shawn 58
British Army
combat refusal in Afghanistan 102–3
performance in Afghanistan 200–1
professionalism 15
see also 7 Platoon, Royal Irish Regiment; British Second Army (1944–45)
British Indian Army
colonially constructed ethnicities 30
Martial Races 30, 31
national culture 30
recruitment of Sikhs 31
Sikhs in Second World War 32–4
uniform regulations for Sikhs 31–2
British Second Army (1944–45) 156
morale and combat cohesion 149–51
morale levels 146, 148
quantitative assessment of morale 143–8
Browne, A. T. A. 142–3
Browning, Christopher 37, 38, 39
Buckley, John 140, 149, 152
buddy system 7
bureaucratic processes, and atrocities 37
Burgaud, Eric 234, 238, 240, 241, 243–4, 245
Bush, George W 295

Cadorno, Luigi 96
Calvert, Mike 33–4
Cameron, Craig 28–9
Canadian Army
professionalism 14–15
in Somalia 9
Special Forces unit (Joint Task Force 2) 110
Cantwell, John 176, 179, 187
capital punishment
decimation 96–7, 101
First World War 95
German forces in Second World War 99–100
Italian Army in First World War 96–7
Second World War 95
see also discipline
careerism 9
Carew, Paul 102
Carrington, Charles 49
censorship reports, and assessment of morale 140, 143–4
citizen armies
cohesion 94
discipline 95–100
city combat, *see* urban combat
Clausewitz, Carl von 49, 122, 153, 154, 156, 157, 182
clinical psychology, and combat 77–8
Cockerham, William 12
cohesion
ad hoc units 85–6
brute force narrative on 135–6, 151–2, 155
buddy system 7

citizen armies 94
collective action problem 6
command structures 236
complexity of 157
core groups 315–16
definition of 135, 235
deviant cohesion 9, 235, 236, 237, 250, 251
divisional headquarters 321
early modern literature on 17–19
early twentieth century understanding of 119–20
efficiency 135
formations 17–20
future research 320–3
historical origins of performative paradigm 13–16
human social cooperation 6
identity 62
ideology 10, 48, 53, 93, 291–2
inadequate attention to dynamics of 236
institutional politics 212
methods of study 312
morale narrative on 137, 151, 155
nineteenth-century literature 19–20
peace and stability operations 237, 242–3
political motivation 10, 53, 311
primary group thesis 6–10, 52, 78, 93, 216, 250, 311
problematic nature of 237
professionalism 11, 61, 79, 88, 94, 312–15
punishment 93
rational self-interest 8
risk aversion 296–7
scale and scope of 236
small unit dynamics 13, 55
strategy 135, 153–6, 157
tactics narrative on 136–7, 152, 155–6
task cohesion 10–11, 13, 78, 174
training 11–12, 93, 202, 312–14
Coker, Christopher 188, 309
collateral damage 58
collective action problem, and cohesion 6
Collins, Randall 12, 29, 50, 51, 205–6
colonial soldiers, *see* subaltern soldiers
columns vs lines 18–19
combat
collective action problem 6
computer-controlled 57–8, 310
conflating individual and group responses to 47
contextual nature of 47
cultural conditioning 62–3
depersonalization of enemy 53–4
difficulties in studying 46–7, 50–1, 312
diversity of individual experiences 51, 54
frightening aspects of 280–2
group responses to 47
individual experience of 47, 49–50
individual responses to 48, 50, 54, 59–60
neutral vocabulary 58
number of soldiers firing weapons 53
patterns in 51
physical and emotional stress 270
popular portrayal of 55
primary groups 6–10, 52
problems with individual testimony 49, 52
religion 57, 61
reluctance to kill 53, 55
small unit dynamics 55
social cohesion 9, 10, 13, 55
war cries 56–7
see also urban combat
combat formations
cohesion and performance 17–20
columns vs lines 18–19
combat motivation debates
military sociology 26–9
national emphasis of 26–9
combat performance 312–15
brute force narrative on 135–6
cultural homogeneity 216
discipline in citizen armies 95–6, 97, 98–9, 100
early modern literature on 17–19
emotional bonds 207–11, 314
formations 17–20
institutional politics 211–12
nineteenth-century literature 19–20
political motivation 10
primary group thesis 6–10, 216
reaction to contact drills 203–5
small unit dynamics 13
sociological explanation of 13
tactics narrative on 136–7
task cohesion 10–11, 13
training 11–12, 105–6, 312–15
combat refusal 101–3
rarity of 106
combat stress reactions 78
command structures, and cohesion 236
commemoration practices 285
communications, and urban combat 81, 82
compensation culture 296, 297
computers, and warfare 57–8, 310
conscription 13–14
contractors 322
core groups 315–16
cowardice
capital punishment 95
discipline in contemporary professional armies 101–2
rarity of 106
see also discipline
Creveld, Martin van 321
Crew, F. A. E. 141

Crichton-Browne, James 124
Crocker, John 98
Crombie, J. E. 50
cultural conditioning 62–3
cultural explanations of behaviour
Chinese communist PoWs in Korean War 34–5
inadequacy of 35
Sikhs, in British Indian army 32–4
cultural homogeneity, and combat performance 216

Davies, Christine 253
Davis, Natalie Zemon 16
De Amicis, Edmondo 259–60
De Saxe, Marshall 17–18
deadly zone, expansion of 120–2
decimation, and discipline 96–7, 101
Deepcut enquiry 299
Delbrück, Hans 121
demonic others 38
der Derian, James 58
desertion
capital punishment 95
discipline in contemporary professional armies 101–2
morale 141–2, 144
rarity of 106
see also discipline
discipline 106
Allied forces in Second World War 98–9
citizen armies 95–100
coercive and normative 28, 62
cohesion 62
combat performance in citizen armies 95–6, 97, 98–9, 100
combat refusal in British forces 102–3
concerns over reputational damage from prosecutions 104–5
courts martial for abuse cases 103–4
decimation 96–7, 101
future prospects 111–12
German forces in Second World War 99–100
impact of training 105–6
informal mechanisms 106, 108–9, 211
Italian Army in First World War 96–7
Italian Army in Second World War 98
lack of courts martial 101, 105–6
leniency of military justice 104
morale 138
new paradigm of 112
peer-group sanctions 106–11, 211, 315
professional armies 101–11
professional shame 109–11, 207, 211, 314–15
professionalism as mechanism of 106–8
rarity of applying military justice to combat refusal 101–2, 105
ridicule 108–9
unprofessional conduct 104
distrust, culture of 296
divisions 321
Dobrofsky, Lynne 8
Dollard, John 271–2
Donaldson, Mark 173, 176, 186, 187
Doubler, Michael 276
Dower, John 38
drill, and cohesion 11, 12, 15, 85, 106, 119–20, 312–14
innovation 83
military conditioning 87
pre-packaged strategy 154
professionalism 79
psychological effects 203
reaction to contact drills 203–5
swift trust 205
use of mnemonics 201–2
drone strikes 57–8
du Picq, Ardant 19–20, 28, 50, 52, 119, 120, 123, 137, 206
Dunlop, Edward 'Weary' 169
duty of care 292, 305
challenging military over 294
human rights 302–3
legal aspects of 292
mental health issues 301–2
moral aspects of 292
negligence 302–3
paradoxical effects of safety concerns 300–1
potentially inhibiting effects of litigation 303
soldier's relationship with officers 298
Dyer, Gwynne 29, 56

East India Company 31
Economic Community of West African States (ECOWAS) 238
Ehrenreich, Barbara 57
Eisenhart, Ronald 7–8
Eisenhower, Dwight D 58
Ellis, John 136
email, and communication with home front 273–4
emotional bonds, and combat performance 207–11, 314
emotions
combat performance 206–7
forward panic 206
underemphasis on 205
in violence 205–6
enemies
attitudes towards 60
stereotyping of 60–1
English, John 14–15
essentialism, in military scholarship 25
Eurocentrism, in military scholarship 24–5, 59
European Convention of Human Rights 303
European Court of Human Rights 295

European Court of Justice 295
Evans, Michael 74–5

Falklands War 103
family, and cohesion 269
  calming effects of family contact 277–8
  competing demands of family and military 279, 281, 283, 285–6
  concerns over personal injury and family welfare 282
  contact with families 273–4
  detrimental effects of family problems 276–7
  father's role in family 283, 284
  forward operating bases 269, 270
  institutional changes in families 283–4, 320
  less contact with family by officers and senior NCOs 274–5
  morale and contact with 273, 277
  as motivating factor in performance 281
  potential distraction by family concerns 279–80, 286
  remote warfare 287
  security and family contact 278–9
  social solidarity and the family 284–5
  soldiers' emotional support for family 277
  support for families 277
  topics discussed with families 275–6
Family Readiness Groups (FRGs) 277
fatalism 281–2
Faulk, Henry 35
fear
  alleviation from 271–3
  concerns over personal injury and family welfare 282
  frightening aspects of combat 280–2
female emancipation 284
female soldiers 316–18
feminisation of public life 294
firepower, and cohesion 135–6, 151–2, 155
First World War 47
  atrocities 48, 54
  Bethune 62
  Bois de Buttes 62
  brute force narrative on cohesion 135–6
  capital punishment 95, 96
  discipline in Italian army 96–7
  humanity 48
  individual experiences of combat 49–50
  infantry assault 128–9
  lack of cohesion at command levels 129–30
  problem of exploiting tactical successes 129
  unit cohesion 129
  war cries 56–7
Flynn, Corporal (changed name) 203, 204, 208–9, 210–11, 212
Folard, Chevalier 17–18
forward operating bases (FOBs)
  adequate sleep for soldiers 271
  calming effects of family contact 277–8
  competing demands of family and military 279, 281, 283, 285–6
  contact with families 269, 270, 273–4
  distraction from fear 272–3
  food provision 271–2
  internet access 274
  less contact with family by officers and senior NCOs 274–5
  mitigation of fear and stress 269, 270–1, 274
  morale and contact with families 273, 277
  potential distraction by family concerns 279–80, 286
  psychological benefits of 271
  recreation and entertainment 272
  as safe havens 270–1, 273
  security and family contact 278–9
  soldiers' emotional support for family 277
  soldiers' expectations 274
  topics discussed with families 275–6
forward panic 50, 87
  role of emotions 206
fragging 27–8
Franco-Prussian War 120–1
French, David 140, 152
French, John 136
French Army, *see* French Foreign Legion; Mahé affair
French Foreign Legion 216, 230, 313
  absence of cultural bonds 216–17
  authoritarianism of 221, 225–6
  availability of legionnaires 224
  cohesion through discipline and hierarchy 225
  composition of 218
  contract system 224–5
  desertion 225
  difficulty in retaining soldiers 225, 226
  disadvantages of extreme adherence to orders 230
  ethnic diversity of 218–19
  fut-fut corporals 226–7
  heterogeneity within 217
  hierarchy within 220–1
  historical background 217–18
  learning French language 222
  legionnaire/officer boundary 221
  legionnaire's dislike of officers 221
  method of study 219–20
  obedience of legionnaires 222–3
  operational experience and status 227–8
  parachute regiment's (2 REP) prestige 228–9
  rank system 220
  rank without status 226–7
  regimental structure 218
  sanctity of orders 221–3, 229

French Foreign Legion (*cont.*)
  status hierarchy 226–9
  superiors are always right 223–4
friendly fire 296, 303
Fritz, Stephen 216
frontline
  changed nature of 3–4
  micro-fronts 4
functionalism 15
Fussell, Paul 293

Gallipoli 55, 57
Gammage, Bill 168
Garfinkel, Harold 16
Geertz, Clifford 16
gender roles 57
  women soldiers 316–18
genocide
  bureaucratic processes 37
  soldiers' motivations 36
Germany Army, debate over tactical doctrine 125–6
Giddens, Anthony 16
Glass, Charles 94
Glenn, Russell W 78–9
Glenton, Joe 101
Goffman, Erving 16, 322
Goldhagen, Daniel 36, 37, 39
Gorrod, Wayne 101, 102–3
Graham, Stephen 309, 310
Granatstein, Jack 14–15
Graves, Robert 48, 54
Gray, Colin 153
Gray, Leon 174
greedy institutions 279, 280
Gregory, R A 141
Griffith, Wyn 50
Grossman, David 54–6, 60
Guadalcanal 29
Guibert, Comte Jacques de 18–19
Gusterson, Hugh 60

Hammett, Jim 172
Hancock, Peter 197n
Handel, Michael 137
Hansen, Liam 176
Hart, Stephen 152
hate, rarity of 60
Healy, Aaron 174, 177–8
Hedges, Chris 63
Hellpach, Willy 124
Helmus, Todd C. 78–9
Henderson, Darryl 9, 14
Henley Management Centre 296
hierarchy, in French Foreign Legion 220–1
Hills, Alice 310
Hirschmann, Albert 225
hit rates 53
Holmes, Richard 54
Holocaust
  bureaucratic processes 37
  foreign soldiers' involvement in 38–9
  German soldiers' motivations 36, 37–8
honour, professional 62, 94, 111, 112, 175, 176, 212, 314–15, 316
Houston, Angus 176
Howard, Michael 154
human rights 292, 300, 302–3
hunger 271–2
Huntington, Samuel 26, 170

identity, and cohesion 62
ideology 291–2
  cohesion 10, 48, 53, 93
immunity, combat 299, 303, 305
improvisation
  professionalism 83
  urban combat 82–3
Improvised Explosive Devices (IEDs) 58, 201
industrialization 122–4
Ingham, Sarah 298
institutions, definition of 217
insurgent warfare 58
  religion 61
internet access, in forward operating bases 274
interpersonal relations 7, 8, 9, 202, 210, 314
Iraq 5
  courts martial for abuse cases 103
ISIS 5
Israel Defense Forces (IDF) 11, 74, 85, 205, 280, 312–13, 315
Italian Army
  after fall of Mussolini 256
  discipline in First World War 96–7
  discipline in Second World War 98
  ineffectiveness of 252–3
  post-war erosion of trust in 257
  see also Alpini Regiment (Italian Army)
Ivory Coast
  Licorne (Unicorn) operation 234, 238–9
  'road blockers' 238–9 *see also* Mahé affair

Jacka, Albert 168–9
Janowitz, Morris 6, 24, 25, 26, 28, 35, 52, 139, 216, 236, 250, 311
Jarvie, Harry 167
Jones, Andrew 176
Judd, Ashley 175, 185
just war principles 291, 295, 296, 304
  accountability for breach of 300

Keegan, John 12, 51–2, 137, 138
Kendall-Smith, Malcolm 101
Khalsa 31
Kier, Elizabeth 10

killing
computer-controlled 57–8, 310
conflating individual and group responses to 47
depersonalization of enemy 53–4
at a distance 56
ease of 47, 55–6
individual experiences of 49–50
individual responses to 48, 50, 54, 59–60
intimate nature of 56
neutral vocabulary 58
popular portrayal of 55
Post-Traumatic Stress Disorder 55
of prisoners 54
religion 57, 61
reluctance to kill 53, 55
soldiers' embrace of 56
Kindsvatter, Peter 270, 272
King, Anthony 61, 62, 79–80, 135, 137, 251, 285
Kingston, Gina 79
Kirkpatrick, John Simpson 169
Korean War
American PoWs 35
Chinese communist PoWs 34–5
cohesion 7
role of families 269
Krulak, Charles C 86
Kuss, Hans 99

Ladurie, Emanuel le Roy 16
Langlois, Bianka 104
Larcombe, Jamie 172
Leahy, Peter 169
Lebanon War, Second 86
LeBon, Gustave 124
Lebrun, Norman 49
LeSuire, Edgar von 128–9
Levi-Strauss, Claude 15
Liddell Hart, Basil 153
Lifton, Robert Jay 56
litigation 305
potentially inhibiting effects of 303
Little, Roger 7, 269, 281
logistics 321
loyalty, and cohesion 61–2, 94, 120, 243, 244, 280, 291
Ludendorff, Erich 129
Lynn, John 30, 62

McCarthy, Sean 176
McCartney, Helen 291, 295
McChrystal, Stanley 14
MacCoun, Robert 10
McCoy, Brian 109–10
McGorman, Ewan 224
Machiavelli, Niccolò 17
McLeish, Kathy 177
MacMunn, George 31
McPherson, A B 138
Mahé, Firmin 104
arrest and killing of 239–40
execution of 234
Mahé affair 234, 235
arrest and killing of Mahé 239–40
belief that killing served a military purpose 242
cause of misconduct 241
character of misconduct 241–2
cohesion 243
detention of personnel 241
dismissal of case against Poncet (2010) 241
group loyalty/disloyalty 243–5
military leadership's reaction 240–1
operational context (Licorne operation) 238–9
Poncet's denial of giving order 244–5
public exposure of information 237
role of political leadership 245
task cohesion 242
Malaussène, Renaud de 238, 241, 244
Manchester, William 49
Market Garden, Operation 149
Marlantes, Karl 212
Marshall, S. L. A. 13, 53, 136, 138, 139, 154–5
Martial Races, and British Indian Army 30, 31
Martin, James 177
masculinity 7, 8, 216
social and military significance of 294
Masters, Chris 172, 176, 177, 178, 179, 180, 182, 183, 185–6
Maxse, Ivor 129
Mazower, Mark 38
media, portrayal of combat 55
memorial practices 285
Menard, Daniel 104, 105
mental health issues 295, 301–2
Mesnil-Durand, Jean de 18
Meyers, Samuel 34
Meyerson, Debra 84
micro-fronts 4
Milgram, Stanley 37
Military Affiliate Radio System (MARS) 273
military contract of unlimited liability 292–3
combat immunity 299, 303, 305
contestation over 299–300
dynamic nature of 293, 297
human rights 302–3
mental health issues 301–2
Military Covenant 297–8
military spaces 299
negligence 302–3
paradoxical effects of safety concerns 300–1
potentially inhibiting effects of litigation 303

military contract of unlimited liability (*cont.*)
public's role 297
reciprocity 297
soldier's dilemma 293
Military Covenant 297
armed forces obligations 297–8
institutionalization of legal culture 298
problematic nature of 298
public's obligation 297
soldier's relationship with officers 298
see also military contract of unlimited liability
military sociology
combat motivation debates 26–9
conceptions of 'society' and 'military' 27
Eurocentrism 24–5
national emphasis of 26–9
towards a non-Eurocentric 40–1
Millet, Allan R. 135
misconduct
'bad apple' explanations 241–2
cohesion 234–5, 243
deviant cohesion 237
dynamics of cohesion 236–7
examples of 235
multiple forms of 241
see also Mahé affair
mission-oriented tactics 126
mnemonics, use in battle drills 201–2
Moltke, Helmuth von 63, 120
Montgomery, Bernard Law 138–9, 141, 152
morale 134
absence without leave 140, 141, 142, 144
affective states 138
assessment of 139–43
attitudinal surveys 139–40
battle exhaustion 141, 144
British Second Army (1944–45) 146, 148
British Second Army (1944–45), quantitative assessment 143–8
British Second Army (1944–45), role in combat cohesion 149–51
censorship reports 140, 143–4
cohesion 137, 151, 155, 156–7
concerns over family 276–7
contact with families 273, 277
definition of 137–9
desertion 141–2, 144
discipline 138
functional conceptualization of 134, 156
historical sources on 140
motivation 138
problems in studying 137
quantification of 134–5
self-inflicted wounds 140, 141, 142–3, 144
sickness rates 140, 141, 144
statistical sources 140–1
morality 13
Morant, Harry 'Breaker' 197n
Moreland, Jacob 176
Morrison, David 182, 187–8
Moshtarrak III, Operation 3–4
Moskos, Charles 8–9, 10, 14, 78, 250, 270, 291–2
motivation
Australian Army in Afghanistan 173–4, 175–7
morale 138
Mousa, Baha 103, 104, 235
Mullins, Nathan 177, 180, 184, 186
Munzu, Simon 238–9
Murray, Williamson 135
mutiny 28

Napoleon I 121
national culture, and shaping of soldiers 30
national essentialism
explanation of soldiers' actions 36–40
military scholarship 24–5
national identity 25
nationalism 28
nation-state, and military sociology 26–9
Natynczyk, Walt 104
Naylor, Jimmy 61
Nazi Germany
foreign soldiers 38–9
Holocaust 36–8
negligence 302–3
Neilson, Glenn 172
Nellis Air Force Base 287
nervousness, nineteenth century concerns over 123–4
Neviaski, Gilles 240
non-firing 53
non-governmental organizations (NGOs) 296
Noy, Shattai 280

Obama, Barack 295
officer corps, nineteenth century changes in 124–5
oral testimony, problems with 49
organizational sociology 217
organizations, definition of 217
Orientalism 34, 35, 60
outsourcing 322

Panetta, Leon 317
paratroopers 8
Parsons, Talcott 15
patriotism 48, 93, 94, 111, 112, 293, 295, 319
Patton, George S 321
peace and stability operations 242–3
cohesion 237
performative paradigm, historical origins of 13–16
performative turn, in military scholarship 12
Perrucchetti, Giuseppe 253, 254

Pitt, Barrie 51
Pogany, Georg 102, 108
policy, and strategy 153–4
political motivation
  cohesion 10, 53, 311
  explanation of soldiers' actions 36–40
Poncet, Henri 234, 238, 240, 241, 244–5
Porter, Patrick 295
post-Fordist military 322
Post-Traumatic Stress Disorder (PTSD) 78, 294, 301
  killing 55
practice, growth of interest in 15–16
precision warfare 58
Predator pilots 287
pressure groups 296
primary groups 52, 216
  cohesion 6–10, 78, 93, 250, 311
prisoners of war (PoWs)
  American soldiers 35
  Chinese communist soldiers in Korean War 34–5
  German soldiers 35
private military and security companies 322
professionalism
  abolition of conscription 13–14
  Australian Army 170
  cohesion 11, 61, 79, 88, 94, 312–15
  development of 14–15
  deviant professionalism 318–19
  as disciplinary mechanism 106–8
  improvisation 83
  reputation 85, 108–9, 110, 111, 209, 210, 211, 315, 316
  shame 109–11, 207, 211, 314–15
  urban combat 79–80
  women soldiers 317, 318
Proud, Matt 175
psychiatry, military 294, 295
punishment 94–5
  cohesion 93
  see also discipline
Putnam, Robert 284–5

Quigley, H. 50

Randle, John 32
Rapid Reaction Corps 321
Raugel, Guy 234, 240, 241, 242, 243, 244
reciprocity, and military contract 297
recovery rate of a unit 121
regimental identity 315, 316
regimental politics, and cohesion 211–12
religion, and combat 57, 61
remote warfare, competing demands of family and military 287
reprisals 60
reputation, professional 85, 108–9, 110, 111, 209, 210, 211, 315, 316
reservists 174, 180, 322
resilience, psychological 119–20, 127–8
revenge, as motivating factor 54, 56, 175–6, 207
Revolution of Military Affairs 14
risk aversion 293, 294, 296, 304
  cohesion 296–7
  public attitudes 295–6
risk culture 296, 297
risk-transfer warfare 309
rites of passage, and combat 57
rituals 57
Robertson, David 174
Roberts-Smith, Ben 175, 179, 186
Rochat, Giorgio 97
Royal Irish Regiment, *see* 7 Platoon, Royal Irish Regiment
Royal Marines 106–7, 108–9
Russo-Turkish War (1877–78) 121

safety concerns, paradoxical effects of 300–1
Saint-Fuscien, Emmanuel 94
Sangin, British Army's withdrawal from 201
Sarbi 187
Sarkesian, Sam C. 135
Schnier, Johannes 234, 241
Schultz, J. H. 54
Second World War
  capital punishment 95
  cohesion 6–7
  discipline in Allied forces 98–9
  discipline in German forces 99–100
  discipline in Italian army 98
  urban operations 74–5
Segal, David 216
Segal, Mady Wechlsler 279
self-inflicted wounds (SIW) 99, 103
  morale 140, 141, 142–3, 144
self-interest, and cohesion 8
7 Platoon, Royal Irish Regiment, in Afghanistan 201, 213, 313
  battle drills 202–3
  emotional bonds and combat performance 207–11
  professionalism of 205
  reaction to contact drills 203–5
  regimental politics and combat performance 211–12
  relations within platoon 209–10
  Sarwan Qala assault 207–8
  Waterloo Day contact 202–5
Severloh, Hein 99
Shalit, Ben 280, 281
shame 109–11, 207, 211, 314–15

Shaw, Martin 309
Shay, Jonathan 277
shell-shock 294
Shephard, Ben 293, 294
Sher, Gregory 176
Shils, Edward 6, 25, 26, 52, 139, 216, 236, 250, 311
sickness rates, and morale 140, 141, 144
Sikhism 30–1
  Khalsa 31
Sikhs, in British Indian army 30
  recruitment of 31
  in Second World War 32–4
  uniform regulations 31–2
Singh, Gobind 30–1
Singh, Ranajit 31
sleep, and combat soldiers 271
Slim, William 61–2
Slovik, Eddie 95
small unit dynamics 55
  cohesion 13
Smith, Chris 172, 174, 183, 184
Smith, Darren 176
Smith, James 176
snipers 59
social capital 284–5
social change, in nineteenth century 122–5
social cohesion 9, 10, 13, 55
social psychology, and combat 77–8
social sciences, change of paradigm 15–16
soldier's dilemma 293
Somalia, Canadian Airborne in 9
Spohn, Nicole 173
sports psychology 110
stacks, and urban combat 80
Stalingrad, Battle of 76
standardized tactics 125–6
Stanski, Keith 61
status hierarchy, French Foreign Legion 226–9
Stouffer, Samuel A 7, 26, 52, 138, 139, 281
Strachan, Hew 11, 93, 111, 136, 153, 155–6, 157
strategy
  balancing ends and means 153–4
  cohesion 135, 153–6, 157
  definition of 153
  as interactive process 155, 156
  levels of war 154–5
  policy 153–4
  tactics and drill 154
structuralism 15
subaltern soldiers 40
  theoretical significance of 26
support troops, vulnerability of 4
survival instinct 48
swift trust 11, 205, 313, 315
  urban combat 84
tactical doctrine
  cohesion 136–7, 152, 155–6
  mission-oriented tactics 126
  pre-First World War developments in 125–7
  pre-packaged strategy 154
  standardized tactics 125–6
Taliban 58, 61
task cohesion 10–11, 13, 78, 174
Taylor, Eleanor 317
technology 309
  combat 57–8
  dangers of 309, 310
  risk-transfer warfare 309
telephones, and communication with home front 273
Thomlinson, Damien 172, 174
Thompson, E. P. 16
Thompson, Mark 95, 96, 97
three block warfare 77, 86
time, compression of 48, 49
training
  cohesion 11–12, 93, 202, 312–14
  embodied repertoires of actions 87–8
  improved combat performance 105–6
  swift trust 205
transitions between tasks 86–8
Truscott, Lucian 98
trust 84
  see also swift trust

United Nations Opération des Nations Unies en Côte d'Ivoire (ONUCI) 238
United States Army
  2nd Division in First World War 129
  buddy system 7
  professionalism 14
United States Marine Corps 28–9, 109–10
  tactical bubble 76
  three block warfare 77
  Vietnam War 7–8
unlimited liability, *see* military contract of unlimited liability
Urban, Mark 296
urban combat 73
  accuracy requirement 79–80
  ad hoc units 83–5
  added pressures of 77
  autonomy 82
  characteristics of 74–5, 79
  combined-arms teams 76
  communications 81, 82
  difficulties with cohesion 73
  fragmentation and dispersion of forces 76, 81
  greater demands of 79
  improvisation at lower level 82–3
  leadership and decision-making 82
  military technology 309, 310

multiple tasks 77, 86
necessity for cohesion 73–4
premodern nature of 310
professionalism 79–80
sociological bubbles 76, 80–3
sociology and psychology of 77–80
special forcification of regular units 88–9
stacks 80
swift trust 84
temporary modular assemblages 76–7
three block warfare 77, 86
transitions between tasks 86–8
urban environments 75
vulnerability 75
urbanization 122–4

Vietnam War 7–8, 28
deviant cohesion 9
violence, role of emotions 205–6
V-mail 273

Wallace, Q. V. B. 142
Walzer, Michael 245
war
autonomy of 57
military technology 309
public attitudes 295–6
risk-transfer warfare 309
as 'spectator sport' 309–10
war cries 56–7
Waterloo 12
Watman, Kenneth H. 135
Watson, Alexander 10, 54, 140
Watterson, D. J. 141
Wavell, Archibald 24
Wehrmacht 6–7, 24, 29, 135, 216, 220, 236, 250, 311
discipline 99–100, 139
Welsh Guards Battle Group 102–3
Wesbrook, Stephen 10, 28, 62, 95, 250–1
Wessely, Simon 216, 294
Williams, Michael 207
Williams, Philip 103
Winslow, Donna 9, 237
Winter, Denis 49
Wittgenstein, Ludwig 15–16
women soldiers 316–18
Wong, Leonard 139, 210
Wood, Brett 179
Wynne, Gavin 103

Yeaman, Stuart 181, 185
Young German League 128
youth groups, pre-First World War period 128

Printed and bound by CPI Group (UK) Ltd, Croydon, CR0 4YY